MATHEMATICS

CONTEMPORARY
TOPICS
AND APPLICATIONS

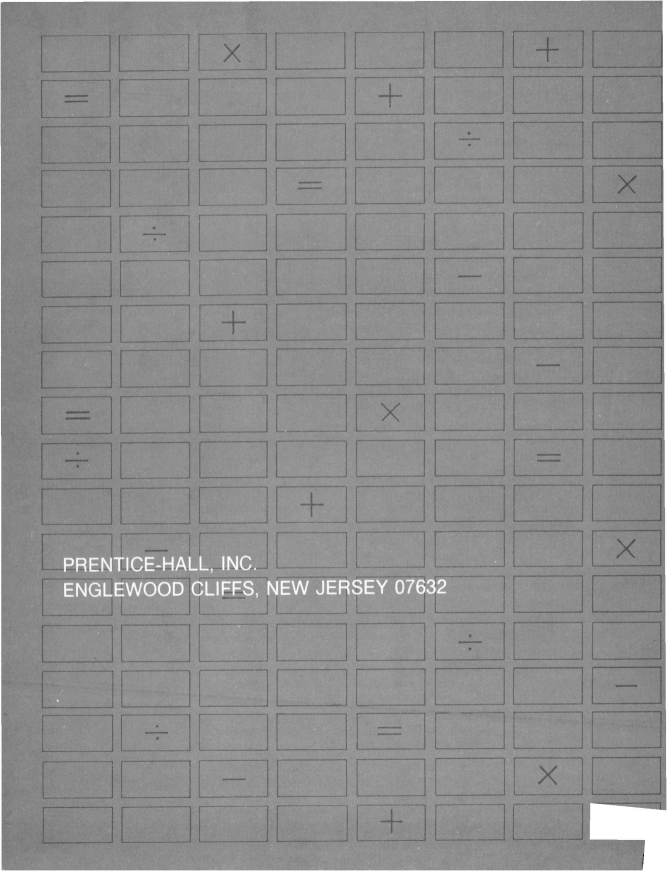

PRENTICE-HALL, INC.
ENGLEWOOD CLIFFS, NEW JERSEY 07632

MATHEMATICS

CONTEMPORARY TOPICS AND APPLICATIONS

Howard A. Silver

Chicago State University

Library of Congress Cataloging in Publication Data

SILVER, HOWARD A   (date)
    Mathematics: contemporary topics and applications.

    Includes index.
    1. Mathematics—1961-   I. Title.
QA37.2.S56   510   78-15133
ISBN   0-13-563304-4

TO BECKY AND LISA
_____

Mathematics: Contemporary Topics and Applications
Howard A. Silver

Printed in the United States of America

10   9   8   7   6   5   4

Cover illustration: "Fractured Hexagon" by Joanna Pinsky

Page layout by Jenny Markus and Suzanne Behnke
Cover and chapter-opening design by Walter Behnke
Manufacturing buyer: Phil Galea

PRENTICE-HALL INTERNATIONAL, INC., *London*
PRENTICE-HALL OF AUSTRALIA PTY. LIMITED, *Sydney*
PRENTICE-HALL OF CANADA, LTD., *Toronto*
PRENTICE-HALL OF INDIA PRIVATE LIMITED, *New Delhi*
PRENTICE-HALL OF JAPAN, INC., *Tokyo*
PRENTICE-HALL OF SOUTHEAST ASIA PTE. LTD., *Singapore*
WHITEHALL BOOKS LIMITED, *Wellington, New Zealand*

CONTENTS

PREFACE

Virtually no one would deny that mathematics is a vital part of our daily life. It is almost impossible to go through a normal twenty-four hour day without meeting situations involving discount percentages, interest, tax, inflation; or the dimensions of rooms, cars, food portions; or the chances of rain, a team's winning, our getting into graduate school; or a computerized billing; and so on.

This book is intended for the general (non-physical science major) college student. The course may have various titles, such as Liberal Arts Math, Introductory Mathematics, Finite Mathematics. My goal in writing this book is not only to expose the student to the beauty of mathematics but, primarily, to equip him or her with the skills and confidence needed to cope with mathematics in the real world. With this purpose in mind I have, by and large, selected topics of a concrete nature. Students will *not* ask the question, "Where will I ever use this stuff?" They will use much of it every day of their lives.

The writing has been kept simple and concise. Every idea is followed by a clearly explained example. Rules and procedures are summarized in boxes. "Hand Calculator Instant Replays" appear throughout the text to familiarize the student with the operation and capabilities of calculators.

Problem sets are also included as a learning tool: there are over 2500 problems, many involving real-life applications. Answers to more than half of the exercises are given at the end of the book. A list of "important words" and a set of review exercises appear at the ends of chapters to help the student integrate the material and prepare for tests.

Prerequisites are minimal: knowing some arithmetic and a little common sense. Chapter 1 is almost all review of arithmetic (except for Section 6 on the hand calculator). This chapter could be skipped. Chapter 2 is on problem solving and may also seem like a review, but it probably should not be skipped. The other chapters are largely independent.

The entire book makes a two-semester sequence. In a one-semester course the instructor would be free to choose the topics that appealed to him or her.

My hope is that instructors will enjoy teaching from this book and that students will enjoy learning from it.

I gratefully acknowledge the help of the following reviewers during various stages in preparation of the manuscript: Elton Beougher, Fort Hays State University; Bruce E. Earnley, Northern Essex Community College; Arthur Kramer, New York City Community College; Albert W. Liberi, Westchester Community College; Ann Miller, Southern Illinois University; Dudley R. Pitt, Northwestern State College of Louisiana; Robert Rapalje, Seminole Community College; Ara Sullenberger, Tarrant County Junior College; and Thomas A. Tredon, Lord Fairfax Community College.

I would like to thank the great Prentice-Hall staff for all their help and encouragement; specifically, Joanne Slivinski, Harry Gaines, Cathy Brenn, Zita de Schauensee, and Ed Lugenbeel. Thanks also go to my typists, Karen Utes (text) and Dolly Gore (instructor's manual); to Carol Olson, and Dan Coffey for their help with the photos; and finally to Mike Sullivan, Pat Blus, Pam Folie, Danny O'Connor, Jenifer Morison, and Fran Silver for their input and feedback along the way.

Park Forest, Illinois   HOWARD A. SILVER

MATHEMATICS

CONTEMPORARY TOPICS AND APPLICATIONS

Massive crowd at rock concert. (Christopher Little/
Camera 5)

1

REVIEW OF REAL NUMBERS

Question What are the real numbers?

ANSWER **Real numbers** are all the numbers we use to express measurements we make in the real world. Real numbers come in different shapes and sizes. We have the following types within the real numbers.

1. **Integers**, such as 10, 215, -31, -27, and 5, are real numbers.
2. **Fractions** (or rational numbers), such as $\frac{1}{2}$, $\frac{17}{20}$, $-\frac{13}{50}$, $\frac{61}{5}$, and $-\frac{21}{4}$, are real numbers.
3. **Decimals**, such as 5.1, 18.75, -4.94, $3.14159\ldots$, and -21.063, are real numbers.

Real numbers can also be pictured on a number line. Every point on the number line stands for a real number.

8:09 a.m. on John Street, New York City. (© 1977 by Jan Lukas from Rapho/Photo Researchers)

$$-4 \quad -3 \quad -2 \quad -1 \quad 0 \quad 1 \quad 2 \quad 3 \quad 4 \quad 5$$

The purpose of this chapter is to review quickly some of the fundamental operations with real numbers. We will not stress the theory of real numbers at all.

1–1 FRACTIONS

Addition and Subtraction

Question   What is the rule for adding and subtracting fractions?

ANSWER   This is the rule that we use.

> To add or subtract fractions:
>
> 1. Find a common denominator, if necessary.
> 2. Rewrite all fractions to have this common denominator.
> 3. Add (or subtract) the numerators; put the answer over the common denominator.

We use the following formulas.

Formula 1-1-1

$$\frac{a}{c} + \frac{b}{c} = \frac{a+b}{c}$$

Formula 1-1-2

$$\frac{a}{c} - \frac{b}{c} = \frac{a-b}{c}$$

Problem   Add $\dfrac{4}{10} + \dfrac{3}{10}$.

ANSWER

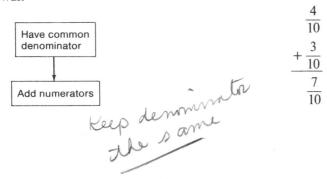

Have common denominator

Add numerators

Keep denominator the same

$$\begin{array}{r} \dfrac{4}{10} \\[6pt] + \dfrac{3}{10} \\[4pt] \hline \dfrac{7}{10} \end{array}$$

Problem   Add $\frac{1}{2} + \frac{2}{3}$.

ANSWER

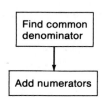

$$\frac{1}{2} = \frac{3}{6}$$
$$+\frac{2}{3} = +\frac{4}{6}$$
$$\frac{7}{6}$$

Problem   Add $\frac{1}{2} + \frac{3}{4} + \frac{5}{6}$.

ANSWER   We can add three or more fractions by finding a common denominator for all of them.

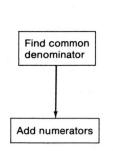

$$\frac{1}{2} = \frac{6}{12}$$
$$\frac{3}{4} = \frac{9}{12}$$
$$+\frac{5}{6} = +\frac{10}{12}$$
$$\frac{25}{12}$$

Problem   Subtract $\frac{7}{9} - \frac{2}{9}$.

ANSWER

Keep denominator the same

$$\frac{7}{9}$$
$$-\frac{2}{9}$$
$$\frac{5}{9}$$

Problem   Subtract $\frac{7}{4} - \frac{1}{6}$.

ANSWER

Find common denominator

Subtract denominators

$$\frac{7}{4} = \frac{21}{12}$$
$$-\frac{1}{6} = -\frac{2}{12}$$
$$\frac{19}{12}$$

Question What is the least common denominator and how do we find it?

ANSWER The **least common denominator** (**LCD**) is the smallest number that can be divided by all the denominators evenly. For example, given a problem such as $\frac{1}{4} + \frac{1}{6}$, it is easy to see that the LCD is 12, since 12 is the smallest number that can be divided by both 4 and 6 evenly.

Sometimes, the LCD isn't quite that easy to find. In such cases, we have a special rule for finding the LCD and then adding (or subtracting) the fractions.

Problem Add $\frac{1}{6} + \frac{3}{10} + \frac{7}{18}$.

ANSWER First, we find the LCD by factoring the denominators. Remember, a **prime** is a number that is divisible only by 1 and itself, such as 2, 3, 5, 7, 11, 13, 17, and 19. We can factor any whole number into a product of primes.

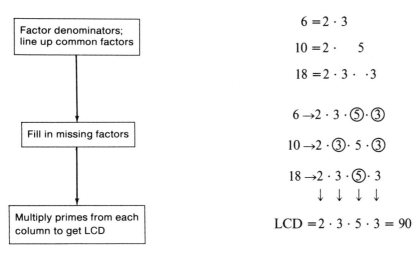

Now we have that the LCD = 90. We can also use the numbers above to help us add the fractions.

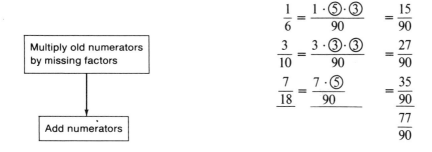

Let us summarize this procedure in the following rule.

To find the LCD and add (or subtract) fractions:

1. Factor all the denominators into primes.

2. Line up the common primes in columns, with unlike primes in different columns.

3. Fill in the missing factors in each column. (Circle these factors for later use.)

4. Bring down a prime from each column. The product is the LCD.

5. For each of the fractions, multiply the old numerator by the missing (circled) factors for that denominator.

6. Add (or subtract) the numerators.

Problem Add $\dfrac{3}{8} + \dfrac{1}{12} + \dfrac{7}{20}$.

ANSWER We use the rule given above.

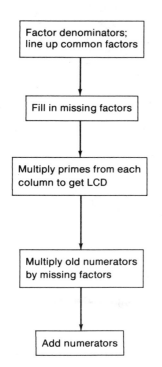

$$8 = 2 \cdot 2 \cdot 2$$
$$12 = 2 \cdot 2 \cdot \quad \cdot 3$$
$$20 = 2 \cdot 2 \cdot \quad \quad \cdot 5$$

$$8 \to 2 \cdot 2 \cdot 2 \cdot ③ \cdot ⑤$$
$$12 \to 2 \cdot 2 \cdot ② \cdot 3 \cdot ⑤$$
$$20 \to 2 \cdot 2 \cdot ② \cdot ③ \cdot 5$$
$$\downarrow \ \downarrow \ \downarrow \ \downarrow \ \downarrow$$

$$\text{LCD} = 2 \cdot 2 \cdot 2 \cdot 3 \cdot 5 = 120$$

$$\frac{3}{8} = \frac{3 \cdot ③ \cdot ⑤}{120} = \frac{45}{120}$$

$$\frac{1}{12} = \frac{1 \cdot ② \cdot ⑤}{120} = \frac{10}{120}$$

$$\frac{7}{20} = \frac{7 \cdot ② \cdot ③}{120} = \frac{42}{120}$$

$$\frac{97}{120}$$

Problem   Subtract $\dfrac{7}{30} - \dfrac{1}{25}$.

ANSWER   We use the same procedure, except we subtract at the end.

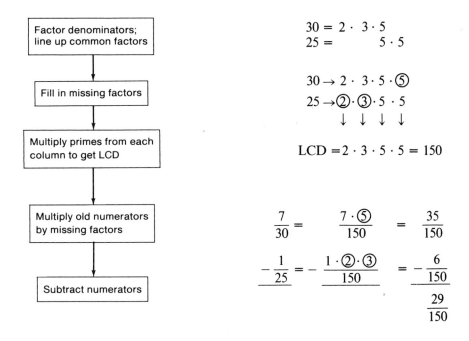

Question   How do we reduce fractions?

ANSWER   We reduce a fraction to its **lowest terms** by dividing out all common factors from the numerator and denominator.

Problem   Reduce $\dfrac{36}{42}$ to lowest terms.

ANSWER

Divide out a 2

Divide out a 3

$$\frac{36}{42} = \frac{18}{21}$$

$$= \frac{6}{7}$$

Since 6 and 7 have no common factors, the fraction is reduced to lowest terms.

Problem   Add $\dfrac{7}{9} + \dfrac{5}{6}$, and reduce to lowest terms.

ANSWER

$$\dfrac{7}{9} = \dfrac{28}{36}$$

$$+\dfrac{5}{6} = +\dfrac{30}{36}$$

$$\dfrac{58}{36}$$

$$= \dfrac{29}{18}$$

Question   How do we add and subtract mixed numbers?

ANSWER   **Mixed numbers** have a whole number part and a fraction part. We just add (or subtract) the whole numbers and add (or subtract) the fraction parts.

  If the sum of the fractions goes over 1, we *carry* the whole number over with the other whole numbers.

Problem   Add $4\frac{1}{2} + 5\frac{2}{5}$.

ANSWER

$$4\frac{1}{2} = \quad 4\frac{5}{10}$$

$$+5\frac{2}{5} = +5\frac{4}{10}$$

$$9\frac{9}{10}$$

Problem   Add $4\frac{2}{3} + 9\frac{3}{4}$.

ANSWER

$$4\frac{2}{3} = \quad 4\frac{8}{12}$$

$$+9\frac{3}{4} = +9\frac{9}{12}$$

$$13\frac{17}{12}$$

$$= 14\frac{5}{12}$$

Notice here that we had $\frac{17}{12}$, which is $1\frac{5}{12}$. Then we carried the 1 over to the 13 to give 14.

Problem   Subtract $7\frac{1}{5} - 3\frac{1}{15}$.

ANSWER

$$7\frac{1}{5} \quad = \quad 7\frac{3}{15}$$

$$-3\frac{1}{15} \quad = \quad -3\frac{1}{15}$$

$$4\frac{2}{15}$$

Problem   Subtract $10\frac{1}{3} - 3\frac{4}{5}$.

ANSWER   Here we will have trouble, since $\frac{1}{3}$ is less than $\frac{4}{5}$. So we will have to *borrow* from the 10.

$$10\frac{1}{3} \quad = \quad 10\frac{5}{15}$$

$$-3\frac{4}{5} \quad = \quad -3\frac{12}{15}$$

$$= \quad 9\frac{20}{15}$$

$$-3\frac{12}{15}$$

$$6\frac{8}{15}$$

We borrowed $1 = \frac{15}{15}$ from the 10 to make the $\frac{5}{15}$ into $\frac{20}{15}$. Then the 10 became a 9.

1. Define or discuss:
   (a) LCD.
   (b) Prime.

Reduce the following fractions as much as possible.

2. $\dfrac{24}{32}$ **3.** $\dfrac{15}{45}$ **4.** $\dfrac{12}{18}$ **5.** $\dfrac{9}{54}$

6. $\dfrac{15}{24}$ **7.** $\dfrac{42}{28}$ **8.** $\dfrac{25}{200}$ **9.** $\dfrac{6900}{1000}$

Add the following fractions and reduce the answer to lowest terms.

10. $\dfrac{1}{4} + \dfrac{2}{4}$ **11.** $\dfrac{2}{7} + \dfrac{4}{7}$ **12.** $\dfrac{1}{2} + \dfrac{3}{4}$

13. $\dfrac{3}{4} + \dfrac{1}{6}$ **14.** $\dfrac{1}{10} + \dfrac{2}{15}$ **15.** $\dfrac{3}{50} + \dfrac{7}{100}$

16. $\dfrac{1}{2} + \dfrac{1}{4} + \dfrac{1}{8}$ **17.** $\dfrac{5}{9} + \dfrac{11}{12} + \dfrac{2}{15}$ **18.** $\dfrac{7}{8} + \dfrac{9}{10} + \dfrac{1}{12}$

Add the following mixed numbers and reduce.

19. $8\frac{1}{2}$ **20.** $15\frac{1}{3}$ **21.** $8\frac{2}{5}$
     $+7\frac{1}{4}$ $+6\frac{1}{9}$ $+4\frac{2}{3}$

22. $79\frac{3}{4}$ **23.** $21\frac{5}{12}$ **24.** $4\frac{13}{15}$
     $+38\frac{5}{6}$ $+15\frac{7}{8}$ $+7\frac{7}{10}$

Subtract the following fractions and reduce.

25. $\dfrac{9}{10} - \dfrac{5}{10}$ **26.** $\dfrac{10}{12} - \dfrac{7}{12}$ **27.** $\dfrac{1}{2} - \dfrac{3}{8}$

28. $\dfrac{3}{4} - \dfrac{5}{16}$ **29.** $\dfrac{1}{8} - \dfrac{1}{12}$ **30.** $\dfrac{17}{10} - \dfrac{4}{15}$

31. $\dfrac{25}{24} - \dfrac{17}{18}$ **32.** $\dfrac{15}{8} - \dfrac{29}{18}$ **33.** $\dfrac{7}{50} - \dfrac{3}{100}$

Subtract the following mixed numbers.

34. $17\frac{1}{2}$ **35.** $14\frac{3}{10}$ **36.** $7\frac{1}{4}$
     $-6\frac{1}{4}$ $-2\frac{1}{5}$ $-3\frac{1}{6}$

37. $8\frac{1}{12}$ **38.** $6\frac{1}{2}$ **39.** $51\frac{2}{7}$
     $-3\frac{3}{4}$ $-\frac{4}{5}$ $-46\frac{3}{11}$

40. Dan bought $6\frac{1}{2}$ gallons of gas on Monday and $11\frac{7}{10}$ on Thursday. How much gas did he buy?

41. Jeanne drove $25\frac{1}{4}$ miles on Monday, $30\frac{1}{2}$ on Tuesday, $18\frac{3}{10}$ on Wednesday, $13\frac{3}{4}$ on Thursday, and $33\frac{1}{10}$ on Friday. How many miles did she drive during the week?

42. Mary mixes $2\frac{1}{2}$ cups of sugar and $3\frac{1}{3}$ cups of flour. How many cups is this?

43. A wall is $12\frac{5}{12}$ feet long. Al and Linda center a $7\frac{1}{2}$-foot couch against the wall. How much total space is left on both sides? $\div 2 = $ Both Sides.-

44. Elmer cuts $4\frac{1}{3}$ feet off an $8\frac{1}{2}$-foot board. What length of board is left?

Multiplication and Division

Question   What is the rule for multiplying fractions?

ANSWER   We use the following rule.

> To multiply fractions:
> 1. Multiply the numerators together.
> 2. Multiply the denominators together.

Formula 1–1–3

$$\frac{p}{q} \cdot \frac{r}{s} = \frac{p \cdot r}{q \cdot s}$$

Problem   Multiply $\dfrac{2}{5} \times \dfrac{7}{9}$.

ANSWER

| Formula 1–1–3 |

$$\frac{2}{5} \times \frac{7}{9} = \frac{14}{45}$$

Problem   Multiply $\dfrac{6}{7} \times \dfrac{3}{4}$.

ANSWER

| Formula 1–1–3 |

| Divide out a 2 |

$$\frac{6}{7} \times \frac{3}{4} = \frac{18}{28}$$

$$= \frac{9}{14}$$

We also could have canceled before we multiplied.

$$\frac{\overset{3}{\cancel{6}}}{7} \times \frac{3}{\underset{2}{\cancel{4}}} = \frac{3}{7} \times \frac{3}{2}$$

$$= \frac{9}{14}$$

Question   How do we convert a mixed number to an improper fraction?

ANSWER   An **improper fraction** is a fraction with a numerator larger than the denominator, such as $\frac{17}{10}$. While the term *improper* seems to delight sixth-grade teachers, it does not mean that something is wrong with the fraction.

> To convert a mixed number to an improper fraction:
> 1. Multiply the whole number by the denominator.
> 2. Add the numerator.
> 3. Place this over the denominator.

Examples   1. $5\frac{2}{3} = \frac{15 + 2}{3} = \frac{17}{3}$

2. $6\frac{1}{5} = \frac{30 + 1}{5} = \frac{31}{5}$

Question   How do we convert an improper fraction to a mixed number?

ANSWER   We use the following rule.

> To convert an improper fraction to a mixed number:
> 1. Divide the denominator into the numerator.
> 2. The quotient is the whole-number part.
> 3. The remainder over the original denominator is the fraction part.

Examples

1. $\dfrac{19}{5} = 5\overline{)19}^{\,3\ \ R4} = 3\dfrac{4}{5}$

2. $\dfrac{37}{4} = 4\overline{)37}^{\,9\ \ R1} = 9\dfrac{1}{4}$

Question   How do we multiply mixed numbers?

ANSWER   We use the following rule.

> To multiply mixed numbers:
> 1. Convert them to improper fractions.
> 2. Use Formula 1–1–3 (with any cancellation).
> 3. Convert the answer back to a mixed number.

Problem   Multiply $4\frac{1}{2} \times 3\frac{1}{5}$.

ANSWER

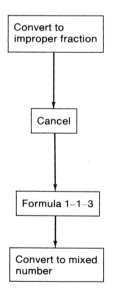

$$4\frac{1}{2} \times 3\frac{1}{5} = \frac{9}{2} \times \frac{16}{5}$$

$$= \frac{9}{\underset{1}{\cancel{2}}} \times \frac{\overset{8}{\cancel{16}}}{5}$$

$$= \frac{72}{5}$$

$$= 14\frac{2}{5}$$

Question What is the rule for dividing fractions?

ANSWER We use the following rule.

> To divide fractions:
> 1. Invert (or flip) the divisor (second number).
> 2. Multiply as fractions.

Formula 1–1–4

$$\frac{a}{b} \div \frac{c}{d} = \frac{a}{b} \cdot \frac{d}{c}$$

Problem Divide $\dfrac{7}{2} \div \dfrac{3}{5}$.

ANSWER

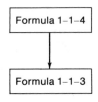

$$\frac{7}{2} \div \frac{3}{5} = \frac{7}{2} \cdot \frac{5}{3}$$

$$= \frac{35}{6}$$

Question How do we divide mixed numbers?

ANSWER The rule is as follows.

> To divide mixed numbers:
> 1. Convert them to improper fractions.
> 2. Use Formula 1–1–4 (invert divisor).
> 3. Use Formula 1–1–3 (multiply).
> 4. Convert back to a mixed number.

Problem   Divide $7\frac{1}{3} \div 1\frac{4}{7}$.

ANSWER

Convert to improper fraction	$7\frac{1}{3} \div 1\frac{4}{7} = \frac{22}{3} \div \frac{11}{7}$
$\downarrow$	
Formula 1–1–4	$= \frac{22}{3} \times \frac{7}{11}$
$\downarrow$	
Cancel	$= \frac{\overset{2}{\cancel{22}}}{3} \times \frac{7}{\underset{1}{\cancel{11}}}$
$\downarrow$	
Formula 1–1–3	$= \frac{14}{3}$
$\downarrow$	
Convert to mixed number	$= 4\frac{2}{3}$

PROBLEM
SET 1–1–2

1. Define or discuss:
   (a) Mixed number.
   (b) Improper fraction.

Write the following mixed numbers as improper fractions.

2. $2\frac{1}{2}$ 3. $6\frac{2}{5}$ 4. $12\frac{1}{8}$

5. $4\frac{6}{11}$ 6. $5\frac{3}{10}$ 7. $18\frac{11}{12}$

Write the following improper fractions as mixed numbers.

8. $\frac{11}{7}$ 9. $\frac{17}{3}$ 10. $\frac{14}{5}$

11. $\frac{108}{15}$ 12. $\frac{691}{100}$ 13. $\frac{172}{13}$

Multiply the following fractions and reduce.

14. $\frac{1}{2} \times \frac{1}{3}$ 15. $\frac{4}{5} \times \frac{6}{7}$ 16. $\frac{6}{5} \times \frac{15}{12}$

17. $\frac{7}{10} \times \frac{5}{14}$ 18. $\frac{5}{8} \times \frac{3}{10} \times \frac{6}{25}$ 19. $\frac{2}{25} \times \frac{5}{12} \times \frac{3}{10}$

Multiply the following mixed numbers, leaving the answers as reduced mixed numbers.

20. $4\frac{1}{2}$ **21.** $6\frac{1}{5}$ **22.** $10\frac{1}{4}$

     $\times 3\frac{1}{3}$ $\times 2\frac{2}{5}$ $\times 8\frac{3}{5}$

23. $2\frac{2}{3}$ **24.** 150 **25.** 280

     $\times 4\frac{5}{8}$ $\times 3\frac{1}{2}$ $\times \frac{2}{5}$

Divide the following fractions and reduce.

26. $\dfrac{11}{3} \div \dfrac{1}{5}$ **27.** $\dfrac{2}{3} \div \dfrac{5}{4}$ **28.** $\dfrac{16}{3} \div 4$

29. $\dfrac{12}{5} \div \dfrac{3}{10}$ **30.** $\dfrac{18}{11} \div \dfrac{4}{5}$ **31.** $\dfrac{6}{13} \div \dfrac{18}{7}$

Divide the following mixed numbers, and leave the answers as reduced mixed numbers.

32. $6\frac{1}{5} \div 2\frac{2}{3}$ **33.** $7\frac{1}{5} \div \frac{2}{3}$ **34.** $14\frac{1}{2} \div 2\frac{1}{5}$

35. $3\frac{1}{2} \div 3\frac{1}{3}$ **36.** $9\frac{1}{4} \div 1\frac{5}{6}$ **37.** $2\frac{1}{10} \div 4\frac{1}{20}$

38. A certain recipe calls for $\frac{2}{3}$ cup sugar, $1\frac{1}{4}$ cups flour, $\frac{1}{2}$ teaspoon salt, $1\frac{3}{4}$ cups water, $\frac{3}{4}$ cup chocolate chips, and $\frac{5}{6}$ tablespoon baking soda. Find the quantities needed to make $2\frac{1}{2}$ times this amount.

39. Bob's shoe is $1\frac{1}{6}$ feet long. He measures a car's length by counting off lengths with his shoes. The car measures $14\frac{1}{2}$ shoes. How many feet is this?

40. Ed stacks up 12 books that are each $1\frac{3}{8}$ inch thick. How high is his pile?

41. Mary's hair grows $\frac{3}{8}$ inch every month. How much will it grow in a year?

In Soybeanville, Illinois, there are 432 first graders. Of these, $\frac{5}{6}$ will graduate from the 8th grade. Of these, $\frac{2}{3}$ will graduate from high school. Of these, $\frac{1}{5}$ will graduate from college. Of these, $\frac{1}{10}$ will get a master's or professional degree.

42. How many will finish eighth grade?

43. How many will finish high school?

44. How many will finish college?

45. How many will get a master's or professional degree?

46. Howie runs $3\frac{1}{2}$ miles in $27\frac{1}{2}$ minutes. How many minutes per mile is this?

47. At a big picnic, 100 pounds of meat are cooked. If each person gets $\frac{3}{8}$ pound, how many people can be fed?

48. Mike plans to own his car 5 years. He will change the oil every $\frac{1}{4}$ year. How many times will he change the oil over the 5 years?

49. A vitamin pill provides $\frac{2}{5}$ the daily need of B_2. How many pills do you need to take to get $1\frac{1}{2}$ times your daily need?

1–2 DECIMALS

Addition and Subtraction

Question   What is the rule for adding or subtracting decimals?

ANSWER   Here is the rule that we use.

> To add (or subtract) decimals:
>
> 1. Line up all the decimal points vertically.
>
> 2. Add (or subtract) as usual.
>
> 3. Put the decimal point in the answer in line, below the decimal points in the problem.

Problem   Add the following decimals:
(a) 61.5 + 9.19 + 1.753
(b) 1.7 + 84.05 + 21.3 + 0.72

ANSWER   We line up the decimal points and add.

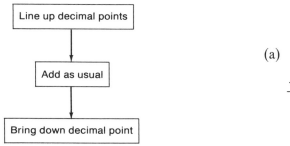

$$
\text{(a)} \quad
\begin{array}{r}
61.500 \\
9.190 \\
+\ 1.753 \\
\hline
72.443
\end{array}
\qquad
\text{(b)} \quad
\begin{array}{r}
1.70 \\
84.05 \\
21.30 \\
+\ 0.72 \\
\hline
107.77
\end{array}
$$

Notice that, after we lined up the decimal points, we filled in extra zeros to give the numbers the same length. This helps many people.

Problem   Subtract the following decimals.
(a) 101.5 − 83.69
(b) 700 − 584.3

ANSWER

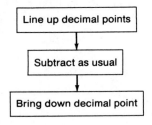

(a)   101.50   (b)   700.0
      − 83.69   −584.3
      ──────   ──────
       17.81   115.7

PROBLEM   Add the following decimals.
SET 1–2–1

1.	2.01	2.	5.878	3.	178.8	4.	80.1
	18.78		23.1		85.66		7.59
	+ 5.42		+54.25		842.72		27.46
					+439.937		+174.532

5. 71.06 + 82.3 + 47.92   **6.** 1.007 + 21.5 + 5.0708

7. 12.3 + 27.47 + 193.7 + 0.357   **8.** 10.031 + 7.85

Subtract the following decimals.

9.	173.01	10.	14.47	11.	60.03	12.	100
	− 85.59		− 8.081		−59.889		− 99.97

13. 19.01 − 8.75   **14.** 841.79 − 58.3

15. 72.3 − 8.553   **16.** 468.93 − 93.2

Balance the following checkbooks.

17. Mar. 1	Balance from February		$319.73
Mar. 3	Coat	−$ 31.55	
Mar. 5	Phone bill	− 16.09	
Mar. 8	Groceries	− 32.75	
Mar. 15	Deposit		+200.00
Mar. 18	Cash	− 25.00	
Mar. 19	Gasoline credit card	− 27.15	
Mar. 21	Groceries	− 21.14	
Mar. 24	Records	− 9.87	
Mar. 27	Rent	−125.00	
Apr. 1	Balance from March		?

Balancing the checkbook. (Peter M. Lerman/Nancy Palmer Photo Agency)

18.	Nov. 1	Balance from October		$523.72
	Nov. 2	Gas bill	−$ 25.17	
	Nov. 3	Groceries	− 57.29	
	Nov. 5	Insurance	− 78.12	
	Nov. 8	Deposit (pay)		+418.33
	Nov. 12	Electric bill	− 35.66	
	Nov. 15	Phone bill	− 25.47	
	Nov. 19	Revolving charge card	−107.53	
	Nov. 21	Clothes	− 83.73	
	Nov. 22	Groceries	− 49.54	
	Nov. 23	Deposit (pay)		+418.33
	Nov. 25	Rent	−275.00	
	Nov. 26	Cash	− 60.00	
	Nov. 28	Car payment	−153.73	
	Dec. 1	Balance from November		?

Multiplication and Division

Question   What is the rule for multiplying decimals?

ANSWER   The rule is as follows.

> To multiply decimals:
>
> 1. Multiply the numbers as if they were whole numbers.
> 2. Count up the total number of decimal places in both factors.
> 3. Move this number of places to the left in the product (filling in zeros, if necessary).

Problem   Multiply 7.13×8.4.

ANSWER

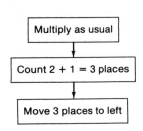

$$
\begin{array}{r}
7.13 \\
\times\ 8.4 \\
\hline
2\ 852 \\
57\ 04 \\
\hline
59.892
\end{array}
$$

Problem   Multiply 0.0007×0.00261.

ANSWER

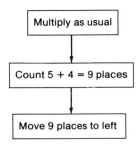

$$
\begin{array}{r}
0.00261 \\
\times\, 0.0007 \\
\hline
0.000001827
\end{array}
$$

Notice that we had to fill in with five extra places to get the nine places we needed.

Question   What is the rule for dividing decimals?

ANSWER   We use the following rule.

> To divide decimals:
>
> 1. Move the decimal point of the divisor (outside number) until it is a whole number.
>
> 2. Move the decimal point of the dividend (inside number) the *same* number of places to the right (fill in zeros, if necessary).
>
> 3. Put the decimal point for the answer above the new decimal point of the dividend.
>
> 4. Divide as usual.

Problem   Divide 44.103 ÷ 24.1.

ANSWER

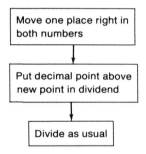

$$
\begin{array}{r}
1.83 \\
24.1\,\overline{)44.1\,03} \\
34\ 1 \\
\hline
20\ 0\ 0 \\
19\ 2\ 8 \\
\hline
7\ 23 \\
7\ 23 \\
\hline
\end{array}
$$

Problem   Divide 1.9 ÷ 0.26.

ANSWER

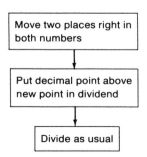

$$
\begin{array}{r}
7.307 \\
0.26\,\overline{)1.90\,000} \\
1.82 \\
\hline
8\ 0 \\
7\ 8 \\
\hline
200 \\
182 \\
\hline
18
\end{array}
$$

Notice that the division does not stop.

Question What do we do when the answer does not stop?

ANSWER We **round off** the answer to a certain number of places.

> To round off a decimal to a certain number of places:
>
> 1. We go out one extra decimal place to the right.
> 2. If this digit is 4 or less (0, 1, 2, 3, or 4), we drop this and all the other digits to the right.
> 3. If this digit is 5 or more (5, 6, 7, 8, or 9), we increase the digit to be rounded by 1, and drop the digits to the right.

Problem Round 7.307 to two decimal places.

ANSWER Since we want to round to two places, we go out to the third place.

$$7.30⑦$$

$$\downarrow$$

$$7.31$$

Problem Round 18.1359 to one decimal place.

ANSWER This time we go to the second place.

$$18.1③59$$

$$\downarrow$$

$$18.1$$

Problem Divide 25 ÷ 0.62 with the answer rounded to two decimal places.

ANSWER

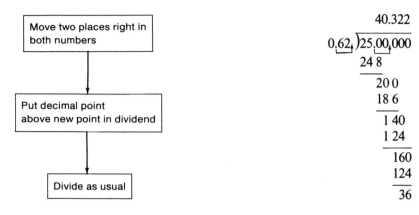

Since we wanted a two-place answer, we carried out the division to a third place. Now we round off the answer.

PROBLEM SET 1–2–2 Multiply the following decimals.

1. 73.18 ×52.57	**2.** 45.03 × 47.6	**3.** 9.035 × 8.2	**4.** 1.008 × 6.02
5. 0.0035 ×0.0008	**6.** 3773.1 × 0.003	**7.** 3.25 × 250	**8.** 7250 ×0.0025

Round off the following decimals.

9. 17.073 to two places   **10.** 325.183 to one place

11. 0.77777 to three places   **12.** 3.14159 to two places

13. 250.078 to one place   **14.** 5.4444 to three places

Divide the following decimals and round the answers to two decimal places.

15. $74.5 \div 1.79$ **16.** $84.01 \div 0.003$

17. $483 \div 17$ **18.** $0.003 \div 0.07$

Divide the following decimals and round the answers to three decimal places.

19. $19 \div 1.1$ **20.** $55 \div 173$

21. $3.7 \div 0.072$ **22.** $4.4 \div 0.19$

23. Mr. N. Vestor owns 250 shares of the Plastic Tree Co. Each share is worth $1.75. How much are Mr. Vestor's shares worth?

24. Joan earns $2.45 an hour at Ye Olde Taco Shoppe. Last week she worked 17.5 hours. How much did she earn?

25. Bob buys 15.3 gallons of gas at 69.4 cents per gallon. How much does the gas cost?

26. Jan and Wayne have $440 in the bank. In 5 years they will have 1.35 times this amount. How much money will they have?

27. Forty five sheets of metal are stacked on top of one another. If each sheet is 0.053 inch, what is the total height?

28. In a class of 26, each student donated an average of $1.26 to charity. How much did the total class donate?

29. Steve agrees to pay $3926 in 24 payments. How much is each payment?

30. Sue cut a 2.8-meter board into 6 equal pieces. How long is each piece?

31. Bonnie buys a gross (144) of pencils for $3.25. How much does each pencil cost?

32. John drove 241.3 miles on 21.7 gallons of gas. How many miles to a gallon does he get?

33. Mary was paid $75 to type a Master's thesis. She worked 17.5 hours typing. How much did she earn per hour?

34. Rich notices that one of his books has 1258 pages (or 629 sheets). He measures the total thickness of these pages to be 4.7 cm. How thick is each sheet?

1–3 NUMERAL CONVERSIONS

Question   How do we change a fraction into a decimal?

ANSWER   We use the following rule.

> To convert a fraction to a decimal:
> 1. Divide the denominator into the numerator.
> 2. Round off, if necessary.

Problem Write $\frac{1}{40}$ as a decimal.

ANSWER

Divide denominator into numerator

$$\begin{array}{r} 0.025 \\ 40\overline{)1.000} \end{array}$$

Problem Write $\frac{4}{7}$ as a three-place decimal.

ANSWER

Divide denominator into numerator
$\downarrow$
| Round off |

$$\begin{array}{r} 0.571④ \\ 7\overline{)4.000\ 0} \end{array}$$

$$\rightarrow 0.571$$

Question How do we change a decimal into a fraction?

ANSWER We have the following rule.

> To convert a decimal to a fraction:
>
> 1. Determine the column value of the last decimal place.
> 2. Write the entire number over this denominator (drop decimal point).
> 3. Reduce the fraction, if necessary.

Question What are the column place values?

ANSWER Every column is a power of 10. To the left of the decimal point are the numbers 1, 10, 100, 1000, and so on. To the right are the fractions $\frac{1}{10}, \frac{1}{100}, \frac{1}{1000}$, and so on.

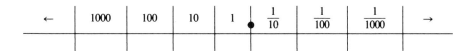

Problem Write 0.07 as a fraction.

ANSWER The last digit, 7, is in the $\frac{1}{100}$ column, so we put the whole number over 100.

$$0.07 = 0.07$$
$$\uparrow \frac{1}{100}$$

$$= \frac{7}{100}$$

Problem Write 16.8 as a fraction.

ANSWER

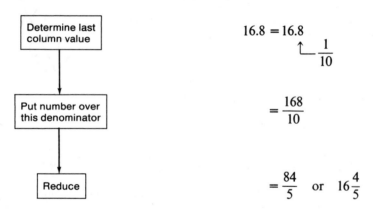

$$16.8 = 16.8$$
$$\uparrow \frac{1}{10}$$

$$= \frac{168}{10}$$

$$= \frac{84}{5} \quad \text{or} \quad 16\frac{4}{5}$$

Question How do we change a decimal to a percent?

ANSWER We use the following rule.

> To convert a decimal to a percent:
> 1. Move the decimal point two places to the right.
> 2. Attach a percent sign (%).

Problem   Write 0.53 as a percent.

ANSWER

$$0.53 = 0.53,$$

$$= 53\%$$

Problem   Write 0.176 as a percent.

ANSWER

$$0.176 = 0.17,6$$

$$= 17.6\%$$

Question   How do we change a percent to a decimal?

ANSWER   The rule is as follows.

> To convert a percent to a decimal:
>
> 1. Move the decimal point two places to the left, filling in zeros if needed.
> 2. Drop the percent sign.

Problem   Write 86% as a decimal.

ANSWER

$$86\% = {}_{,}86\%$$

$$= 0.86$$

Problem Write 5.5% as a decimal.

ANSWER

$$5.5\% = 05.5\%$$

$$= 0.055$$

Problem Write 1000% as a decimal.

ANSWER

$$1000\% = 10\,00\%$$

$$= 10$$

Question How do we change a fraction to a percent?

ANSWER We use the following rule.

> To convert a fraction to a percent:
> 1. Convert the fraction to a decimal.
> 2. Then convert the decimal to a percent.

We can think of the following diagram.

$$\text{Fractions} \longleftrightarrow \text{decimals} \longleftrightarrow \text{percents}$$

Decimals are in the middle, and we always travel through them to convert between fractions and percents.

Problem Write $\dfrac{4}{13}$ as a percent.

ANSWER

$$13 \overline{)\,4.0000}^{0.3076}$$

$$\rightarrow 0.308$$

$$= 30.8\%$$

Question   How do we convert a percent to a fraction?

ANSWER   Here is the rule we use.

> To convert a percent to a fraction:
> 1. Convert the percent to a decimal.
> 2. Then convert the decimal to a fraction.

Problem   Write 23.7% as a fraction.

ANSWER

| Convert to a decimal |

$$23.7\% = 0.237$$

$$\uparrow \quad \frac{1}{1000}$$

| Convert to a fraction |

$$= \frac{237}{1000}$$

PROBLEM SET 1-3-1   Write the following fractions as three-place decimals.

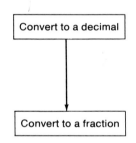

1. $\frac{1}{2}$ 2. $\frac{1}{3}$ 3. $\frac{1}{7}$

4. $\frac{3}{11}$ 5. $\frac{9}{13}$ 6. $\frac{37}{50}$

7. $\frac{24}{15}$ 8. $\frac{1}{200}$ 9. $\frac{22}{7}$

Write the following decimals as reduced fractions.

10. 0.4   11. 0.39   12. 1.6

13. 0.25   14. 0.170   15. 0.0002

16. 0.1   17. 0.0525   18. 0.22

Write the following decimals as percents.

19. 0.29   20. 0.85   21. 0.07

22. 0.308   23. 0.9944   24. 1.25

25. 0.0005   26. 3.7   27. 1.02

Write the following percents as decimals.

28. 32%   **29.** 91%   **30.** 2%

31. 6.25%   **32.** 0.2%   **33.** 4.1%

34. 300%   **35.** 0.001%   **36.** 71.4%

37–45. Write Problems 1–9 as percents.

46–54. Write Problems 28–36 as reduced fractions.

1–4 SIGNED NUMBERS

Question   What are signed numbers?

ANSWER   **Signed numbers** are numbers (whole, decimal, or fraction) that have a positive (+) or negative (−) sign in front of them. The number 0 has no sign in front of it. On the number line, the negative numbers are always to the left of 0; the positive numbers are always to the right. (A number with no sign is considered positive.)

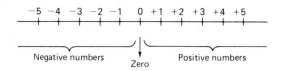

Question   What do the signed numbers mean?

ANSWER   Signed numbers are used when a quantity has two opposite possibilities. For example, money has two opposite possibilities: being in debt or having money. We say −$5 means being $5 in debt, and +$5 means having $5. Table 1-1 gives the most common uses of positive and negative numbers.

Table 1–1   Uses of Negative and Positive

Situation	Use of negative	Use of positive
Money	Debt (liabilities)	Having money (assets)
Weight	Losing	Gaining
Time	Past	Future
Space	Down	Up
	Backward	Forward
Temperature	Below 0°C	Above 0°C
Personal mood	Sad	Happy

Question   What is the absolute value of a signed number?

ANSWER   The **absolute value** is the distance of the number from 0 on the number line. It is easiest to remember as the signed number without its sign. The absolute value of a number x is written $|x|$.

Examples   1. $|+7| = 7$
2. $|-8| = 8$
3. $|0| = 0$
4. $|+12.1| = 12.1$
5. $\left|-\dfrac{13}{4}\right| = \dfrac{13}{4}$

Notice that the absolute value is always 0 or positive.

Question   How do we add signed numbers?

ANSWER   We use the following rule.

> To add signed numbers:
>
> 1. If the numbers have the *same sign*,
>   (a) Add the absolute values (ignore the signs; add as usual).
>   (b) Give the answer the common sign.
> 2. If the numbers have *different signs*,
>   (a) Subtract the absolute values (larger minus smaller).
>   (b) Give the answer the sign of the larger absolute value.

Examples   If the signs are the same, we add the absolute values.

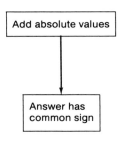

1. $(+5) + (+3) = +8$

2. $(+12) + (+14) = +26$

3. $(-4) + (-5) = -9$

4. $(-21) + (-13) = -34$

Examples   If the signs are different, we subtract the absolute values.

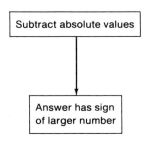

1. $(-5) + (+9) = +4$

2. $(-13) + (+6) = -7$

3. $(+20) + (-18) = +2$

4. $(+12) + (-21) = -9$

Problem   Add $(-8.1) + (-3.7)$.

ANSWER

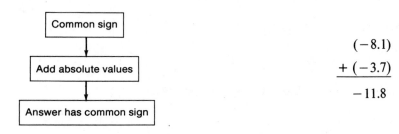

$$(-8.1)$$
$$+ (-3.7)$$
$$\overline{ -11.8}$$

Problem   Add $-\dfrac{7}{8} + \dfrac{5}{6}$.

ANSWER

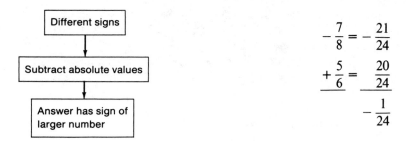

$$-\frac{7}{8} = -\frac{21}{24}$$
$$+\frac{5}{6} = \frac{20}{24}$$
$$\overline{\phantom{+\frac{5}{6}=} -\frac{1}{24}}$$

PROBLEM SET 1–4–1

1. Define or discuss:
   (a) Positive numbers.
   (b) Negative numbers.
   (c) Absolute value.

Give a signed number that is the same as the following expressions.

2. Win $7.20

3. Eight weeks ago

4. Drive 10 miles per hour in reverse

5. Three years in the future

6. Lose $15.50

7. One hundred feet above the ground

8. Two hundred fifty dollars in debt

9. Fourteen degrees below zero

10. Run 7 miles per hour forward

Add the following signed numbers.

11. $(+4)$
    $+ (+5)$

12. $(+9)$
    $+ (-3)$

13. $(+11)$
    $+ (-16)$

14. (-11.2)
    $+ (-183.6)$

15. (4.02)
    $+ (18.5)$

16. (-15.77)
    $+ (-4.3)$

17. $\left(+\frac{1}{2}\right)$
    $+ \left(-\frac{2}{3}\right)$

18. $\left(-\frac{5}{6}\right)$
    $+ \left(-\frac{3}{8}\right)$

19. $\left(+\frac{1}{12}\right)$
    $+ \left(-\frac{7}{10}\right)$

Use the rules for adding signed numbers to find the results of the following.

20. Win $5.20
    Win $7.25
    Lose $4.64
    Lose $0.78
    Win $3.25
    Lose $7.08
    Lose $4.54

21. 10.5 miles backward
    8.6 miles forward
    5.4 miles forward
    4.7 miles backward
    9.4 miles backward
    11.7 miles forward
    1.9 miles backward

22. Gain 6 yards
    Gain 10 yards
    Gain 4 yards
    Lose 5 yards
    Gain 2 yards
    Gain 14 yards
    Lose 3 yards
    Gain 4 yards

23. Lose 3 pounds
    Lose $1\frac{1}{2}$ pounds
    Gain $\frac{1}{2}$ pound
    Gain 1 pound
    Lose $\frac{3}{4}$ pound
    Lose 2 pounds
    Gain $\frac{1}{4}$ pound
    Lose $1\frac{1}{2}$ pounds

Subtraction

Question   How do we subtract signed numbers?

ANSWER   We use the following rule.

> To subtract signed numbers:
>
> 1. Change the sign of the second (or bottom) number.
> 2. Add, using the rules for signed numbers.

Formula 1–4–1

$$a - b = a + (-b)$$

Notice in Formula 1–4–1 that we make two changes. First, we change b to $-b$. Then we change subtraction to addition.

Problem   Subtract $14 - (-12)$.

ANSWER

$$\begin{array}{c}14 \\ \underline{-\,(-12)} \end{array} \quad = \quad \begin{array}{c} 14 \\ \underline{\oplus(\oplus 12)} \\ 26 \end{array}$$

Problem   Subtract $(-18.1) - (-2.5)$.

ANSWER

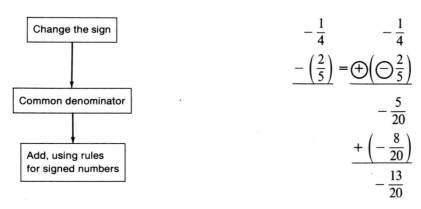

$$\begin{array}{c}-18.1 \\ \underline{-\,(-2.5)} \end{array} \quad = \quad \begin{array}{c} -18.1 \\ \underline{\oplus(\oplus 2.5)} \\ -15.6 \end{array}$$

Problem   Subtract $-\dfrac{1}{4} - \left(\dfrac{2}{5}\right)$.

ANSWER

Change the sign

Common denominator

Add, using rules for signed numbers

$$\begin{array}{c} -\dfrac{1}{4} \\[4pt] \underline{-\left(\dfrac{2}{5}\right)} \end{array} \quad = \quad \begin{array}{c} -\dfrac{1}{4} \\[4pt] \underline{\oplus\left(\ominus\dfrac{2}{5}\right)} \end{array}$$

$$\begin{array}{c} -\dfrac{5}{20} \\[6pt] \underline{+\left(-\dfrac{8}{20}\right)} \\[6pt] -\dfrac{13}{20} \end{array}$$

PROBLEM SET 1–4–2   Subtract the following signed numbers.

1. $\begin{array}{c} 5 \\ \underline{-\,(-2)} \end{array}$ **2.** $\begin{array}{c} -18 \\ \underline{-\,(12)} \end{array}$ **3.** $\begin{array}{c} -17 \\ \underline{-\,(-24)} \end{array}$

4. $\begin{array}{c} 1.17 \\ \underline{-\,(-8.4)} \end{array}$ **5.** $\begin{array}{c} -6.3 \\ \underline{-\,(-4.45)} \end{array}$ **6.** $\begin{array}{c} 20.9 \\ \underline{-\,35.7} \end{array}$

7. $\begin{array}{c} \frac{1}{10} \\ \underline{-\,\frac{3}{20}} \end{array}$ **8.** $\begin{array}{c} \frac{3}{2} \\ \underline{+\left(+\frac{2}{5}\right)} \end{array}$ **9.** $\begin{array}{c} -\frac{7}{4} \\ \underline{-\left(\frac{5}{6}\right)} \end{array}$

Multiplication and Division

Question How do we multiply and divide signed numbers?

ANSWER The rule is as follows.

> To multiply (or divide) signed numbers:
> 1. If the numbers have the same sign,
>   (a) Multiply (divide) the absolute values.
>   (b) Answer is always *positive*.
> 2. If the numbers have *different signs*,
>   (a) Multiply (divide) the absolute values.
>   (b) Answer is always *negative*.

Examples If the signs are the same, the answer is positive.

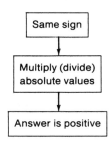

1. $(+4) \times (+5) = +20$

2. $(-4) \times (-9) = +36$

3. $(36) \div (6) = +6$

4. $(-44) \div (-4) = +11$

Examples If the signs are different, the answer is negative.

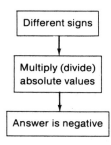

1. $(-7) \times (+3) = -21$

2. $(+10) \times (-2) = -20$

3. $(-50) \div (+5) = -10$

4. $(+42) \div (-6) = -7$

Examples Multiply $\left(-\dfrac{7}{9}\right) \times \left(\dfrac{3}{8}\right)$.

$$\frac{-7}{9} \times \frac{3}{8} = \frac{-7}{\underset{3}{\cancel{9}}} \times \frac{\overset{1}{\cancel{3}}}{8}$$

$$= \frac{-7}{24}$$

Question Why does positive × negative = negative, and negative × negative = positive?

ANSWER Without trying to prove it formally, let us look at an example. Suppose that we are filming a man playing cards. We have two quantities to consider: his chips (winning or losing) and time (forward and backward).

1. Suppose that the man is *losing* $4 an hour (-4); 5 hours *from now* $(+5)$, he will lose $20; $(-4) \times (+5) = -20$.

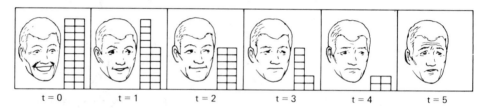

2. Suppose that the man is *winning* $4 an hour $(+4)$. If we run the movie *backward* for 5 hours (-5), he will appear to *lose* $20; $(+4) \times (-5) = -20$.

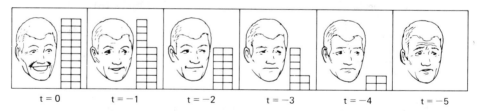

3. Suppose that the man is *losing* $4 an hour (-4), and we run the movie in *reverse* for 5 hours (-5). Now it will appear that he is *winning* $20; $(-4) \times (-5) = +20$.

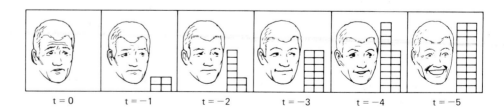

Problem Divide $-24.5 \div (-1.1)$.

ANSWER

| Same signs |
| Divide absolute values |
| Answer is positive |

$$\begin{array}{r} +\ 22.\ 272 \\ -1.1\overline{)-24.5\,000} \end{array}$$

We can round this to 22.27.

PROBLEM SET 1-4-3 Multiply the following signed numbers.

1. $\begin{array}{r}(-8)\\ \times\,(-4)\end{array}$
2. $\begin{array}{r}(+5)\\ \times\,(-6)\end{array}$
3. $\begin{array}{r}(-13)\\ \times\,(12)\end{array}$

4. $\begin{array}{r}(-2.2)\\ \times\,(8.5)\end{array}$
5. $\begin{array}{r}-21.2\\ \times\,(-1.65)\end{array}$
6. $\begin{array}{r}(-0.0071)\\ \times\,(8.3)\end{array}$

7. $\begin{array}{r}(-2\frac{1}{2})\\ \times\,(1\frac{1}{2})\end{array}$
8. $\begin{array}{r}-5\frac{1}{4}\\ \times\,(-\frac{1}{2})\end{array}$
9. $\begin{array}{r}-\frac{2}{3}\\ \times\,(-\frac{11}{12})\end{array}$

Divide the following signed numbers.

10. $(-24) \div (4)$
11. $(36) \div (-9)$
12. $-210 \div (-7)$
13. $(-100) \div 13$
14. $29.1 \div (-64.5)$
15. $-101.5 \div 34.7$
16. $-\frac{3}{5} \div (2\frac{1}{2})$
17. $(-5\frac{1}{4}) \div (-1\frac{1}{5})$

1-5 EXPONENTS AND SQUARE ROOTS

Question What is an exponent?

ANSWER When we first learn about multiplication, we are told that it is "repeated addition"; for example, $5 \times 7 = 7 + 7 + 7 + 7 + 7$. In the same way, we use exponents to do "repeated multiplication"; for example, $2^6 = 2 \cdot 2 \cdot 2 \cdot 2 \cdot 2 \cdot 2$.

The number that we wish to multiply is called the **base**. The number of times that the base appears as a factor is called the **exponent** or **power**. We write this as follows.

Formula 1–5–1

$$b^p = \underbrace{b \cdot b \cdot b \cdot \ldots \cdot b}_{p \text{ times}}$$

where b is the base, and p is the power (or exponent).

Examples
1. $2^5 = 2 \cdot 2 \cdot 2 \cdot 2 \cdot 2 = 32$
2. $5^2 = 5 \cdot 5 = 25$
3. $7^1 = 7$
4. $1^6 = 1 \cdot 1 \cdot 1 \cdot 1 \cdot 1 \cdot 1 = 1$
5. $8.1^4 = (8.1) \cdot (8.1) \cdot (8.1) \cdot (8.1) = 4304.6721$
6. $(0.75)^3 = (0.75) \cdot (0.75) \cdot (0.75) = 0.421875$

Notice from Examples 1 and 2 that $b^p \neq p^b$, usually.

Question How do we use exponents?

ANSWER Exponents are used to simplify many of the formulas that we will use later. Here are a few examples that we will see later in the book.

1. The area of a square is $A = s^2$. For instance, if the side s of a square is 8, the area is $8^2 = 64$.
2. The volume of a cube is $V = s^3$. For instance, if the side s of a cube is 5, the volume is $5^3 = 125$.
3. The value of a depreciating car might be given by the formula $V = P(0.75)^n$. For instance, suppose that the original price P is \$5500, and the car is $n = 6$ years old. Then its value is about $5500 (0.75)^6 = \$978.88$.
4. For money receiving compound interest in the bank, the formula is $P = P_0(1 + r)^n$. For instance, suppose that the original deposit P_0 is \$2000, the interest rate r is 6%, and the number of years n is 8. Then the total is $2000(1.06)^8 = \$3187.70$.

PROBLEM SET 1–5–1

1. Define or discuss:
   (a) Exponent (or power).
   (b) Base.

Write the following numbers in exponent form.

2. $6 \cdot 6 \cdot 6 \cdot 6$ 3. $8 \cdot 8 \cdot 8 \cdot 8 \cdot 8 \cdot 8 \cdot 8 \cdot 8 \cdot 8 \cdot 8$ 4. $10 \cdot 10 \cdot 10$

5. $(0.75) \cdot (0.75) \cdot (0.75) \cdot (0.75) \cdot (0.75)$ 6. $(1.07) \cdot (1.07) \cdot (1.07)$

Write the following as a single number.

7. 4^5 8. 2^6 9. $(1.065)^2$ 10. 1.1^3

11. $(0.8)^4$ 12. 3^4 13. 40^3 14. 10^8

15. The area of a certain square is 46^2. What is the area?

16. The volume of a certain cube is 31^3. What is the volume?

17. The volume of a certain sphere is $\frac{4}{3} \cdot (3.14) \cdot 8^3$. What is the volume?

18. The value of a car is $\$4400 \, (0.8)^5$. What is the value?

19. Judy's bank account has grown to $\$500(1.05)^4$. What is the bank balance?

20. Complete the following table.

10^1	10^2	10^3	10^4	10^5	10^6	10^7	10^8
10	100	1000					

Scientific Notation

Question What is scientific notation?

ANSWER **Scientific notation** is the way scientists write very large and very small numbers. The idea is to separate the "digits" from the power of 10. For instance, $20,000 = 2 \times 10^4$. By moving the decimal point, we can always write a number in the following form:

$$\boxed{\text{Any number}} = \boxed{\text{number between 1 and 10}} \times \boxed{\text{power of 10}}$$

Question What is the rule for writing a number in scientific notation?

ANSWER We have the following rule.

> To write a number in scientific notation:
>
> 1. If number is 10 or more,
>   (a) Move the decimal point p places to the left until the number is between 1 and 10.
>   (b) The power of 10 is 10^p.
>
> 2. If number is less than 1,
>   (a) Move the decimal point q places to the right until the number is between 1 and 10.
>   (b) The power of 10 is 10^{-q}.
>
> 3. If number is already between 1 and 10, leave it alone.

Question   What does 10^{-q} mean?

ANSWER   The term 10^{-q} means the fraction $\dfrac{1}{10^q}$. For example, $10^{-3} = \dfrac{1}{10^3} = \dfrac{1}{1000}$.

Examples

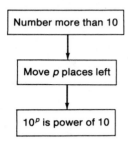

1. $5900 = 5\,900. = 5.9 \times 10^3$
2. $210,000 = 2\,10,000. = 2.1 \times 10^5$
3. $870 = 8\,70. = 8.7 \times 10^2$
4. $50,000,000 = 5\,0,000,000 = 5 \times 10^7$
5. $217 = 2\,17. = 2.17 \times 10^2$

Examples

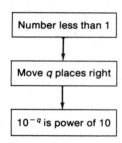

1. $0.04 = 0.04 = 4 \times 10^{-2}$
2. $0.017 = 0.01\,7 = 1.7 \times 10^{-2}$
3. $0.00063 = 0.0006\,3 = 6.3 \times 10^{-4}$
4. $0.159 = 0.1\,59 = 1.59 \times 10^{-1}$
5. $0.0077 = 0.007\,7 = 7.7 \times 10^{-3}$

Examples   1. In 1972, Richard Nixon spent $60,000,000 to get 47,000,000 votes. We can write this as 6×10^7 and 4.7×10^7 votes.
2. There are about 220,000,000 Americans. We can write this as 2.2×10^8.
3. In a recent year, Americans drove 1,040,000,000,000 miles. We can write this as 1.04×10^{12}.
4. The thin layers wrapped around nerve axons are about 0.00000009 meter thick. This can be written as 9×10^{-8}.
5. The chance of winning a car in a certain radio giveaway is about 0.0000033. This can be written as 3.3×10^{-6}.

This notation is also important because many electronic calculators and computers use it. The calculators and computers do not use exactly the same notation that we do. They have their own shorthand notation. For example,

Standard notation	625,000	0.00027
Scientific notation	6.25×10^5	2.7×10^{-4}
Computer (FORTRAN)	6.25E5	2.7E − 4
Hand calculator	6.25 + 05	2.7 − 04

Notice in the last three that the idea is always the same. Separate the "digits part" (6.25 or 2.7) from the "power-of-ten part" (10^5 or 10^{-4}). In the last two, only the power is given, but we know that it means 10 to that power.

PROBLEM SET 1–5–2

1. Define or discuss scientific notation.

Write the following numbers in scientific notation.

6.2×10^1

2. 71,500 3. 87,000,000 4. 5,100. 5. 62.

6. 0.007 7. 0.00015 8. 565,000 9. 200,000,000 2×10^8

10. 0.0000021 11. 1,400,000,000 12. 600 13. 0.00000725

7×10^{-3}

Write the following calculator numbers in both scientific notation and standard notation.

14. 6.18 + 07 15. 7.05 + 02 16. 6.0 − 05

17. 8.1 + 12 18. 9.8 − 10 19. 3.34175 − 07

Write the numbers in the following statements in scientific notation.

20. The sun is 93,000,000 miles from the earth.

21. India has an area of 3,287,590 square kilometers.

22. The population of the world is about 4,400,000,000.

23. The speed of light is 30,000,000,000 centimeters per second.

24. The weight of the earth is 6,586,000,000,000,000,000,000 tons.

25. The earth is believed to be 4.7 billion years old.

26. In 1941, Ted Williams batted 0.406.

27. In 1973, 105,944,521,000 gallons of gasoline were used in American cars.

28. In 1970, 26.9 people out of every 100,000 died in an automobile accident.

29. In 1970, there were 0.0000000007 deaths for every mile of railroad traffic.

30. The chance of winning a certain lottery is 0.00000004.

31. The wavelength of green light is 0.0000005 meters.

32. A modern computer can do an addition problem in about 0.0000001 seconds.

Square Root

Question What is $\sqrt{25}$, and what does it mean?

ANSWER $\sqrt{25}$ is called the **square root** of 25. We are asking what number times itself is 25. The answer is 5. We write this as $\sqrt{25} = 5$.

Even though $(-5)(-5) = 25$, we do *not* say $-5 = \sqrt{25}$. This is because we want to avoid any possible confusion. Some people might say $\sqrt{25} = +5$; others might say $\sqrt{25} = -5$. To avoid this, we say $\sqrt{25}$ is only $+5$.

Question What is $\sqrt{-25}$?

ANSWER There is no such real number. Recall that (negative) × (negative) = positive, and (positive) × (positive) = positive. So it is impossible to multiply a real number by itself and get a negative number. We can only take the square root of zero and positive numbers.

In some cases, it is easy to find the square roots. For example,

$$\sqrt{0} = 0 \qquad \sqrt{1} = 1$$

$$\sqrt{4} = 2 \qquad \sqrt{9} = 3$$

$$\sqrt{16} = 4 \qquad \sqrt{25} = 5$$

$$\sqrt{36} = 6 \qquad \sqrt{49} = 7$$

$$\sqrt{64} = 8 \qquad \sqrt{81} = 9$$

These numbers, 1, 4, 9, 16, 25, and so on, are called **perfect squares**. As we see, their square roots are easy. Unfortunately, it is not so easy to find the square roots of other numbers.

Question How do we find the square root of any number, for example $\sqrt{381}$?

ANSWER We will give two of the easier methods (without a calculator).

1. *Trial and error*. Here, we guess different numbers and square them to see how close we are. We then make adjustments, up or down, and square again.

   For example, take $\sqrt{381}$. We will try to find a number whose square is close to 381. Let us start by guessing 17.

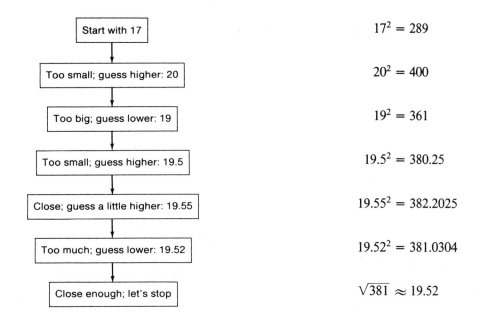

$17^2 = 289$

$20^2 = 400$

$19^2 = 361$

$19.5^2 = 380.25$

$19.55^2 = 382.2025$

$19.52^2 = 381.0304$

$\sqrt{381} \approx 19.52$

We could continue this on and on to more accurate answers. But at this point, we are satisfied, and we say that $\sqrt{381} \approx 19.52$. ($\approx$ means "approximately equal to.") Notice how we kept closing in on 381. If our guess was too high, we guessed lower the next time, and vice versa.

2. *Square-root formula.* Sometimes we can use a simple formula. To use the formula, we must first find a perfect square (1, 4, 9, 16) close to the number. We can call this number A^2, since it is the square of some number A.

Formula 1–5–2

$$\sqrt{N} \approx \frac{A^2 + N}{2A}$$

where A^2 is close to N. For example, 381 is close to 400, a perfect square equal to 20^2. We let $A = 20$ and $A^2 = 400$.

$$\sqrt{N} \approx \frac{A^2 + N}{2A}$$

$$= \frac{400 + 381}{2(20)}$$

$$= \frac{781}{40} = 19.525$$

This is very close to the trial-and-error answer.

Problem   Find $\sqrt{18}$ using both methods.

ANSWER   1. *Trial and error*. We start with 4.

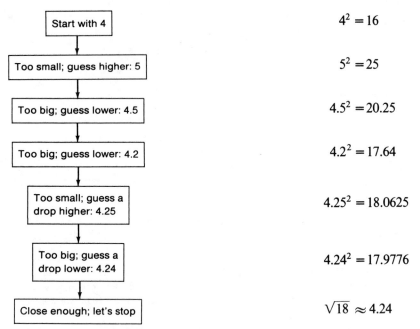

Start with 4	$4^2 = 16$
Too small; guess higher: 5	$5^2 = 25$
Too big; guess lower: 4.5	$4.5^2 = 20.25$
Too big; guess lower: 4.2	$4.2^2 = 17.64$
Too small; guess a drop higher: 4.25	$4.25^2 = 18.0625$
Too big; guess a drop lower: 4.24	$4.24^2 = 17.9776$
Close enough; let's stop	$\sqrt{18} \approx 4.24$

We say that $\sqrt{18} \approx 4.24$.

Notice how we adjusted our next guess. If we were close but high, we guessed a drop lower. If we were much too low, we guessed much higher.

2. *Formula 1–5–2*. 18 is closest to the perfect square $16 = 4^2$. So $A = 4$ and $A^2 = 16$.

Formula 1–5–2 $\sqrt{N} \approx \dfrac{A^2 + N}{2A}$

Substitute $= \dfrac{16 + 18}{2(4)}$

Simplify $= \dfrac{34}{8} = 4.25$

So $\sqrt{18} \approx 4.25$.

PROBLEM   **1.** Define or discuss:
SET 1–5–3   (a)   Square root of a number.
                     (b)   Perfect square.

Find the following square roots using both the trial-and-error and formula method.

2. $\sqrt{10}$ **3.** $\sqrt{29}$ **4.** $\sqrt{99}$ **5.** $\sqrt{1900}$

6. $\sqrt{87}$ **7.** $\sqrt{12.52}$ **8.** $\sqrt{5000}$ **9.** $\sqrt{2}$

10. $\sqrt{75,000}$ **11.** $\sqrt{43.5}$ **12.** $\sqrt{1.1}$ **13.** $\sqrt{60}$

1–6 HAND CALCULATORS

Question Should I buy a hand calculator?

ANSWER Why not? They're cheap, fast, and very accurate. But they are not a substitute for knowing the basic skills. After all, even though you drive a car, you still must know how to walk.

Question Which calculator should I buy?

ANSWER We will not recommend any brand but will suggest certain features to look for. These features are on many calculators selling for under $10.

    1. *Eight-digit display.* You should get an eight-digit calculator. To check, punch the digits 1, 2, 3, 4, 5, 6, 7, 8. All 8 should appear on the display.
    2. *Usual arithmetic.* Your calculator should do simple arithmetic in the usual order. Before you buy any calculator, try this simple problem: 2 + 3 = 5.
    Punch $\boxed{2}$; punch $\boxed{+}$; punch $\boxed{3}$; punch $\boxed{=}$. If you do not get 5, put the calculator back. (This may sound silly, but some calculators do arithmetic very strangely.)
    3. *Memory.* A memory is necessary for doing long calculations. Look for keys marked $\boxed{M+}$, $\boxed{M-}$, $\boxed{MR}$, and $\boxed{MC}$. If the calculator calls the keys something else, be sure it has a memory. Some use the keys $\boxed{STO}$ and $\boxed{RCL}$.
    4. *Square root.* This is suggested. It is only a dollar or two more. It will be worth your while to get a square root key, $\boxed{\sqrt{}}$, which is very handy.
    5. *Other options.* There are many other features, such as logarithm functions, trigonometric functions, scientific notation, and reciprocals. You can get most of these features in addition to the necessary ones for under $20. I honestly don't think you will need them, but they certainly do make the calculator look more impressive.

Question How do we use the hand calculator? What do the keys mean?

ANSWER The sketch on the following page shows a typical $10 calculator with the features mentioned. We will explain the meaning and use of these keys as we go along.

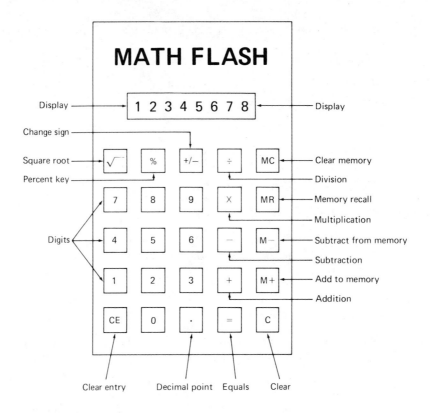

Simple Arithmetic

Question   Where do we start?

ANSWER   We will start with very simple problems just to get the feel of the machine. The keys we will use are the following:

1. $\boxed{\text{C}}$ This key clears the entire machine and sets the display at 0. Always push this before starting a problem.

2. $\boxed{\text{CE}}$ This key will erase or clear a mistaken entry but not the other numbers in a problem.

3. $\boxed{1}$, $\boxed{2}$, $\boxed{3}$, $\boxed{4}$, $\boxed{5}$, $\boxed{6}$, $\boxed{7}$, $\boxed{8}$, $\boxed{9}$, $\boxed{0}$. These are the usual digits used in our numbers.

4. $\boxed{\cdot}$ This is the decimal point.

5. $\boxed{\div}$, $\boxed{\times}$, $\boxed{-}$, $\boxed{+}$ These are the usual operations of arithmetic: division, multiplication, subtraction, and addition.

6. $\boxed{=}$ The equal sign is usually used to get our final answer.

7. $\boxed{+/-}$ This key changes the sign of the number from + to −, or from − to +. This key varies from machine to machine. On other machines, it may be $\boxed{\text{CS}}$, $\boxed{+/-}$, and so on.

When we go through some sample problems, we will have three columns:

1. PUNCH: These are the keys you actually punch.
2. DISPLAY: This will show what is being displayed after the last punch. This will help you to check to see that you are on the right track.
3. MEANING: In this column, we will comment on what we are doing.

A note of caution: different calculators sometimes get slightly different answers after a long calculation.

Problem Display 14.702.

ANSWER We punch the keys just as the numbers appear.

PUNCH	DISPLAY	MEANING
C	0.	Clear
1 4 · 7 0 2	14.702	This is 14.702

Problem Add 8 + 4.

ANSWER This is a simple warm-up problem.

PUNCH	DISPLAY	MEANING
C	0.	Clear
8	8.	8
+ 4	4.	Plus 4
=	12.	Answer

Notice that we punch this problem in the usual order 8 + 4. We then punch = to get our answer.

Problem Subtract 61.69 − 9.053.

ANSWER We punch this just as it reads.

PUNCH	DISPLAY	MEANING
C	0.	Clear
6 1 · 6 9	61.69	61.69
− 9 · 0 5 3	9.053	Minus 9.053
=	52.637	Answer

Problem Use the hand calculator to balance the following checkbook.

Mar. 1	Balance from February	$219.20
Mar. 6	Frank's Foods	−21.79
Mar. 8	The Jeans Shop	−31.85
Mar. 10	The Spill Oil Co.	−15.16
Mar. 11	Cash	−30.00
Mar. 15	Deposit	+215.72
Mar. 20	Jo's Records	−11.34
Mar. 25	Frank's Foods	−19.48
Mar. 29	Rent	−115.00
Apr. 1	Balance from March	?

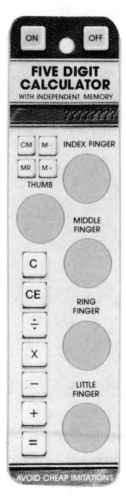

(© 1977 Antioch Bookplate Company)

ANSWER   We enter the balance first. Then we subtract all checks and add all deposits.

PUNCH	DISPLAY	MEANING
C	0.	Clear
2 1 9 · 2 0	219.20	Enter balance
− 2 1 · 7 9	21.79	
− 3 1 · 8 5	31.85	
− 1 5 · 1 6	15.16	
− 3 0 · 0 0	30.00	Subtract checks,
+ 2 1 5 · 7 2	215.72	add deposits
− 1 1 · 3 4	11.34	
− 1 9 · 4 8	19.48	
− 1 1 5 · 0 0	115.00	
=	190.30	Balance (answer)

The April 1 balance is $190.30.

Problem   Multiply $(7.02) \cdot (218.5) \cdot (0.057) \cdot (-6.8)$.

ANSWER   We punch these just as they appear. We must punch $\boxed{+/-}$ to get our minus sign in − 6.8.

PUNCH	DISPLAY	MEANING
C	0.	Clear
7 · 0 2	7.02	
× 2 1 8 · 5	218.5	
× · 0 5 7	0.057	Multiply numbers
× +/− 6 · 8	−6.8	
=	−594.528	Answer

Problem   Divide $14.3 \div 2.72$.

ANSWER   We punch the keys just as they appear.

PUNCH	DISPLAY	MEANING
C	0.	Clear
1 4 · 3	14.3	14.3
+ 2 · 7 2	2.72	Divided by 2.72
=	5.2573529	Answer

Problem Simplify $\dfrac{(14.7)(253.04)}{(0.42)(82.9)(1.09)}$.

ANSWER We could multiply out the numerator, then multiply out the denominator, and then divide the two. But there is an easier way. Watch carefully.

PUNCH	DISPLAY	MEANING
$\boxed{\text{C}}$	$\boxed{0.}$	Clear
$\boxed{1}\ \boxed{4}\ \boxed{.}\ \boxed{7}$	$\boxed{14.7}$ $\Big\}$	Multiply each
$\boxed{\times}\ \boxed{2}\ \boxed{5}\ \boxed{3}\ \boxed{.}\ \boxed{0}\ \boxed{4}$	$\boxed{253.04}$	numerator term
$\boxed{\div}\ \boxed{.}\ \boxed{4}\ \boxed{2}$	$\boxed{0.42}$ $\Big\}$	Divide by each
$\boxed{\div}\ \boxed{8}\ \boxed{2}\ \boxed{.}\ \boxed{9}$	$\boxed{82.9}$	denominator term
$\boxed{\div}\ \boxed{1}\ \boxed{.}\ \boxed{0}\ \boxed{9}$	$\boxed{1.09}$	
$\boxed{=}$	$\boxed{98.011302}$	Answer

> To simplify (using a calculator) a fraction containing *only multiplications* (no addition or subtraction):
>
> 1. Multiply all terms in the numerator.
> 2. Divide by each term in the denominator.

To put this in a formula,

Formula 1-6-1

$$\frac{A \cdot B \cdot C}{D \cdot E \cdot F} = A \cdot B \cdot C \div D \div E \div F$$

where the operations on the right side are carried out from left to right in that order.

Problem Simplify $\dfrac{(81.53)(17.2)(0.04)}{(106.9)(0.0029)}$.

ANSWER Since there are only multiplications in this fraction, we multiply the numerator terms and divide by each denominator term.

PUNCH	DISPLAY	MEANING
C	0.	Clear
8 1 · 5 3	81.53	Multiply numerator
× 1 7 · 2	17.2	terms
× · 0 4	0.04	
÷ 1 0 6 · 9	106.9	Divide by each
÷ · 0 0 2 9	0.0029	denominator term
=	180.93813	Answer

We will get to mixed operations of addition and subtraction with multiplication and division later.

Question Suppose that we hit the wrong key by mistake halfway through a problem.

ANSWER This is where CE or the clear entry key is used. For example, add 14.8 + 29.3 + 43.7.

PUNCH	DISPLAY	MEANING
C	0.	Clear
1 4 . 8	14.8	
+ 2 9 . 3	29.3	
+ 4 2 . 7	42.7	Mistake!
CE	0.	Erase 42.7
4 3 . 7	43.7	Enter 43.7
=	87.8	Answer

Notice that we did not have to start over. After a mistake, we punch CE to erase the mistake; then we punch the correct number and continue.

In the following problem set, you will be given practice problems to get you comfortable with the calculator. Later there will be many applications and uses of the calculator.

PROBLEM SET 1–6–1 Use the calculator to do the following. (Do the easy ones, too, just to get the feel of the machine and the keys.)

1. 2 + 3

2. 4 + 7

3. 49 + 17.8

4.
```
   5,123
   4,913
     682
     483
+ 51,089
```

5.
```
   411.2
   60.73
  193.18
 254.333
+698.714
```

6.
```
  0.003
 0.0216
 0.5103
 0.8034
+0.0076
```

7. 6 − 2

8. 12 − 5

9. 27.8 − 9.9

10.
```
  41,782
− 19,814
```

11.
```
  106.73
− 695.88
```

12.
```
 14.007
−  9.93
```

13.
```
  518.27
+ 125.75
−  14.46
−  25.88
−  53.92
− 176.84
+   2.50
```

14.
```
  291.04
− 105.96
−  17.81
−  25.23
+ 150.00
−  39.95
−  87.91
```

15.
```
 1717.77
− 851.42
− 119.54
− 202.59
−  59.43
+ 210.00
+  57.82
```

16. 2 × 4

17. 7 × 12

18. 27 × 73

19.   41.07
   $\times 683.9$

20.   1754.2
   $\times 0.0039$

21.   0.00751
   $\times 0.00385$

22. $12 \div 6$

23. $45 \div 9$

24. $2 \div 7$

25. $14.5 \div 2.86$

26. $14{,}592 \div 19.8$

27. $0.0074 \div 0.00271$

28. $\dfrac{(14.7)(19.2)}{(27.8)(0.53)}$

29. $\dfrac{(175)(20.8)(-64)}{4.93}$

30. $\dfrac{2.98}{(157)(.0079)(0.52)}$

The Memory

Question   How do we use the memory?

ANSWER   The memory keys allow you to do many complex problems. Here is a brief description of the memory and the memory keys.

1. *The memory.* Your calculator should have a memory. (Pay the extra $2 and get one.) Have you ever been working a problem and said to yourself (or wrote down on paper), "Remember the number 6.2. I'll need it later." This is just what the memory does: It remembers one number while you're working with some other numbers. Think of the memory as if it were a piece of paper with a number written on it for later use.

2. $\boxed{\text{MC}}$ This is the *memory clear* key. This key says to the calculator, "Erase (or forget) whatever you were remembering. Start over again at 0." We punch this whenever we start a problem just in case something is left over in the memory from another problem.

3. $\boxed{\text{M+}}$ This key will *add* to the memory. Suppose that the calculator is remembering the number 7. If we punch $\boxed{4}$ and then $\boxed{\text{M+}}$, we are telling the calculator, "Add 4 to the memory (7), and now remember the sum (11)." When the memory is blank or zero, punching $\boxed{\text{M+}}$ enters a new number in the memory. For example, if we punch $\boxed{8}$ and then $\boxed{\text{M+}}$, this stores the number 8 in the memory for later use.

4. $\boxed{\text{M}-}$ This key will *subtract* from the memory. Suppose that 15 is in the memory. If we punch $\boxed{7}$ and then $\boxed{\text{M}-}$, we are telling the calculator, "Subtract 7 from the memory (15) and now remember the difference (8)."

5. $\boxed{\text{MR}}$ This is the *memory recall.* By punching this key, we are asking the calculator, "What is in the memory now?"

We use the memory for complex problems where we have additions, subtractions, multiplications, and divisions.

Problem   Simplify $7 \cdot 4 + 8 \cdot 2$.

ANSWER   This is so easy we should do it first by hand so that we know what answer to expect. Recall that we do all multiplications first.

$$7 \cdot 4 + 8 \cdot 2 = 28 + 16 = 44$$

Suppose that we tried to do this on a calculator in the same order.

PUNCH	DISPLAY	MEANING
C	0.	
7	7.	
×	7.	
4	4.	Work in same order
+	28.	
8	8.	
×	36.	
2	2.	
=	72.	Wrong answer!

What happened? We got 72 instead of 44. The problem is that we *cannot* do addition and multiplication together in the order that they appear, because multiplication has priority over addition. (Recall that we first multiplied 7 · 4 and 8 · 2 to get 28 and 16; then we added.)

Here is where we use our memory. We do a multiplication, add it to the memory, do another multiplication, add it to the memory, and so on. At the end, we use [MR] to get the answer.

PUNCH	DISPLAY	MEANING
C MC	0.	Clear everything
7 × 4 =	28.	Do 7 · 4
M+	28.	Put in memory
8 × 2 =	16.	Do 8 · 2
M+	16.	Add to memory
MR	44.	Recall answer

Notice that we did each multiplication separately. We did 7 · 4 = 28, then added it to memory with [M+]; we did 8 · 2 = 16, then added it to the memory with [M+]. Finally, we get the answer out of the memory by punching [MR].

Problem Sam is paneling a basement room in his house; he is buying the following materials. Use the calculator to compute his cost.

Item	Quantity	Cost per unit ($)
Ten-penny nails	3 lb	0.59 per lb
Finishing nails	2 lb	1.19 per lb
2-in. × 4-in. × 8-ft studs	55	1.09 per stud
4-ft by 8-ft panels	11	9.50 per panel
Floor molding	50 ft	0.14 per ft
Corner molding	32	0.17 per ft

ANSWER If we were working this with pencil and paper, we would multiply each quantity by its cost. We would write down each of these and then add up the total.

   With our calculator, we multiply each quantity by its cost, then punch [M+] to add it to the memory. Finally, we punch [MR] to get the total.

PUNCH	DISPLAY	MEANING
[C] [MC]	0.	Clear everything
[3] [×] [.] [5] [9]	0.59	
[=]	1.77	$1.77 for nails
[M+]	1.77	Put in memory
[2] [×] [1] [.] [1] [9]	1.19	
[=]	2.38	$2.38 for nails
[M+]	2.38	Add to memory
[5] [5] [×] [1] [.] [0] [9]	1.09	
[=]	59.95	$59.95 for studs
[M+]	59.95	Add to memory
[1] [1] [×] [9] [.] [5] [0]	9.50	
[=]	104.5	$104.50 for panels
[M+]	104.5	Add to memory
[5] [0] [×] [.] [1] [4]	0.14	
[=]	7.	$7.00 for floor molding
[M+]	7.	Add to memory
[3] [2] [×] [.] [1] [7]	0.17	
[=]	5.44	$5.44 for corner molding
[M+]	5.44	Add to memory
[MR]	181.04	Recall answer

Thus the whole job is $181.04.

   Notice that we blocked the steps together. Multiply quantity by cost, punch [=], and add to memory with [M+]. We repeated the procedure with each item. Finally, we use [MR] to get the answer out.

Problem Add $\frac{2}{7} + \frac{5}{6}$ with the calculator.

ANSWER Let us first do this the usual way by finding a common denominator.

$$\frac{2}{7} + \frac{5}{6} = \frac{12}{42} + \frac{35}{42} = \frac{47}{42}$$

With the calculator, we *cannot* write fractions, only decimals. We cannot enter $\frac{2}{7}$ as such. We can enter [2] [÷] [7] and let the calculator change this to a decimal. Here, too, we enter each decimal and add it to the memory.

PUNCH	DISPLAY	MEANING
[C] [MC]	0.	Clear
[2] [÷] [7] [=]	0.2851742	Enter $\frac{2}{7}$ as a decimal
[M+]	0.2851742	Put into memory
[5] [÷] [6] [=]	0.8333333	Enter $\frac{5}{6}$ as a decimal
[M+]	0.8333333	Add to memory
[MR]	1.1190475	Recall answer

Thus the answer is 1.1190475 when written as a decimal.

Problem Simplify $\dfrac{(1501)}{(70.1)(2.8)} - \dfrac{(429)(8.4)}{(0.54)(294.7)}$.

ANSWER This expression has two simpler terms. We simplify the first term using the rule from the last section, and put it in the memory with [M+]. Then we simplify the second the same way. Since this is a subtraction problem, we punch [M−] to subtract the second term from the memory.

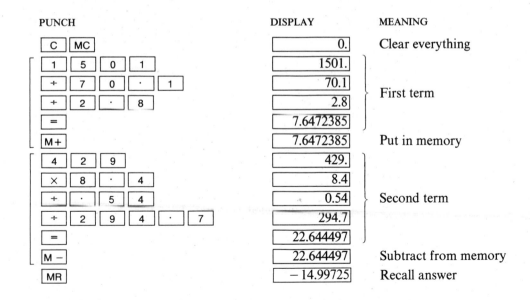

PUNCH	DISPLAY	MEANING
[C] [MC]	0.	Clear everything
[1] [5] [0] [1]	1501.	
[÷] [7] [0] [.] [1]	70.1	
[÷] [2] [.] [8]	2.8	First term
[=]	7.6472385	
[M+]	7.6472385	Put in memory
[4] [2] [9]	429.	
[×] [8] [.] [4]	8.4	
[÷] [.] [5] [4]	0.54	Second term
[÷] [2] [9] [4] [.] [7]	294.7	
[=]	22.644497	
[M−]	22.644497	Subtract from memory
[MR]	− 14.99725	Recall answer

Simplify the following expressions into a single number.

1. $2 \cdot 9 + 3 \cdot 8$ **2.** $4 \cdot 10 + 5 \cdot 9 + 12 \cdot 9$

3. $(2.1)(6.2) + (8.9)(0.6)$ **4.** $(0.51)(7.9)(12.5) + (16.8)(6.3)$

5. $(1.9)(3.9)(7.5) + (10.4)(3.3) - (8.6)(21.3) + (14.2)(3.06)$

6. $\frac{1}{2} + \frac{1}{3} + \frac{1}{4}$ **7.** $\frac{2}{7} + \frac{5}{9} + \frac{8}{15} + \frac{10}{17}$

8. $\frac{1}{4} - \frac{1}{7}$ **9.** $\frac{7}{6} - \frac{8}{7} - \frac{9}{8}$ **10.** $\frac{1}{20} + \frac{1}{30} - \frac{1}{40}$

11. $\frac{2}{11} - \frac{3}{14} + \frac{8}{19} - \frac{51}{101}$ **12.** $\frac{16}{17} + \frac{25}{12} - \frac{18}{13}$

13. $\frac{(1.7)(8.9)}{18.3} + \frac{14.2}{29.3}$ **14.** $\frac{(16.2)(21.7)}{38.67} - \frac{108}{(2.6)(5.3)}$

15. $\frac{66.1}{18.2} - \frac{402}{(1.6)(23)} + \frac{1178}{29.1}$ **16.** $\frac{(102)(0.9)}{57.3} - \frac{18.3}{6.5} - \frac{257}{(38.1)(0.53)}$

17. Find the total cost of the following Little League team.

Item	Quantity	Cost per unit ($)
Baseballs	20	1.59
Bats	8	3.49
Bases	4	4.50
Shoes (baseball)	25 pairs	12.95
Uniforms	25	19.95

(Ann Chwatsky/Editorial Photocolor Archives)

18. Find the total cost of the following party.

Item	Quantity	Cost per unit ($)
Paper plates	2 dozen	1.29 per dozen
Paper cups	2 dozen	1.19 per dozen
Potato chips	4 bags	0.79 per bag
Fritos	3 bags	0.69 per bag
Dip	3 tubs	0.89 per tub
Beer	10 six-packs	1.59 per six-pack
Coke	4 six-packs	1.49 per six-pack
Cole slaw	6 pounds	0.85 per pound
Potato salad	6 pounds	0.79 per pound

19. Find the total cost of the following gift.

Item	Quantity	Cost per unit ($)
Drummers drumming	12	30 per drummer
Pipers piping	11	25 per piper
Lords a-leaping	10	50 per lord
Ladies dancing	9	15 per lady
Maids a-milking	8	4 per maid
Swans a-swimming	7	25 per swan
Geese a-laying	6	10 per goose
Golden rings	5	100 per ring
Calling birds	4	20 per bird
French hens	3	8 per hen
Turtle doves	2	9 per bird
Partridge	1	15 per partridge
Pear tree	1	50 per tree

Exponents

Question   Can we do exponents on a calculator?

ANSWER   Yes, very quickly.

Problem   Find 2^6.

ANSWER   We could do this as $2 \times 2 \times 2 \times 2 \times 2 \times 2$. But this is the slow way, especially if the problem were like $(2.0152)^{10}$. A simpler way is to use the $\boxed{\times}$ and $\boxed{=}$ keys together. Watch carefully.

PUNCH	DISPLAY	MEANING
C	0.	Clear
2	2.	Enter base, 2
×	2.	Start multiplying
=	4.	2^2
=	8.	2^3
=	16.	2^4
=	32.	2^5
=	64.	2^6 (answer)

Notice that we punched ⎡2⎤, then ⎡×⎤, and then ⎡=⎤ 5 times.

> To find an exponent, B^n:
>
> 1. Clear calculator.
> 2. Enter base, B.
> 3. Punch ⎡×⎤.
> 4. Punch ⎡=⎤ $(n-1)$ times.

Problem Find $(2.79)^8$.

ANSWER This is a problem that we certainly would not want to do by hand. But it is easy with a calculator. We enter the base, 2.79. We punch ⎡×⎤. Then we punch ⎡=⎤ 7 times (remember, $n-1$ times for ⎡=⎤).

PUNCH	DISPLAY	MEANING
C	0.	Clear
2 · 7 9	2.79	Enter base, 2.79
×	2.79	Start multiplying
=	7.7841	2.79^2
=	21.717639	2.79^3
=	60.592212	2.79^4
=	169.05227	2.79^5
=	471.65583	2.79^6
=	1315.9197	2.79^7
=	3671.4159	2.79^8 (answer)

Problem Find $700 \cdot (1.06)^{20}$.

ANSWER This problem might come up in a bank interest situation: $700 at 6% interest for 20 years. We first compute $(1.06)^{20}$; then multiply by 700.

PUNCH	DISPLAY	MEANING
C	0.	Clear
1 · 0 6	1.06	Enter base, 1.06
×	1.06	Start multiplying
=	1.1236	$(1.06)^2$
=	1.191016	$(1.06)^3$
$\vdots$ } 19 times	$\vdots$	$\vdots$
=	3.0255983	$(1.06)^{19}$
=	3.2071341	$(1.06)^{20}$
× 7 0 0	700.	Times 700
=	2244.9938	Answer

Problem Find $(4500)(0.75)^6$.

ANSWER This problem might represent the value of a $4500 car after 6 years of 25% depreciation.

PUNCH	DISPLAY	MEANING
C	0.	Clear
· 7 5	0.75	Enter base, 0.75
×	0.75	Start multiplying
=	0.5625	$(0.75)^2$
=	0.421875	$(0.75)^3$
=	0.3164062	$(0.75)^4$
=	0.2373046	$(0.75)^5$
=	0.1779784	$(0.75)^6$
× 4 6 0 0	4500.	Times 4500
=	800.9028	Answer

Some expensive calculators do exponents differently. They have a $\boxed{y^x}$ key. This works as follows.

Problem   Find $(1.07)^{25}$.

ANSWER

PUNCH	DISPLAY	MEANING
[C]	0.	Clear
[1] [·] [0] [7]	1.07	Enter base, 1.07
[y^x] [2] [5]	25.	Raise to 25th power
[=]	5.4274326	Answer

Remember, this [y^x] key is very useful, but it is not on most cheap calculators. It is probably not worth the extra money to most students.

Question   Can we do square roots on our calculators?

ANSWER   Yes. If your calculator has a [$\sqrt{}$] key this is the easiest way. If not, we have Formula 1–5–2.

Problem   Find $\sqrt{51.2}$.

ANSWER   First, if our calculator has a [$\sqrt{}$] key, the solution is simple.

PUNCH	DISPLAY	MEANING
[C]	0.	Clear
[5] [1] [·] [2]	51.2	Enter 51.2
[$\sqrt{}$]	7.1554175	Take square root

If we do not have a [$\sqrt{}$] key, we use Formula 1–5–2, $\sqrt{N} \approx (A^2 + N)/2A$. Here 51.2 is closest to $49 = 7^2$. So $A = 7$.

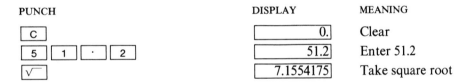

PUNCH	DISPLAY	MEANING
[C]	0.	Clear
[7] [×] [7]	7.	A^2
[+] [5] [1] [·] [2]	51.2	Plus N
[+] [2]	2.	} Divide by 2A
[+] [7]	7.	
[=]	7.1571425	Answer

This answer is pretty close to the real answer, so we will stop. Sometimes we can take this answer and use it for A. Then we redo the whole procedure.

Simplify the following with a hand calculator.

1. 5^2 2. 4^3 3. 10^4

4. 5.8^4 5. 16.9^5 6. $(0.62)^{10}$

7. $(1.065)^{15}$ 8. $(1.07)^6$ 9. $(1.5)^{12}$

10. $5000(0.85)^5$ 11. $3500(1.08)^7$ 12. $1600(0.5)^{10}$

13. $650(1.065)^{12}$ 14. $250(2.5)^4$ 15. $15,000(0.95)^{20}$

16. $\sqrt{2}$ 17. $\sqrt{57}$ 18. $\sqrt{91.62}$

19. $\sqrt{1784}$ 20. $\sqrt{526}$ 21. $\sqrt{0.0781}$

22. $\sqrt{0.00071}$ 23. $\sqrt{1.71}$ 24. $\sqrt{1,818,192}$

IMPORTANT WORDS

Absolute value (1–4)   Negative number (1–1)
Base (1–5)   Perfect square (1–5)
Exponent (1–5)   Positive number (1–4)
Improper fraction (1–1)   Prime (1–1)
Least common denominator (LCD) (1–1)   Scientific notation (1–5)
Mixed number (1–1)   Square root (1–5)

REVIEW EXERCISES

Perform the indicated operations on the following fractions, and reduce all answers.

1. $\frac{1}{12} + \frac{2}{15} + \frac{3}{8} =$ 2. $2\frac{7}{10} + 5\frac{8}{15} =$

3. $\frac{3}{10} - \frac{1}{18} =$ 4. $12\frac{3}{16} - 4\frac{7}{8} =$

5. $\frac{2}{5} \cdot \frac{15}{16} \cdot \frac{2}{7} =$ 6. $5\frac{1}{5} \cdot 1\frac{7}{13} =$

7. $\frac{7}{20} \div \frac{13}{30} =$ 8. $3\frac{3}{5} \div 2\frac{7}{10} =$

9. Write $7\frac{1}{4}$ as an improper fraction.

10. Write $\frac{62}{7}$ as a mixed number.

Perform the indicated operations on the following decimals.

11. 124.09
     82.126
   $+$ 49.8

12.   60.05
    $- 42.963$

13.   4.05
    $\times$ 6.3

14. $0.37 \overline{)\, 2.8157}$

15. Add $0.63 + 5.2 + 8.346$.

16. Subtract $100 - 64.82$.

17. Multiply 0.002 by 0.00073.

18. Divide 0.0482 by 0.61 and round answer to two decimal places.

19. Round 17.38142 to three decimal places.

20. Write $\dfrac{9}{14}$ as a three-place decimal.

21. Write 0.45 as a reduced fraction.

22. Write 0.316 as a percent.

23. Write 0.1% as a decimal.

24. Write $\dfrac{3}{11}$ as a percent.

25. Write 44% as a reduced fraction.

Perform the indicated operations on the following signed numbers.

26. $(+11) + (-18) =$

27. $(-4.2) + (-6.93) =$

28. $\left(\dfrac{1}{2}\right) - \left(-\dfrac{1}{4}\right) =$

29. $(-3.7) - (-4.5) =$

30. $(-6) \times (-4.2) =$

31. $\left(\dfrac{3}{8}\right)\left(\dfrac{-2}{9}\right) =$

32. $(-20) \div (-2) =$

33. $(0.64) \div (-0.08) =$

34. Write 2^5 as a single number.

35. Write $(1.3)(1.3)(1.3)(1.3)$ in exponent form.

36. Write 24,500,000,000 in scientific notation.

37. Write 0.00000082 in scientific notation.

38. What is $\sqrt{49}$?

39. Approximate $\sqrt{61}$.

Work the following problems with a hand calculator.

40. $(0.0006743)(53.86) =$

41. $15.7 \div 0.691 =$

42. $\dfrac{(18.2)(6.31)}{(65.8)(1.28)(12.532)} =$

43. $\dfrac{46.2}{1.7} + \dfrac{18.9}{(2.9)(0.778)} =$

44. $(1.065)^{10}$

45. $\sqrt{3.7654}$

46. Janet has $173.02 in her checking account. If she writes a check for $28.44, what is her new balance?

47. A course at Carroll State College transfers as $2\frac{2}{3}$ credit hours at another school. If Randy has 12 such courses, how many credit hours will be transferred?

48. Ruth drives 207.6 miles on 12.3 gallons of gas. How many miles per gallon is this?

Building a community center. (Paul Conklin/ Monkmeyer Press Photo Service)

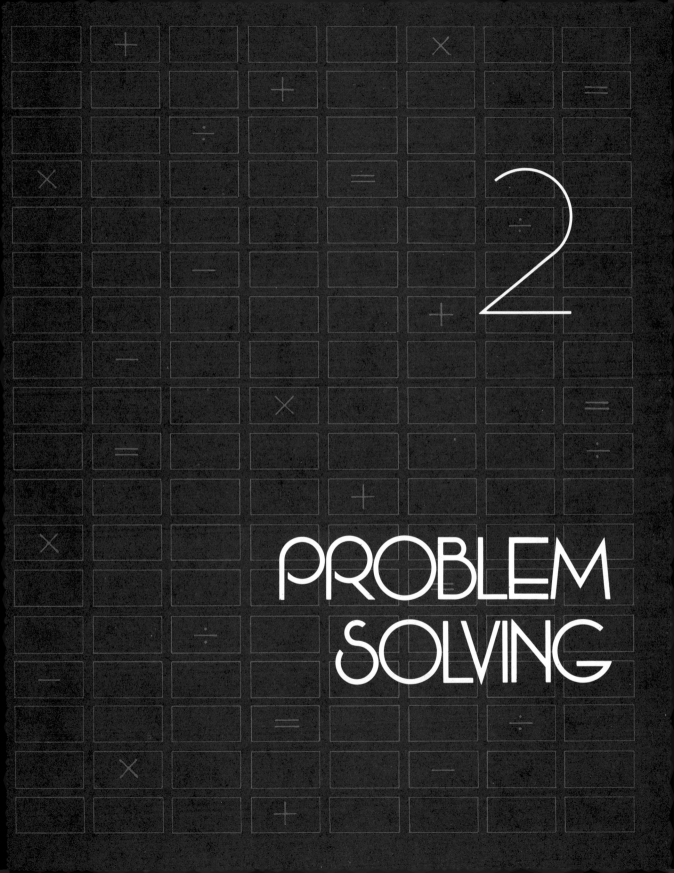

2

PROBLEM SOLVING

Question Is there a simple rule for solving mathematics problems?

ANSWER Sad to say, the honest answer would have to be no; there is no simple rule. Every problem is different. Fortunately, there are a few special types of problems that can be solved with special rules. We will study word problems, percent problems, proportion problems, and units conversion.

We will give a rule for solving each type of problem. However, as is usually the case, the best way to learn is by following through the worked examples. After seeing numerous examples worked out, then you will begin to see the pattern.

2–1 ALGEBRA

Question What is $x + y$?

ANSWER $x + y = x + y$, since we don't know what x and y are. This seems to panic students in math courses. We are taught throughout much of our schooling that $+$ is an operation that takes two numbers and magically rolls them into one number. Also, with the new electronic calculators, we punch two numbers and get a nice answer. But there is nothing wrong with the expression $x + y$. We are adding two numbers. We just don't know what they are.

Question Why are letters sometimes used instead of numbers?

ANSWER There are several reasons.

1. *Unknowns.* The letter might stand for a number that we don't know yet. Then we solve for the letter. For instance, in $x + 7 = 12$, the x stands for an unknown. We can of course solve this and find $x = 5$.
2. *Patterns.* We can use letters to show general patterns for all numbers. For example, we know $1 + 2 = 2 + 1$; $2 + 4 = 4 + 2$; $5 + 3 = 3 + 5$; and $8 + 11 = 11 + 8$. We can never begin to write out all such equations. So we use letters and say that $a + b = b + a$, for all numbers a and b.
3. *Formulas.* Letters stand for quantities in a formula. For example,
   (a) $A = l \cdot w$ (area = length $\cdot$ width)
   (b) $I = P \cdot r \cdot t$ (interest = principal $\cdot$ rate $\cdot$ time)
   (c) $F = \frac{9}{5}C + 32$ (Fahrenheit = $\frac{9}{5} \cdot$ (Celsius) + 32) are formulas that use letters to stand for the terms. Suppose we were told that $l = 5$ and $w = 3$. Then we can *substitute* into (a) and get $A = 15$.

Question Do we treat the letters the same as numbers?

ANSWER Yes. We add, subtract, multiply, and divide the letters just as if they were numbers that we knew. This section will review some of the rules for working with letters and numbers together.

The Distributive Law

Question   What is the distributive law?

ANSWER   The **distributive law**, or **distributivity**, is the following rule.

Formula 2–1–1

$$a \cdot (b + c) = a \cdot b + a \cdot c$$

for any numbers a, b, c. (This is an example of letters being used for a general pattern of numbers.)

Examples   1. Consider $7 \cdot (2 + 9)$. We can simplify this two ways.

$$7 \cdot (2 + 9)$$

$$= 7 \cdot 11 \qquad\qquad = 7 \cdot 2 + 7 \cdot 9$$

$$= 77 \qquad\qquad = 14 + 63$$

$$= 77$$

The method on the right used distributivity.

2. $5 \cdot (a + 4) = 5a + 20$
3. $p \cdot (q + 6) = pq + 6p$
4. $x \cdot (y + z) = xy + xz$

Notice that the outside number *distributes* itself over the inside sum.

Formula 2–1–2

$$a \cdot (b - c) = a \cdot b - a \cdot c$$

where a, b, and c are any numbers.

Examples   1. $7(10 - 2) = 70 - 14 = 56$
2. $8(x - 5) = 8x - 40$
3. $s(t - 9) = st - 9s$
4. $u(v - w) = uv - uw$

We can put these two formulas together.

Examples   1. $5(a + b - 2c - 3) = 5a + 5b - 10c - 15$
2. $x(t + 2u + x - 4) = xt + 2xu + x^2 - 4x$

Notice that when we use distributivity the outside number multiplies all the terms on the inside. Sometimes we can do this in reverse by pulling a common term out of all the terms.

Examples   1. $3x + 3y = 3(x + y)$

2. $7a + 8a + a = (7 + 8 + 1)a = 16a$

3. $xa - xb + x = x(a - b + 1)$

In each of the problems above we found a number or letter common to all the terms on the left. We then pulled this out and left the other terms together. This is another use of distributivity.

PROBLEM SET 2-1-1

1. Define or discuss distributivity.

Use distributivity to remove the parentheses.

2. $5(a + b)$	3. $7(x + y)$	4. $8(x + 10)$
5. $10(p - q)$	6. $11(a - 6)$	7. $a(6 + r)$
8. $s(10 - t)$	9. $x(y - z)$	10. $a(b - 12)$
11. $10(a + b + c)$	12. $7(x - y - 4)$	13. $x(x + y - 12)$

Simplify by pulling out the common term.

14. $4a + 5a$	15. $9x - 3x$	16. $5x + 5y$
17. $6x - 12y$	18. $x^2 + x$	19. $st - 7t$
20. $x7 - xz + x$	21. $6a + 3b - 9c$	22. $5x - 10x^2 + 15x^3$

$3(2a + b - 3c)$ $5x(1 - 2x^2 + 3x^3)$

Solving Equations

Question   How do we solve simple equations?

ANSWER   We must keep in the back of our minds our goal in solving problems.

Goal

| Get the unknown quantity by itself. |

To do this, there are some simple rules we can use to help us solve equations. These rules tell us what we can do to both sides of an equation and make sure it will still be an equation.

Rule AS: The same number can be *added* to or *subtracted* from both sides of an equation, and the results are still equal.

Rule MD: Both sides of an equation can be *multiplied* or *divided* by the same number, and the results are still equal. (We cannot divide by zero.)

Rule I: We can *invert* (or flip) both sides of an equation, and the results are still equal. (We cannot invert zero.)

Rule SR: We can always take the *square root* of both sides of an equation, and the results are still equal.

An equation is like a delicate balance. These rules tell us what we can do to both sides and still maintain this balance.

Greek vase painting of merchants weighing (c. 540 B.C.). The Metropolitan Museum of Art, Purchase, 1947, Pulitzer Bequest)

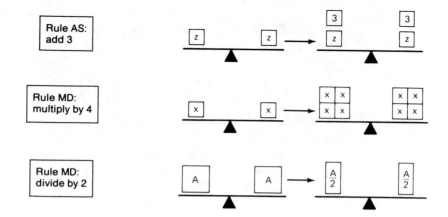

The accompanying figure gives a few examples of how these rules preserve the balance of the equation. The best way to show these rules, however, is just to begin using them in problems.

Remember, our goal is to isolate the unknown. Even though we do not know what the unknown is, we treat it like any other number.

Problem Solve $x - 9 = 15$ for x.

ANSWER

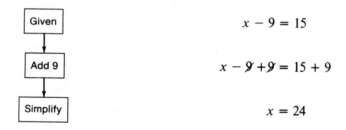

$$x - 9 = 15$$

$$x - \cancel{9} + \cancel{9} = 15 + 9$$

$$x = 24$$

Problem Solve $x + 7 = 12$ for x.

ANSWER

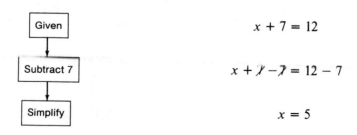

$$x + 7 = 12$$

$$x + \cancel{7} - \cancel{7} = 12 - 7$$

$$x = 5$$

In these two problems, notice that addition and subtraction undo each other.

Problem   Solve $5y = 17$ for y.

ANSWER

$$5y = 17$$

$$\frac{\cancel{5}y}{\cancel{5}} = \frac{17}{5}$$

$$y = \frac{17}{5} = 3.4$$

Problem   Solve $\dfrac{x}{8} = 9$ for x.

ANSWER

$$\cancel{8} \cdot \frac{x}{8} = 9 \cdot \cancel{8}$$

$$\cancel{8} \cdot \frac{x}{\cancel{8}} = 8 \cdot 9$$

$$x = 72$$

In these two problems, notice that multiplication and division undo each other.

Problem   Solve $\dfrac{12}{z} = \dfrac{5}{7}$ for z.

ANSWER

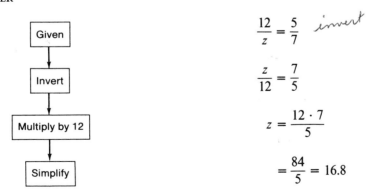

$$\frac{12}{z} = \frac{5}{7} \quad \textit{invert}$$

$$\frac{z}{12} = \frac{7}{5}$$

$$z = \frac{12 \cdot 7}{5}$$

$$= \frac{84}{5} = 16.8$$

We had to invert the equation since we always want x in the numerator.

Problem Solve $x^2 + 4 = 13$ for positive x.

ANSWER

$$x^2 + 4 = 13$$

$$x^2 = 9$$

$$x = 3$$

Notice that, if x were allowed to be negative, $x = -3$ would also have been an answer. We can check this by substituting 3 or -3 into the given equation: $3^2 + 4 = 13$ and $(-3)^2 + 4 = 13$.

Problem Solve $\frac{2}{15}x + 16 = 22$ for x.

ANSWER

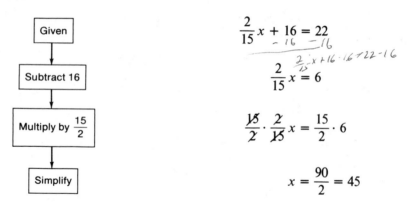

$$\frac{2}{15}x + 16 = 22$$

$$\frac{2}{15}x = 6$$

$$\frac{\cancel{15}}{2} \cdot \frac{2}{\cancel{15}}x = \frac{15}{2} \cdot 6$$

$$x = \frac{90}{2} = 45$$

Notice that we multiplied both sides by $\frac{15}{2}$ to cancel the $\frac{2}{15}$ on the left side.

Problem Solve $6x + ax - 10 + b = 14$ for x.

ANSWER We treat the letters a and b just as though they were numbers. We still group all the x terms together.

$$6x + ax - 10 + b = 14$$

$$(6 + a)x - 10 + b = 14$$

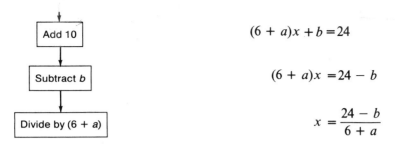

$$(6 + a)x + b = 24$$

$$(6 + a)x = 24 - b$$

$$x = \frac{24 - b}{6 + a}$$

Notice how we carefully get all the x terms on one side of the equation and all the other terms on the other side. Then, at the end, we divide by the term $(6 + a)$.

Problem   Solve $5x - 17 = 3x + 9$ for x.

ANSWER   Here we first get all the x terms together.

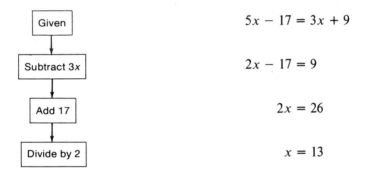

$$5x - 17 = 3x + 9$$

$$2x - 17 = 9$$

$$2x = 26$$

$$x = 13$$

Let us check this answer by substituting it back into the given equation.

| Left side |

$$5(13) - 17 = 65 - 17 = 48$$

| Right side |

$$3(13) + 9 = 39 + 9 = 48$$

Therefore, $x = 13$ is correct. Many times it is not possible to check an answer because the numbers of the problem are somewhat messy, and this makes substitution painful. In these cases we must be careful to work the problem correctly.

Problem   Solve $3(x + 5) = 5x + 2(x - 3)$ for x.

ANSWER   Here we must first use distributivity, and then combine terms to solve for x.

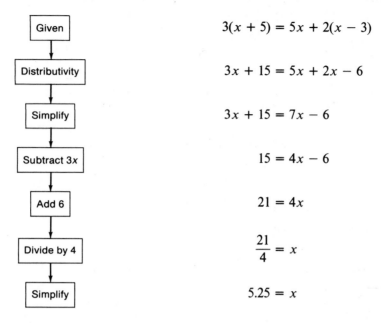

Given	$3(x + 5) = 5x + 2(x - 3)$
Distributivity	$3x + 15 = 5x + 2x - 6$
Simplify	$3x + 15 = 7x - 6$
Subtract 3x	$15 = 4x - 6$
Add 6	$21 = 4x$
Divide by 4	$\dfrac{21}{4} = x$
Simplify	$5.25 = x$

We also can check this by substitution.

Left side	$3(5.25 + 5) = 3(10.25) = 30.75$
Right side	$5(5.25) + 2(5.25 - 3) = 26.25 + 2(2.25)$
	$= 26.25 + 4.5 = 30.75$

So $x = 5.25$ checks out as the answer.

To solve these simple equations for x,
1. Use distributivity to remove any parentheses.
2. Simplify by combining common terms on the same side of the equation.
3. Gather all x terms to one side of the equation, and all number terms to the other (use rule AS).
4. Get x by itself (use rule MD).

For some problems, not all these steps will be needed. If a step is not needed, skip it.

Problem   Solve $4(x - 7) + 6 = 3(2x + 1) - 5x$.

ANSWER

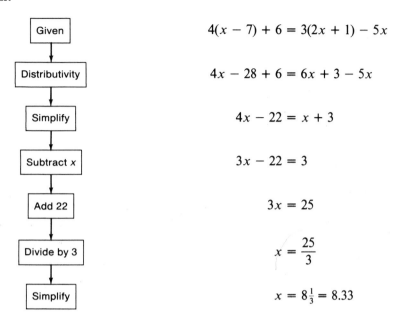

Given	$4(x - 7) + 6 = 3(2x + 1) - 5x$
Distributivity	$4x - 28 + 6 = 6x + 3 - 5x$
Simplify	$4x - 22 = x + 3$
Subtract x	$3x - 22 = 3$
Add 22	$3x = 25$
Divide by 3	$x = \dfrac{25}{3}$
Simplify	$x = 8\frac{1}{3} = 8.33$

PROBLEM SET 2–1–2   Solve for the unknown.

1. $x - 18 = 31$ 2. $y + 3.1 = 5.6$

3. $z - \dfrac{3}{4} = \dfrac{7}{8}$ 4. $t + 15 = 8$

5. $w + 23 = 101$ 6. $a - 13.9 = 21.2$

7. $b + \dfrac{1}{5} = \dfrac{9}{10} - \dfrac{1}{5}$ 8. $3c = 36$

9. $\dfrac{p}{4} = 10$ N_0 10. $6q = 47$

11. $\dfrac{23}{5}x = 46$ 12. $\dfrac{16}{9}x = \dfrac{32}{27}$

13. $8x + 12 = 20$ 14. $4x - 6 = -3$

15. $2.6y + 1.1 = 4.9$ 16. $-4.6z - 15.8 = 3.9$

17. $\dfrac{3}{11}y - 6 = \dfrac{1}{2}$ 18. $\dfrac{12}{13}t + 7 = 3$

19. $6x + 9 = 8x - 3$ 20. $10x - 13 = 3x + 5$

21. $12x + 3 - 5x = 4x - 7$ 22. $18t - 9 - t = 8 - 3t - 2$

23. $5x + tx = 2x - 3$
(solve for x)

24. $3x - ax + bx - 8 = 12 - x$
(solve for x)

25. $6(a + 7) = 5(2a - 4)$

26. $4(3b - 5) = 7(2b + 9)$

27. $\frac{1}{2}(m + 8) = \frac{3}{2}(m - 6)$

28. $4.1(2r + 5.1) = 6.9(r + 7.5)$

Inequalities

Question   What is an inequality?

ANSWER   An **inequality** is a mathematical relation which states that one number is larger than another. In symbols, we have

> $<$ **is less than**
> $>$ **is greater than**

For example,

$$5 < 7, \quad 7.9 < 8.1, \quad -3 < 2, \quad -5 < -1$$
$$18 > 11, \quad 56 > 55.9, \quad 5 > -8, \quad -6 > -11$$

On a number line, we can picture $<$ as "to the left of," and $>$ as "to the right of." For example,

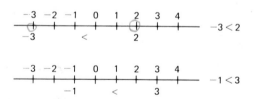

We also have the combined symbols

> $\leqslant$ **is less than or equal to**
> $\geqslant$ **is greater than or equal to**

For example,

$$10 \leqslant 12, \quad 60 \leqslant 60, \quad 90 \leqslant 100, \quad -5 \leqslant -5$$
$$14 \geqslant 13, \quad 7 \geqslant 7, \quad 119 \geqslant 83, \quad -12 \geqslant -12$$

Notice that the numbers may be equal with this symbol.

Question   What does $x \geqslant 2$ look like on a number line?

ANSWER   The relation $x \geqslant 2$ can be pictured as shown. This tells us that x can be 2 or any number to the right of 2.

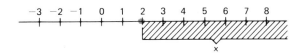

Question   What does $5 \leqslant x \leqslant 9$ mean?

ANSWER   We are making two statements about x. First, we are saying that x is 5 or more. Also, we are saying that x is 9 or less. Together, we are saying that x is *between* 5 and 9, including 5 and 9.

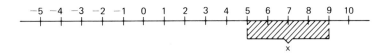

PROBLEM SET 2-1-3

1. Define or discuss inequality.

State in words what the following inequalities mean. Then indicate each on a number line.

2. $x < 7$ 3. $x \geqslant 9$ 4. $x < -3$ 5. $x > -6$

6. $-2 < x$ 7. $10 > x$ 8. $6 < x < 10$ 9. $-4 \leqslant x \leqslant 4$

10. $6.5 < x < 7.5$ 11. $10.1 \leqslant x \leqslant 10.3$

2-2   TRANSLATION AND VERBAL PROBLEMS

Question   Why are we studying translation in a mathematics book? Is mathematics a foreign language?

ANSWER   Yes, mathematics is like a foreign language to most people, and to solve many problems we must learn how to translate a problem written in English into a solvable mathematical equation.

Every word problem has two basic parts:

1. Translate the English problem into mathematics.
2. Solve the mathematical equation.

Usually, step 2 (solving) is much easier. We did this in the last section where we solved for x.

The harder part of the problem is step 1 (translating). In this step we must inventory all our information and put it all into a neat equation or inequality. One thing that will help is Table 2–1, in which we have listed the most common English terms with their mathematical translations.

Table 2–1 English to Mathematics Translations

English	Mathematics	Comment
More than, over, above, and	+	Addition
Less	–	Subtraction
Less than, under	– (reverse order)	Subtraction
Times, of	× or ·	Multiplication
Out of, per, for, to	÷	Division
Is, was, should be, same as	=	Equals (like a verb)
Is greater than	>	Inequality (like a verb)
Is less than	<	Inequality (like a verb)
What	x	Unknown
Quantity	a, b, c	Any letter will do
Percentage	÷ 100	Change to decimal

The term *quantity* in Table 2–1 means any quantity in the problem, such as Mary's age, the price of a radio, or the percentage of men in a room. We usually use a single letter to represent the quantity.

Examples In the following examples, we will translate an English phrase or sentence into mathematics.

1. English: Five more than Gary's age.
   Mathematics: 5 + g

2. English: Eighteen dollars less than the cost of a radio. —watch order
   Mathematics: c –   18   (terms are reversed)

3. English: Half of the people in the room.
   Mathematics: $\frac{1}{2}$ · p

4. English: Five out of nine.
   Mathematics: 5 ÷ 9

5. English: The price of a Ford is half of the price of a Cadillac.
   Mathematics: F $= 1/2$ · C

6. English: The price of a house is less than $2\frac{1}{2}$ times his income.
   Mathematics: H $<$ $2\frac{1}{2}$ · I

of = mult.

7. English: Ten thousand dollars more than 5 times her income.

   Mathematics: $10{,}000 \quad + \quad 5 \quad \cdot \quad I$

8. English: Twenty-four is four-fifths of what?

   Mathematics: $24 \quad = \quad \dfrac{4}{5} \quad \cdot \quad x$

9. English: What is 17% of 29?

   Mathematics: $x \quad = (0.17) \cdot (29)$

10. English: Ed's weight is 15 pounds over average.

    Mathematics: $E \quad = \quad 15 \quad + \quad A$

11. English: Eighty-three is what percent of 117?

    Mathematics: $83 \quad = \quad \dfrac{x}{100} \quad \cdot \quad 117$

    or $83 \quad = \quad p \quad \cdot \quad 117$

12. English: Joe's income is more than twice Harry's.

    Mathematics: $J \quad > \quad 2 \cdot \quad H.$

     Notice that $>$ is the verb "*is* more than," whereas $+$ is just "more than." These are very similar. The key difference is the word "is" for $>$. This makes it a verb. The same is true for $<$, which is the verb "*is* less than," whereas $-$ means "less."

PROBLEM SET 2-2-1   Translate the following phrases into mathematics. (Make up a letter symbol for the quantities.)

1. Three under par $3 - p$ *then* $3 < x$

2. Nine degress more than usual $9 > w$

3. Thirteen times the price of a toaster $13 \cdot x$ $3x > w$

4. One sixth of every professor on campus

5. Nine out of 10 doctors $9 \div 10$ Doc.

6. Two degrees above normal

7. One out of 3 adults $\dfrac{1}{3} \times A$

8. Two hundred dollars for 4 tires

9. Fourteen percent of Dan's income $0.14 \times Dan$

10. Five times the daily minimum requirement

11. Ten more than twice Fran's cookie intake $10 + 2F$

Translate the following sentences into mathematics. (Make up letter symbols for the quantities.)

12. Sixty-seven is 6 under par.

13. One-fifth of all the students is 650.

14. Twenty-two percent of Al's monthly income is $205.

15. The discount is 20% of 95.

16. One hundred seventeen is what percent of 443?

17. Area is length times width.

18. Bob's height is 4 inches less than Ralph's.

19. The men are 80% of the people in the room.

20. Three hundred dollars is 85% of the list price.

21. A groom's age should be seven less than twice the bride's age. (Old advice for marriage.)

22. Mary's height is less than Connie's age.

23. The money spent on education and medicine is less than the money spent on defense.

24. The sales tax is 5% of the price.

Verbal Problems

Problem Three-fifths of a certain group are women. There are 117 women. How many total members are there?

ANSWER Let us first identify the terms (or quantities) of the problem. Then we will translate the three sentences into mathematics and solve.

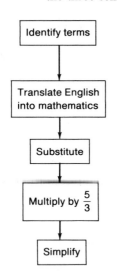

w = number of women = 117

t = total group membership = ?

Three fifths of a group are women:

$$\frac{3}{5} \cdot t = w$$

$$\frac{3}{5} \cdot t = 117$$

$$t = \frac{5}{3} \cdot 117$$

$$t = 195$$

Notice that we translated all three sentences into information. The first gave us the basic equation of the problem, $\frac{3}{5} \cdot t = w$. The second told us that $w = 117$, and the third that t was the unknown.

Problem An antique bit of advice was "A groom's age should be seven less than twice the bride's age." Chuck is 25. By this advice, he should marry a woman of what age?

ANSWER We will first determine the terms of the problem, translate the sentences, and then solve.

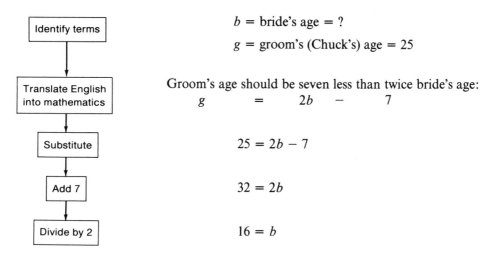

b = bride's age = ?

g = groom's (Chuck's) age = 25

Groom's age should be seven less than twice bride's age:

$$g \quad = \quad 2b \quad - \quad 7$$

$$25 = 2b - 7$$

$$32 = 2b$$

$$16 = b$$

By this maxim, Chuck should marry a 16-year-old girl. (Needless to say, this match up would not be too common today.)

Notice again how we use each sentence to extract all the information we can from the problem. Also, when we translated "seven less than twice the groom's age," we got $2g - 7$, since we reverse order for "less than."

To solve verbal problems:

1. *Identify* all the terms (or quantities) of the problem. Assign each of them a letter.

2. *Translate* each sentence (or question) in a mathematics equation or inequality.

3. *Substitute* any simple information (such as $w = 117$) into the longer equation or inequality.

4. *Solve* for the unknown.

Problem Banks suggest that families buy a house that is less than two times their family income. The Turners like a $50,000 house. How much income do they need?

ANSWER We first identify the terms. Notice that family and Turners are considered the same.

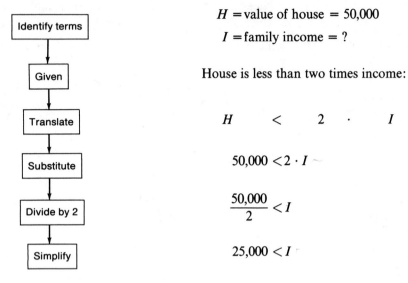

$H =$ value of house $= 50,000$

$I =$ family income $= ?$

House is less than two times income:

$$H \quad < \quad 2 \quad \cdot \quad I$$

$$50,000 < 2 \cdot I$$

$$\frac{50,000}{2} < I$$

$$25,000 < I$$

The Turners must earn at least $25,000 to afford the house.

Problem Helene earns $80 a week plus a commission of $\frac{1}{25}$ of her sales. Last week she earned $210. What were her sales that week?

ANSWER We first identify the terms as E, the earnings, and S, her sales.

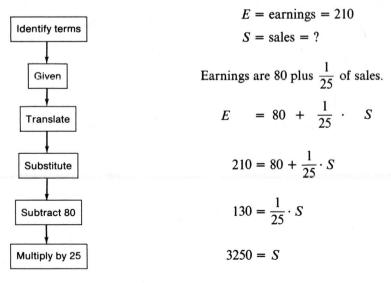

$E =$ earnings $= 210$

$S =$ sales $= ?$

Earnings are 80 plus $\frac{1}{25}$ of sales.

$$E \quad = 80 + \frac{1}{25} \cdot S$$

$$210 = 80 + \frac{1}{25} \cdot S$$

$$130 = \frac{1}{25} \cdot S$$

$$3250 = S$$

Problem   Rob and Candy sell their house and lot for $50,000. They figure that the house was worth $2\frac{1}{2}$ times the lot. How much is the house worth; how much is the lot worth?

ANSWER   Here we have two unknowns, the worth of the house and the worth of the lot. We will call the worth of the lot x. Since the total worth was $50,000, then the value of the house was $50,000 - x$.

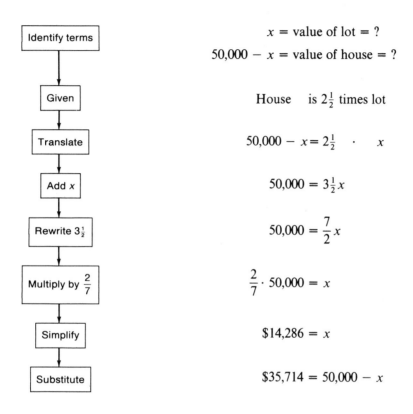

Identify terms	x = value of lot = ?
	$50,000 - x$ = value of house = ?
Given	House is $2\frac{1}{2}$ times lot
Translate	$50,000 - x = 2\frac{1}{2} \cdot x$
Add x	$50,000 = 3\frac{1}{2}x$
Rewrite $3\frac{1}{2}$	$50,000 = \dfrac{7}{2}x$
Multiply by $\dfrac{2}{7}$	$\dfrac{2}{7} \cdot 50,000 = x$
Simplify	$\$14,286 = x$
Substitute	$\$35,714 = 50,000 - x$

The lot was worth about $14,286, and the house worth $35,714.

Problem   There is a simple formula, distance is rate times time. If Bob can average 55 miles per hour, how long will it take to drive 375 miles?

ANSWER   We first translate the formula, and then we substitute.

Identify terms	d = distance = 375
	r = rate = 55

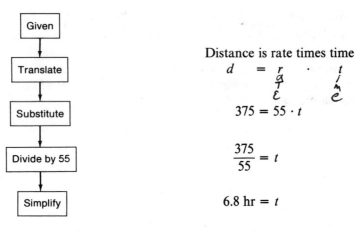

Distance is rate times time

$$d = r \cdot t$$

$$375 = 55 \cdot t$$

$$\frac{375}{55} = t$$

$$6.8 \text{ hr} = t$$

Bob can figure on it taking just under 7 hours to make the trip.

PROBLEM SET 2-2-2

1. Ralph earns $1\frac{1}{2}$ times what Ed makes. Ralph makes $16,000. What does Ed make?

2. Four-fifths of a fraternity plays touch football; 28 men play football. How many are in the fraternity?

3. A clothing store is having a sale. Everything is $\frac{1}{3}$ off list price. Jan is going to buy a $35 outfit. What will it cost her?

4. Steve receives 38 more votes than he needed to be class president. He received 752 votes. How many did he need?

5. Kathy is $\frac{3}{4}$ of the age of her sister Judy. Two years ago, Kathy was 13. How old is Judy now?

6. Bill and Carol spend $\frac{1}{6}$ of their income on travel. Last year they spent $2200 traveling. How much did they earn?

7. The price of a car is 1.7 times what it was 10 years ago. How much would a $5000 car have cost 10 years ago?

Use the relation, distance is rate times time, to solve Problems 8–11.

8. Sherri drives 1156 miles from Detroit to Denver in 22 hours. What was her average speed?

9. Tony drives 1019 miles from Philadelphia to Miami at a speed of about 45 miles per hour. How long will the drive be?

10. Chuck and Donna are planning a drive from Boston to Los Angeles (3049 miles). They can average 45 miles per hour. How many hours will it take? If they can drive 12 hours a day, how many days will it take?

11. Mike jumps in his car and starts driving. He drives 4 hours at 55 miles per hour. How far is he from home?

12. There are 317,000,000 cubic miles of ocean water. This represents about $\frac{39}{40}$ of all the water in the world. How much water is there in the world?

13. Old man Arbuckle leaves his entire estate to his nephews, Clarence and Nigel. Clarence is to get $1\frac{1}{2}$ what Nigel gets. The entire estate is finally settled at $175,000. How much does each nephew get?

2–3 PERCENTAGES

Question Why do we study percentages?

ANSWER For some reason, we are used to thinking in terms of 100. We know that $\frac{3}{4}$, 0.75, and 75% are the same number, but most people feel more comfortable with 75%. This means 75 out of 100.

But we must be careful; 75% is really an English term. When we want to do a mathematics problem, we usually have to use $\frac{3}{4}$ or 0.75. This is, of course, translation from English to mathematics.

Question How do we solve percentage problems?

ANSWER Many times, we can merely translate the percentage problem directly into mathematics and solve it.

$$14 = x \cdot 33$$

Problem Fourteen is what percent of 33?

ANSWER We simply translate into mathematics and solve.

Translate	$14 = x \cdot 33$
Divide by 33	$\frac{14}{33} = x$
Simplify	$0.424 = x$
Convert to percent	$42.4\% = x$

Since the problem asked for a percent, we convert the decimal to a percent.

(Mimi Forsyth/Monkmeyer Press Photo Service)

Problem What is 17% of 93?

ANSWER

$$x = (0.17) \cdot 93$$

$$= 15.81$$

Problem Two hundred twenty-seven is 18% of what?

ANSWER

$$227 = (0.18) \cdot x$$

$$\frac{227}{0.18} = x$$

$$1261.1 = x$$

Notice that we had to first translate the percent to decimal to do these problems.

Problem   Twenty-seven out of 53 is what percent?

ANSWER

| Translate | $27 \div 53 = x$ |

| Simplify | $0.509 = x$ |

| Convert to percent | $50.9\% = x$ |

$$0.9 \\ x = .\cancel{89} \times 10 =$$

PROBLEM SET 2–3–1

1. What is 9% of 10?

2. Eighteen is what percent of 59?

3. Seventy-five is 68% of what?

4. What is 5% of 350?

5. Eighty-three is what percent of 271?

6. One hundred seventy-five is 95% of what?

7. Fifteen is what percent of 3375?

8. What is 68% of 952?

9. One hundred ninety-three is 6% of what?

10. Two hundred forty-one out of 629 is what percent?

11. Two hundred twenty-five is 12% of what?

12. What is 5.5% of 650?

13. Seven hundred fifty-nine is what percent of 641?

14. What is 1000% of 25?

More Percents

Question   What do we do if the problem isn't so easy to translate?

ANSWER   We notice that just about all the percentage problems of the last section had the form A is some percent of B. Sometimes A was unknown, sometimes the percent, and sometimes B. But the form is always the same.
   We can translate A *is some percent of* B into

Formula 2–3–1

$$\boxed{A = P \cdot B}$$

To solve the problems, we must first identify the three parts:

> A = amount = a *part* of the whole, change
> P = percent = a *part* of 100%
> B = base = *total*, whole, original, list price

To use Formula 2–3–1, *A and P must represent the same quantity*. For example,

1. Amount of *men*, percent of *men*.
2. Amount of *discount*, percent of *discount*.
3. Amount of *increase*, percent of *increase*.

To solve a percentage problem:

1. Identify the terms A, P, and B. Determine what is known and what is unknown.

2. Substitute into Formula 2–3–1, $A = P \cdot B$.

3. Solve for the unknown.

4. If the unknown is P, be sure to convert the answer to a percentage.

Problem Twenty-seven percent of a group smoke. There are 500 people in the group. How many people smoke?

ANSWER Here we know the total (B) and the percent (P). We want to find the part or amount (A).

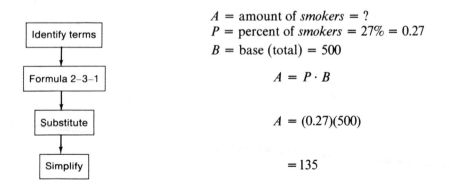

Identify terms

A = amount of *smokers* = ?
P = percent of *smokers* = 27% = 0.27
B = base (total) = 500

Formula 2–3–1

$$A = P \cdot B$$

Substitute

$$A = (0.27)(500)$$

Simplify

$$= 135$$

Notice that A and P were both the same quantity, *smokers*.

Problem Karyn is doing a survey. She contacts 173 women. She finds that 81 of them hold jobs. What percent is this?

ANSWER We first determine the terms. Obviously, the percent (P) is the unknown. We do know that the whole (B) is 173, and the part (A) is 81.

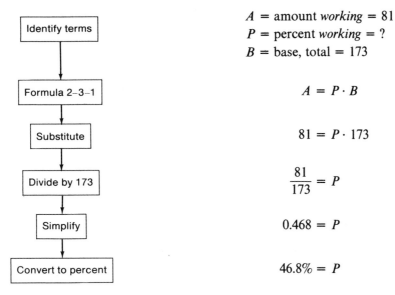

A = amount *working* = 81
P = percent *working* = ?
B = base, total = 173

$$A = P \cdot B$$

$$81 = P \cdot 173$$

$$\frac{81}{173} = P$$

$$0.468 = P$$

$$46.8\% = P$$

Problem Experts say that a person's rent should be about 25% of his or her monthly income. Becky is thinking of renting a $235 apartment. How much should her monthly income be?

ANSWER Here we know the percent (P) that goes toward rent and the amount (A) that goes toward rent. We want to find the total income (B).

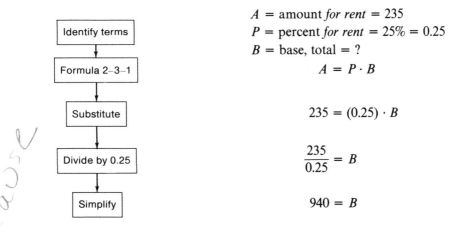

A = amount *for rent* = 235
P = percent *for rent* = 25% = 0.25
B = base, total = ?

$$A = P \cdot B$$

$$235 = (0.25) \cdot B$$

$$\frac{235}{0.25} = B$$

$$940 = B$$

Becky should earn $940 a month to afford the apartment.

Problem Ron earned $11,200 last year. This year he earned $11,900. What is the percent of increase in pay?

ANSWER Here we must be careful. Since we are asked to find percent of *increase*, we must use amount of *increase*. (Remember, the percent and amount must always be the same thing.) The increase in salary is $11,900 - 11,200 = 700$.

Furthermore, we use as the base (B) the *original* salary, 11,200. In all increase–decrease problems, we always use the original price, salary, size, or whatever as the base.

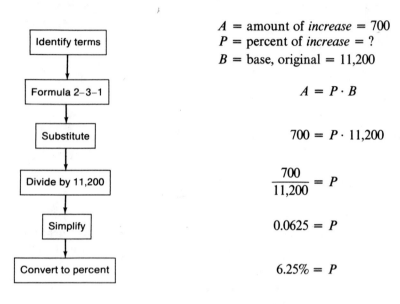

A = amount of *increase* = 700
P = percent of *increase* = ?
B = base, original = 11,200

Identify terms	
Formula 2–3–1	$A = P \cdot B$
Substitute	$700 = P \cdot 11,200$
Divide by 11,200	$\dfrac{700}{11,200} = P$
Simplify	$0.0625 = P$
Convert to percent	$6.25\% = P$

Hand Calculator Instant Replay

PUNCH						DISPLAY	MEANING
C						0.	Clear
7	0	0				700.	700
÷	1	1	2	0	0	11200.	Divided by 11,200
=						0.0625	Answer

Problem Joanne buys a $210 stereo. She is told that this is 30% off list price. What was the list price?

ANSWER The unknown is the original or list price. This is the base, *B*. To find the amount (*A*) and percent (*P*), we again have to be a bit careful. The $210 is the amount she *paid*. So we must know the percent she *paid*. Since the discount was 30%, she paid 70%. (If we used 30%, it would be wrong!) The amount (*A*) and percent (*P*) *must* be the same thing.

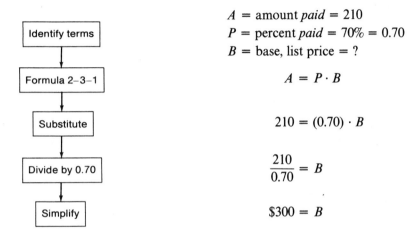

A = amount *paid* = 210
P = percent *paid* = 70% = 0.70
B = base, list price = ?

$$A = P \cdot B$$

$$210 = (0.70) \cdot B$$

$$\frac{210}{0.70} = B$$

$$\$300 = B$$

Problem Danny and Dianne put $850 into a bank account paying $5\frac{1}{2}\%$ interest. How much interest will they earn? What will their new balance be in a year?

ANSWER First we compute the amount of interest.

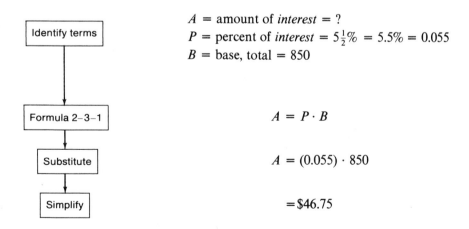

A = amount of *interest* = ?
P = percent of *interest* = $5\frac{1}{2}\%$ = 5.5% = 0.055
B = base, total = 850

$$A = P \cdot B$$

$$A = (0.055) \cdot 850$$

$$= \$46.75$$

They will earn $46.75 in interest. Then their balance will be $850.00 + 46.75 = $896.75.

Problem Lee and Marcy are selling their home through a realtor who charges 7% commission. They want to clear $32,000 *after* the commission. What must they ask to get the $32,000?

ANSWER We must be careful here. The 7% is *not* charged on the $32,000. The 7% is charged on a higher amount, and $32,000 is the amount ($A$) left *after commission*. To make P the same, we must take P as 93% *after commission* (100% − 7%). The total price B is the unknown.

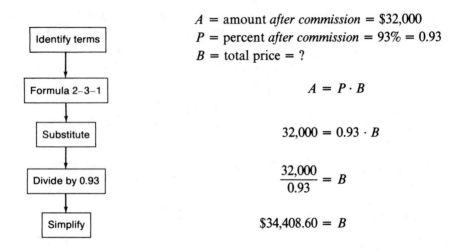

A = amount *after commission* = $32,000
P = percent *after commission* = 93% = 0.93
B = total price = ?

$$A = P \cdot B$$

$$32,000 = 0.93 \cdot B$$

$$\frac{32,000}{0.93} = B$$

$$\$34,408.60 = B$$

To ensure making $32,000, they must actually sell their house for about $34,400.

Hand Calculator Instant Replay

PUNCH	DISPLAY	MEANING
C	0.	Clear
3 2 0 0 0	32000.	32,000
÷ . 9 3	0.93	Divided by 0.93
=	34408.602	Answer

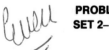

PROBLEM SET 2-3-2 **1.** An IQ test is given to 5793 people; 68% score between 85 and 115. How many people are between 85 and 115?

2. The price of bacon was $1.25 a pound. The price goes up 12%. What is the amount of the increase? What is the new price?

3. Gene earns $1050 a month. He spends $210 a month on his car. What percent is this? *20%*

4. In an election, Senator Clay got 1,218,429 votes out of 1,944,802 cast. What percent is this?

5. Peter pays $250 for a $350 stereo. What percent is this discount?

6. A shirt cost $21. Bob gets a 20% discount. How much does he pay?

7. A bank pays 6.5% interest on certain accounts. Mary and Mike would like to earn $10,000 a year in interest when they retire. How much money must they save to put in this account to earn that interest?

8. A group is 80% men. There are 82 women. How many are in the group (total)? *102.5*

9. Sid pays $4800 for a new car. A year later it is only worth $3750. By what percent has the value decreased?

10. Fran earns $1050 a month. Next year she will make $1190 a month. What percent increase is this? *13.3*

11. It takes a 20% down payment to buy a used home. Jeff and Debbie have saved $6500 to use as a down payment. How much can they spend on a house?

12. A certain new housing development requires only 5% down on a home. What down payment is needed to buy a $44,500 house?

13. In a recent governor's race, Jefferson defeated Adams by 728,684 to 661,072. There were 15,249 votes for other people. What was Jefferson's percent of the vote?

14. Ted pays 22% income tax on every dollar he earns. He needs $1000 (after taxes) to buy a sound system. How much does he have to earn so that he will have $1000 left after taxes?

15. Mary buys an antique rocking chair for $35. Later she sells it for $55. What percent profit is this?

16. On a box of oatmeal, it is claimed that each serving has 4 grams of protein. It is also claimed that this is 6% of one's daily recommended need. Using this, what is one's daily recommended need for protein?

Experts say that the monthly cost of a house (mortgage, taxes, repairs, and so on) is about 1% of the total cost. Gary and Toni are interested in a $53,000 house.

17. About what will it cost per month? *530*

18. If this cost should be about 25% of their income, what must their monthly income be to afford the house? *2120*

19. If the bank requires 20% down, how much of a down payment do they need?

In 1958, Folie State College had 151 on the faculty, 22 of whom were women. In 1978, there were 339 on the faculty, 56 of whom were women.

20. What was the percent of women in 1958?

21. What was the percent of women in 1978?

22. What was the percent of increase in faculty in 20 years?

23. What was the percent of increase in women in 20 years?

24. What was the percent of increase (or decrease) in the percent of women? (Use Problems 20 and 21.)

Jason and Jennifer want to clear $28,000 on the sale of their house.

25. How much should the sale price be so that they make $28,000 after a 6% commission?

26. What should their asking price be so that they can appear to "discount" their asking price by 10% and get their sale price (Problem 25)?

27. George Gonoff owns a store. He is trying to sell a $300 washer. He first raises the price 20%, then takes 20% off the new price. What is the final price?

28. Copperblat's is a department store that gives a 20% discount to its employees. There is also a 5% sales tax. Which is better: to charge the tax first, and then give the discount, or vice versa? (*Hint:* solve the problem for an imaginary $100 purchase.)

2–4 RATIO AND PROPORTION

Ratio

Question What is a ratio?

ANSWER A **ratio** is a comparison of two quantities by division. The quotient may be left as a fraction or a decimal, whichever is more meaningful. Usually, there will be some units involved that tell what the quantities are.

Examples The following are ratios.

1. $\dfrac{240 \text{ miles}}{15 \text{ gallons}} = 16 \dfrac{\text{miles}}{\text{gallons}}$

2. $\dfrac{105 \text{ men}}{70 \text{ women}} = \dfrac{3 \text{ men}}{2 \text{ women}} = \dfrac{1.5 \text{ men}}{1 \text{ woman}}$

3. $\dfrac{3 \text{ cans of water}}{1 \text{ can of orange juice concentrate}}$

4. $\dfrac{160 \text{ pounds}}{72 \text{ inches}} = 2.2 \dfrac{\text{pounds}}{\text{inches}}$

5. $\dfrac{42 \text{ miles}}{0.75 \text{ hour}} = 56 \dfrac{\text{miles}}{\text{hours}}$

6. $\dfrac{17 \text{ births}}{1000 \text{ population}}$

7. $\dfrac{\$3.52}{2.7 \text{ pounds of meat}} = 1.30 \dfrac{\text{dollars}}{\text{pounds}}$

8. $\dfrac{6500 \text{ students}}{315 \text{ faculty}} = 20.6 \dfrac{\text{students}}{\text{faculty}}$

9. $\dfrac{1 \text{ inch}}{2.54 \text{ centimeters}} = 0.39 \dfrac{\text{inch}}{\text{centimeters}}$

Notice how all these examples have retained their units, such as miles per gallon, dollars per pound, or students per faculty. The importance of ratios is that they tell us how the two quantities compare; furthermore, we can use ratios to compute what other quantities should be. We will see this in the next section on proportion.

PROBLEM SET 2–4–1

1. Define or discuss ratio.

Simplify the following ratios. Write them either as a simple fraction or as a decimal with a 1 in the denominator. Keep the units.

2. $\dfrac{\$225 \text{ saved}}{8 \text{ weeks}}$

3. $\dfrac{\$5000 \text{ from Joe}}{\$7000 \text{ from Ralph}}$

4. $\dfrac{35 \text{ pounds of meat}}{91 \text{ people}}$

5. $\dfrac{273 \text{ liberals}}{314 \text{ conservatives}}$

6. $\dfrac{144 \text{ grams of protein}}{\$1.29}$

7. $\dfrac{120 \text{ grams of protein}}{\$0.83}$

8. $\dfrac{75 \text{ calories}}{6 \text{ ounces}}$

9. $\dfrac{353 \text{ miles}}{24 \text{ gallons}}$

10. $\dfrac{353 \text{ miles}}{6.5 \text{ hours}}$

11. $\dfrac{21 \text{ games won}}{17 \text{ games lost}}$

12. $\dfrac{129 \text{ pounds}}{65 \text{ inches}}$

13. $\dfrac{100 \text{ kilometers}}{62 \text{ miles}}$

14. $\dfrac{132,000 \text{ cents spent on car}}{11,000 \text{ miles}}$

15. $\dfrac{\$210 \text{ cost in 1978}}{\$200 \text{ cost in 1977}}$

Proportion

Question   What is a proportion, and what is it used for?

ANSWER   **A proportion** is an equality of two ratios. Let's look at this statement. Remember that a ratio is a comparison of two quantities. A proportion tells us that two different comparisons come out the same. For example, if steak is $2 per pound and $4 for 2 pounds, this is a proportion.

$$\frac{\$2}{1 \text{ lb}} = \frac{\$4}{2 \text{ lb}}$$

Notice that the ratios are the same, or equal. Proportions are used to solve problems where we are missing one of the quantities. This will become clearer in the examples.

Problem   Bob drove 212 miles on 8.3 gallons of gas. How much gas does he need to go 750 miles?

ANSWER   We are given one ratio of miles to gallon, 212 miles per 8.3 gallon. We assume that this will be the same ratio as 750 miles to whatever amount of gas is needed.

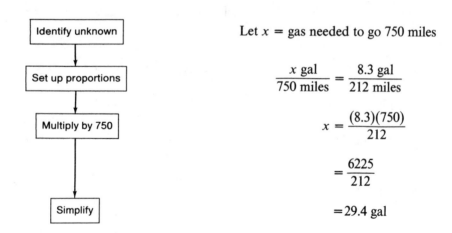

Let x = gas needed to go 750 miles

$$\frac{x \text{ gal}}{750 \text{ miles}} = \frac{8.3 \text{ gal}}{212 \text{ miles}}$$

$$x = \frac{(8.3)(750)}{212}$$

$$= \frac{6225}{212}$$

$$= 29.4 \text{ gal}$$

Notice that we inverted the ratio so that x would be in the numerator. This is perfectly correct to do, and it makes the problem simpler.

To solve proportion problems:

1. Label the unknown by x.

2. With x in the numerator, write down the ratio comparing x to the other quantity (in the denominator).

3. Make a proportion by setting the first ratio equal to the second ratio of the given quantities. Be sure to put the same quantities in both numerators and the same quantities in both denominators. For example,

$$\frac{\$}{lb} = \frac{\$}{lb}$$

In fact, the best way to work these is to actually write in the units when setting up the proportion.

4. Solve for x by multiplying across.

Problem If Jerry can earn $8200 in 5 months, how much will he earn in 1 year?

ANSWER To do this problem, we assume that Jerry will earn at the same rate. If he gets a raise or stops working, we can't do the problem.

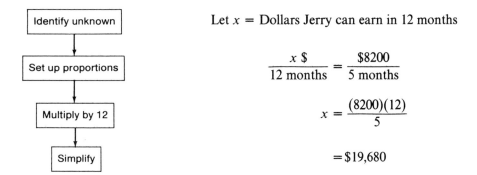

Let x = Dollars Jerry can earn in 12 months

$$\frac{x\ \$}{12\ \text{months}} = \frac{\$8200}{5\ \text{months}}$$

$$x = \frac{(8200)(12)}{5}$$

$$= \$19,680$$

Notice that units were the same in both ratios, dollars per month = dollars per month.

Problem Five years ago, Becky and Lisa opened an art store. Becky put up $5000, and Lisa put up $3500. Now the business is worth $22,000. What is Becky's share of the business?

ANSWER   Remember that the units must be the same. Since $22,000 is the *total* worth, we must use $5000 + $3500 = $8500 as the *total* investment. Now we can solve the problem.

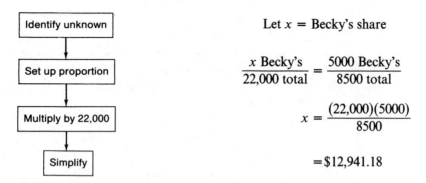

Let x = Becky's share

$$\frac{x \text{ Becky's}}{22{,}000 \text{ total}} = \frac{5000 \text{ Becky's}}{8500 \text{ total}}$$

$$x = \frac{(22{,}000)(5000)}{8500}$$

$$= \$12{,}941.18$$

Problem   Doug is the manager of a townhouse group interested in aluminum siding. He learns that it will cost $212,000 to do 135 units of another group. From this, about what will it cost to do Doug's 85 units?

ANSWER   We will assume that the costs are proportional.

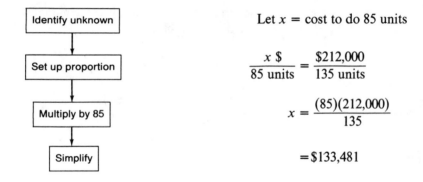

Let x = cost to do 85 units

$$\frac{x \text{ \$}}{85 \text{ units}} = \frac{\$212{,}000}{135 \text{ units}}$$

$$x = \frac{(85)(212{,}000)}{135}$$

$$= \$133{,}481$$

Hand Calculator Instant Replay

PUNCH	DISPLAY	MEANING
C	0.	Clear
8 5	85.	85
× 2 1 2 0 0 0	212000.	Times 212000
÷ 1 3 5	135.	Divided by 135
=	133481.48	Answer

Problem Bonnie was able to save $265 in 3 months. At this rate, how long will it take her to save $1200?

ANSWER Again, we assume that Bonnie will continue to save at the same rate.

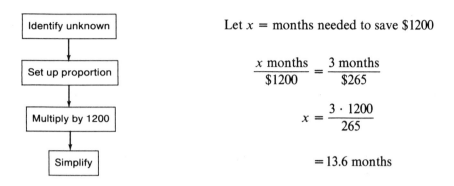

Let x = months needed to save $1200

$$\frac{x \text{ months}}{\$1200} = \frac{3 \text{ months}}{\$265}$$

$$x = \frac{3 \cdot 1200}{265}$$

$$= 13.6 \text{ months}$$

Problem Fran, a social worker, spends 180 hours with 50 clients. She has 600 hours to spend with clients. How many clients can she see?

ANSWER We assume she sees her clients at the same rate.

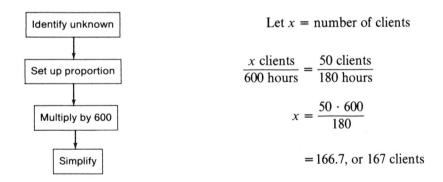

Let x = number of clients

$$\frac{x \text{ clients}}{600 \text{ hours}} = \frac{50 \text{ clients}}{180 \text{ hours}}$$

$$x = \frac{50 \cdot 600}{180}$$

$$= 166.7, \text{ or } 167 \text{ clients}$$

Since 0.7 clients makes no sense, we rounded the answer up to the next highest number.

Problem One kilogram is 2.2 pounds. How many kilograms does a 160-pound man weigh?

ANSWER   We set this problem up as a proportion.

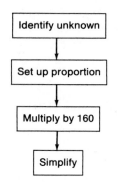

Let x = weight in kg

$$\frac{x \text{ kg}}{160 \text{ lb}} = \frac{1 \text{ kg}}{2.2 \text{ lb}}$$

$$x = \frac{160}{2.2}$$

$$= 72.7 \text{ kg}$$

Proportions can be used to go from one unit system to another. In the next section, we will see another method for doing this.

Problem   If 50,000 Americans out of 220,000,000 are killed every year in traffic accidents, how many are killed per 100,000?

ANSWER   Rates are frequently given per 100,000.

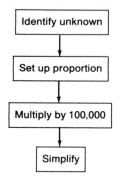

Let x = number killed per 100,000

$$\frac{x \text{ killed}}{100,000 \text{ total}} = \frac{50,000 \text{ killed}}{220,000,000 \text{ total}}$$

$$x = \frac{100,000 \cdot 50,000}{220,000,000}$$

$$= 22.73$$

Then about 23 out of every 100,000 Americans will die on the road in a year.

$$\frac{x}{1.50} \times \frac{110}{63} \div = 2.62 \text{ nails}$$

**PROBLEM
SET 2–4–2**

1. Define or discuss proportion.

2. A bag of 110 nails costs 63 cents. How many nails can one get for $1.50?

3. Thirty miles per hour is equivalent to 44 feet per second. If a man is driving 55 miles per hour, how fast is he going in feet per second?

4. In her luxury car, Pam drove 123 miles on 9.5 gallons of gas. How far can she go on 25 gallons of gas?

5. In his subcompact, Howie drove 204 miles on 8.2 gallons of gas. How many gallons of gas does he need to drive 490 miles?

Even
2–16

6. It cost Judy $100 to carpet 12 square yards of floor. What will it cost her to carpet 45 square yards of floor?

7. It cost $8 for enough wallpaper to cover 55 square feet. How much will it cost to wallpaper a room that is 320 feet square?

8. If a $10,000 life insurance policy costs $180 a year, how much will a $25,000 policy cost?

9. If 1 inch is 2.54 centimeters, how many inches are in 100 centimeters?

10. If 2 cups of flour will make 15 pancakes, how much flour is needed to make 250 pancakes?

11. Out of a group of 128, 17 are joggers. At this rate, how big should a group be to have 75 joggers?

12. The ratio of men to women in a certain field is 5 to 2. Hojo, Inc., employs 500 people. How many women should be on their staff to be in proportion to the total ratio in the field?

13. A 2-meter stick casts a 1.3-meter shadow. How tall is a tree that casts a 21-meter shadow?

14. Art, Bob, and Charlie form an investment partnership. Art puts up $3000, Bob puts up $4000, and Charlie puts up $5000. Their investments are now worth $19,000. What is Art's share? Bob's share? Charlie's share?

(Sam Falk/Monkmeyer Press Photo Service)

15. A survey indicates that out of 150 people interviewed, 9 are unemployed. Using this, how many are unemployed in a labor force of 85,000,000 people?

16. Fred Frugal is a runner, but is too cheap to buy a stop watch. So he counts as he runs: one-one-thousand, two-one-thousand, three-one-thousand, and so on. He knows that he makes 42 counts in 60 seconds. If he runs a mile in 208 counts, how many seconds is this; how many minutes and seconds?

17. If there are 1,000,000 divorces a year out of 220,000,000 people, what is the rate per 1000?

2–5 UNITS CONVERSIONS AND THE METRIC SYSTEM

Problem One kilogram is 2.2 pounds. How many kilograms does a 125-pound woman weigh?

ANSWER Earlier we saw how to use proportions to solve this type of problem. Here we will use a different method. We put what we want (kilograms) on the right, and what we have (pounds) on the left. We also save space for a conversion factor.

$$\text{lb} \times \boxed{?} = \text{kg}$$

We want the conversion factor to somehow cancel out the pounds and leave us with kilograms. So we set up a fraction to do that.

$$\cancel{\text{lb}} \times \frac{\text{kg}}{\cancel{\text{lb}}} = \text{kg}$$

Notice that pounds will cancel pounds and kilograms will remain. Now that the units are right, we fill in the numbers and multiply.

$$125 \, \cancel{\text{lb}} \times \frac{1 \, \text{kg}}{2.2 \, \cancel{\text{lb}}} = \frac{125}{2.2} \, \text{kg}$$

$$= 56.8 \, \text{kg}$$

So 125 pounds is the same as 56.8 kilograms.

Question How do we convert a measurement from one unit to another unit?

ANSWER We use the *multiplying by 1 trick*. Why is it called that? For example, since 1 kilogram = 2.2 pounds, the fraction $\frac{1 \, \text{kilogram}}{2.2 \, \text{pounds}}$ is really just the number 1 (since the numerator and denominator are equal.)

The secret is to set up the units so that everything cancels except the units that we want.

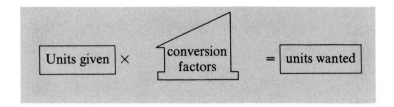

Table 2–2 Common Units Conversions

Length		
1000 mm = 1 m	5280 ft = 1 mi	mm = millimeter
100 cm = 1 m	1 in. = 2.54 cm	cm = centimeter
1000 m = 1 km	1 ft = 0.305 m	m = meter
10 mm = 1 cm	1 mi = 1.61 km	km = kilometer
12 in. = 1 ft	1 m = 1.09 yd	in. = inch
		ft = foot
		yd = yard
		mi = mile
Weight (or mass)		
1000 mg = 1 g	2000 lb = 1 ton	mg = milligram
1000 g = 1 kg	1 oz = 28.4 g	g = gram
1000 kg = 1 MT	1 kg = 2.2 lb	kg = kilogram
16 oz = 1 lb		MT = metric ton
		oz = ounce
		lb = pound
Liquid Capacity		
1000 ml = 1 L	4 qt = 1 gal	ml = milliliter
2 cups = 1 pt	1 qt = 0.95 L	L = liter
2 pt = 1 qt		pt = pint
		qt = quart
		gal = gallon
Time		
60 sec = 1 min	7 days = 1 week	sec = second
60 min = 1 hr	365 days = 1 yr	min = minute
24 hr = day		hr = hour
		yr = year
Temperature		
$°F = 32 + \dfrac{9}{5}°C$	$°C = \dfrac{5}{9}(°F - 32)$	°F = degrees Fahrenheit
		°C = degrees Celsius

To convert units:

1. Put the units we *want* to the right of the equal sign, and the units we *have* on the left.

2. Use Table 2–2 to find the conversion factors that will connect what we have to what we want. Each fact becomes a fraction equal to 1. (We may have to use extra units as middle steps in the process.)

3. Arrange the units so that numerators and denominators will cancel out unwanted units.

4. Fill in all numbers and multiply as fractions; then simplify.

Problem Betty is 5 feet, 5 inches tall. How tall is she in meters?

ANSWER First we must write Betty's height in feet alone. Since there are 12 inches in a foot,

$$5 \text{ ft, 5 in.} = 5\tfrac{5}{12} \text{ ft} = 5.42 \text{ ft}$$

We now use Table 2–2 to find the relation between feet and meters: 1 foot = 0.305 meter.

Converting feet to meters.

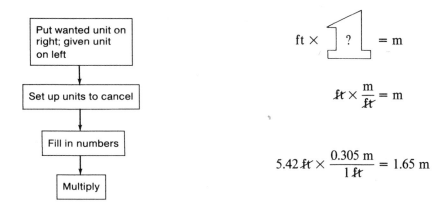

Betty is 1.65 meters tall.

Problem   Erika is in France and buys 30 liters of gas. How many gallons is this?

ANSWER   Table 2–2 tells us that 4 quarts = 1 gallon and 1 quart = 0.95 liter. We will need both of these facts. We have two conversions, liters → quarts → gallons.

We put what we want (gallons) on the right, and what we have (liters) on the left. We leave room for conversion factors.

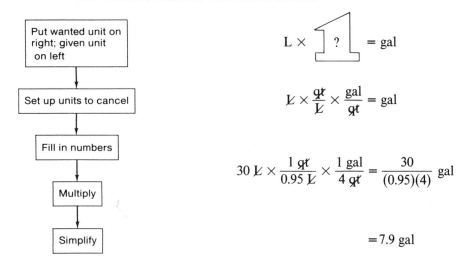

So 30 liters is 7.9 gallons.

Problem   Billy Edd works in a mine and digs 16 tons of no. 9 coal. How many kilograms is this?

ANSWER   The *no. 9* term has nothing to do with the mathematics. It is just a type of coal. (Not every number in a mathematics problem is useful.)

There is no direct connection between tons and kilograms. So we use pounds as a middle step. We have tons → pounds → kilograms.

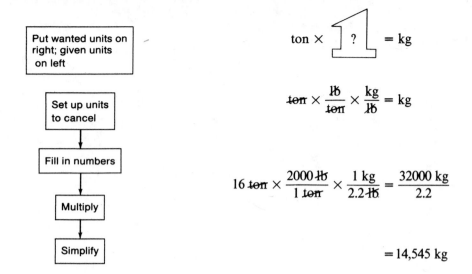

Put wanted units on right; given units on left	$\text{ton} \times \boxed{?} = \text{kg}$
Set up units to cancel	$\cancel{\text{ton}} \times \dfrac{\cancel{lb}}{\cancel{\text{ton}}} \times \dfrac{\text{kg}}{\cancel{lb}} = \text{kg}$
Fill in numbers Multiply	$16\,\cancel{\text{ton}} \times \dfrac{2000\,\cancel{lb}}{1\,\cancel{\text{ton}}} \times \dfrac{1\,\text{kg}}{2.2\,\cancel{lb}} = \dfrac{32000\,\text{kg}}{2.2}$
Simplify	$= 14{,}545\,\text{kg}$

Billy Edd must dig 14,545 kilograms of no. 9 coal each day.

Problem Bill can run 440 yards in 50 seconds. How fast is this in miles per hour?

ANSWER Here we have a *double conversion*: yards to miles, and seconds to hours. It will take several conversion factors, but it is not difficult.

We convert as follows: yards → feet → miles, and seconds → minutes → hours.

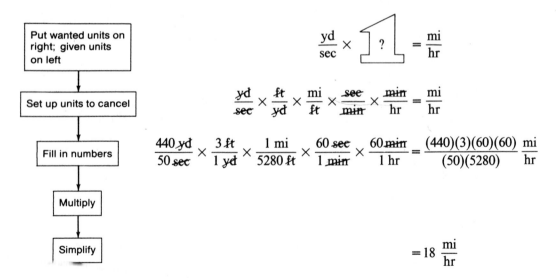

Put wanted units on right; given units on left	$\dfrac{\text{yd}}{\text{sec}} \times \boxed{?} = \dfrac{\text{mi}}{\text{hr}}$
Set up units to cancel	$\dfrac{\cancel{\text{yd}}}{\cancel{\text{sec}}} \times \dfrac{\cancel{\text{ft}}}{\cancel{\text{yd}}} \times \dfrac{\text{mi}}{\cancel{\text{ft}}} \times \dfrac{\cancel{\text{sec}}}{\cancel{\text{min}}} \times \dfrac{\cancel{\text{min}}}{\text{hr}} = \dfrac{\text{mi}}{\text{hr}}$
Fill in numbers Multiply	$\dfrac{440\,\cancel{\text{yd}}}{50\,\cancel{\text{sec}}} \times \dfrac{3\,\cancel{\text{ft}}}{1\,\cancel{\text{yd}}} \times \dfrac{1\,\text{mi}}{5280\,\cancel{\text{ft}}} \times \dfrac{60\,\cancel{\text{sec}}}{1\,\cancel{\text{min}}} \times \dfrac{60\,\cancel{\text{min}}}{1\,\text{hr}} = \dfrac{(440)(3)(60)(60)}{(50)(5280)}\,\dfrac{\text{mi}}{\text{hr}}$
Simplify	$= 18\,\dfrac{\text{mi}}{\text{hr}}$

Thus a very fast sprinter, like Bill, can run about 18 miles per hour.

MATH FLASH

12345678

√ % +/- ÷ MC
7 8 9 X MR
4 5 6 − M−
1 2 3 + M+
CE 0 . = C

Hand Calculator Instant Replay

PUNCH	DISPLAY	MEANING
C	0.	Clear
4 4 0	440.	
× 3	3.	Multiply terms
× 6 0	60.	in numerator
× 6 0	60.	
÷ 5 0	50.	Divide by terms
÷ 5 2 8 0	5280.	in denominator
=	18.	Answer

The Metric System

Question   What is the metric system?

ANSWER   In the world today there are two major systems of measurement. The **English system** is the system that most Americans use in their daily lives: inches, feet, miles; ounces, pounds; quarts, gallons; and so on. Although we are accustomed to these measures, they are really very clumsy to use. (For instance, there are 1728 cubic inches in a cubic foot.) Historically, the units in the English system developed without much plan or coordination.

The **metric system** is a very carefully planned system of measurement. It was devised about 1795 and has gained almost universal acceptance since then. The United States is the only major nation in the world that does not use the metric system exclusively. Table 2–3 shows us the major differences between these two systems.

The beauty of the metric system is that it is based on a decimal system. That is, all conversions involve only multiplying or dividing by powers of 10. This means moving the decimal point to the right or left. We will see this shortly in the examples of metric conversions that we will work.

Since the rest of the world is metric, American industry is committed to going metric in order to compete in world trade markets. Congress and the president have also committed the American people to going metric. No doubt, this will be easier for children than for adults.

Table 2–3 Standard Units in English and Metric Systems

Quantity	English system	Metric system or SI (le Système International d'Unités)
Length	Inch (in.) Foot (ft) Yard (yd) Mile (mi)	Millimeter (mm) Centimeter (cm) Meter (m) Kilometer (km)
Mass (weight)	Ounce (oz) Pound (lb) Ton (t)	Milligram (mg) Gram (g) Kilogram (kg) Megagram (Mg) (metric ton)
Liquid capacity	Teaspoon (tsp) Cup (c) Quart (qt) Gallon (gal)	Milliliter (ml) Liter (L)
Time	Second (sec) Minute (min) Hour (hr)	Second (s) Minute (min) Hour (h)
Temperature	Degree Fahrenheit (°F)	Degree Celsius (°C)

Just as when learning a foreign language, it is suggested that people try not to translate from English into metric, but rather to try to *think metric*. For instance,

A meter is little longer than a yard.
A liter is a little bigger than a quart.
A kilogram is a little heavier than 2 pounds.
A paper clip weighs about 1 gram.
A dime is about 1 millimeter thick.
A teaspoon is about 5 milliliters.
The driving distance from Boston to San Francisco is just over 5000 kilometers.
Fifty miles per hour is about 80 kilometers per hour.
Room temperature is about 18 to 20°C.
Body temperature is 37°C.

A dime is about 1 millimeter thick.

The author has chosen to include both systems of measurement in the book since the metric system is the system of the future, although most American readers, at present, can relate to only the English system.

Problem   A can of coffee weighs 1.5 kilograms. How many grams is this?

ANSWER   We see in Table 2–2 that 1 kilogram = 1000 grams. Thus

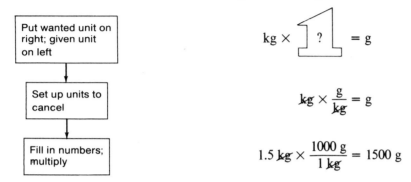

Notice that this is the same as moving the decimal point three places to the right: 1.500.

Problem   Tina is 173 centimeters tall. How many meters is this?

ANSWER   We see in Table 2–2 that 1 meter = 100 centimeters. So

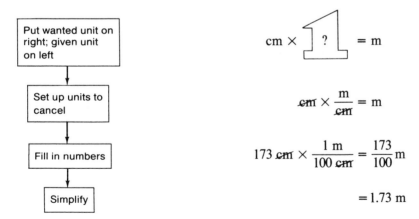

Again we obtain the answer by simply moving the decimal point two places to the left: 1.73.

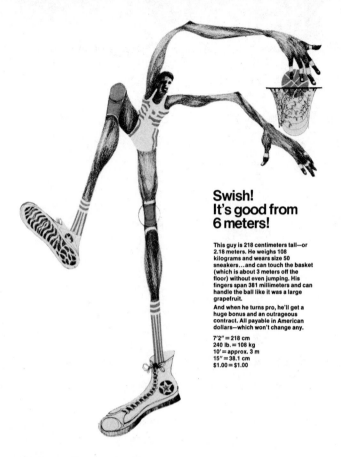

**Swish!
It's good from
6 meters!**

This guy is 218 centimeters tall—or
2.18 meters. He weighs 108
kilograms and wears size 50
sneakers...and can touch the basket
(which is about 3 meters off the
floor) without even jumping. His
fingers span 381 millimeters and can
handle the ball like it was a large
grapefruit.

And when he turns pro, he'll get a
huge bonus and an outrageous
contract. All payable in American
dollars—which won't change any.

7'2" = 218 cm
240 lb. = 108 kg
10' = approx. 3 m
15" = 38.1 cm
$1.00 = $1.00

(Nekoosa Papers, Inc.)

Problem   A can of soda pop is about 0.355 liter. How many milliliters is this?

ANSWER   We see in Table 2–2 that 1 liter = 1000 milliliters. Therefore,

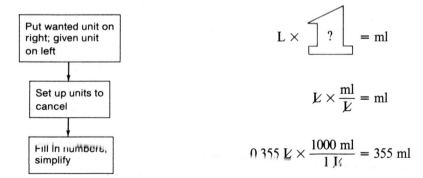

Put wanted unit on right; given unit on left	$$L \times \boxed{} = ml$$
Set up units to cancel	$$\not{L} \times \frac{ml}{\not{L}} = ml$$
Fill in numbers, simplify	$$0.355 \not{L} \times \frac{1000\ ml}{1\ \not{L}} = 355\ ml$$

Once again we see how simple conversion really is. We multiplied by 1000, or
moved the decimal point three places to the right: 0.355.

Problem   The driving distance from Chicago to New York is about 1350 kilometers. How many centimeters is this?

ANSWER   This is a double conversion: kilometers → meters → centimeters. We see in Table 2–2 that 1 kilometer = 1000 meters, and 1 meter = 100 centimeters. Thus we get

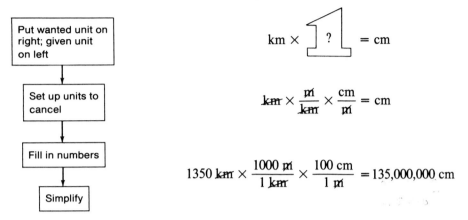

Even though this was a double conversion, it was still only a decimal shift: three places right for the 1000 meters and two places right for the 100 centimeters. Together this is five places right: 135000000.

By now, the appeal of the metric system must be clear. In fact, it will become even clearer when we do areas and volumes in Chapter 3. These are downright ugly in the English system, but they require just a decimal shift in the metric system. In Chapter 4, we will look at the relation between the Celsius and Fahrenheit temperature scales.

Question   Can we use this conversion method to solve other problems?

ANSWER   Yes, as long as we are careful to arrange the units to cancel properly, we can use this method.

Problem   Ron gets 13 miles to a gallon on his car. Gasoline costs 71 cents per gallon. How much does it cost Ron for gas for each mile that he drives?

ANSWER   Here the desired units are cents per mile. We are given the values 13 miles per gallon and 71 cents per gallon. We must arrange these so that cents ends up on top and miles on botton.

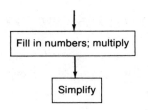

Fill in numbers; multiply

Simplify

$$\frac{71¢}{1 \text{ gal}} \times \frac{1 \text{ gal}}{13 \text{ mi}} = \frac{71¢}{13 \text{ mi}}$$

$$= 5.46 \frac{¢}{\text{mi}}$$

PROBLEM SET 2–5–1

1. Define or discuss:
   (a) Metric system.
   (b) English system.

Complete the following table. (You need not do a box with an × in it.)

	mm	cm	m	km	in.	ft	yd	mi
2.				×	18			×
3.			0.4	×				×
4.	×	×			×		660	
5.	35			×			×	×
6.	×	×		5	×			
7.	×			×		14		×
8.	×	×			×			100
9.				17	×	×		

Complete the following table. (You need not do a box with an × in it.)

	mg	g	kg	MT	oz	lb	ton
10.	500			×			×
11.				×	8		×
12.	×					185	
13.			75				
14.		300		×			×

Complete the following table. (You need not do a box with an × in it.)

	ml	L	cup	pt	qt	gal
15.			2			
16.		5				
17.	×		×			1000
18.	100					×
19.					12	
20.				2		

(© Cary Herz/Nancy Palmer Photo Agency)

21. Fred is 5 feet, 10 inches tall. How tall is he in meters?

22. A certain canoe is 4 meters long. How many feet is this?

23. Bob used 35-millimeter film. How many inches is this?

24. There are 6500 students on a certain campus. The average height is 5.5 feet. If everyone on campus were laid end to end, how long would this be in miles? in kilometers?

25. The distance from Chicago to St. Louis is 314 miles. How many kilometers is this?

26. An Olympic event is the 1500-meter run. How many yards is this? How many miles is this?

27. Pat weighs 145 pounds. How many kilograms is this?

28. Ed orders 3 metric tons of dirt for his yard. How many pounds is this?

29. A certain pill has 500 milligrams of vitamin C. How many ounces is this?

30. A car weighs 3500 pounds. How many kilograms is this? How many metric tons?

31. A hamburger is made with $\frac{1}{4}$ pound of meat. How many grams is this?

32. Mary put 35 liters of gasoline in her car. How many gallons is this?

33. A medicine dropper holds 10 milliliters. How many liters is this? How many cups?

34. A swimming pool holds 70,000 gallons of water. How many liters is this?

35. A coffee maker makes 20 cups of coffee. How many liters is this?

36. A tablespoon is $\frac{1}{16}$ cup. How many milliliters is 1 tablespoon?

37. For liquids, 1 pint = 1 pound. Bob buys 24 cans of beer, each of which is 12 ounces. How many gallons is this? How many liters?

38. Two towns are 24 kilometers apart. How many meters is this?

39. A certain bottle of wine holds 750 milliliters. How many liters is this?

40. Most adults need about 70 grams of protein per day. How many kilograms is this?

41. Bronco Cyzinski weighs 112 kilograms. How many metric tons is this?

42. Sandy is 162 centimeters tall. How many meters is this?

43. A bottle holds 0.4 liter. How many milliliters is this?

IMPORTANT WORDS

Distributive law (distributivity) (2–1)
English system (2–5)
Inequality (2–1)

Metric system (SI) (2–5)
Proportion (2–4)
Ratio (2–4)

REVIEW EXERCISES

Use distributivity to remove the parentheses.

1. $5(a - 3)$ **2.** $x(6 - x - y)$

Simplify by pulling out the common term.

3. $7a + ma$ **4.** $ab + b^2 - 2b$

Solve for x.

5. $x + 12 = 25$ **6.** $6x = 31$

7. $7x - 11 = 13$ **8.** $\frac{2}{3}x - 5 = 12$

9. $4x - 2 = 6x + 11$ **10.** $3(x + 5) = 2(6x + 1) - 3$

Indicate the following inequalities on a number line.

11. $x < 6$ **12.** $x \geqslant -2$ **13.** $-3 \leqslant x \leqslant 12$

Translate the following English into mathematics.

14. Five percent of the people in the world.

15. Two hundred twenty-five dollars is $\frac{2}{11}$ of his monthly income.

Solve the following problems.

16. Bob spent $\frac{2}{3}$ of his paycheck for a car repair. He spent $186 on the repair. How much was his paycheck?

17. Tony's college is 384 miles from his home. If he can average 45 miles per hour, how long will it take him to drive this distance?

18. What is 17% of 850?

19. Two hundred fifty is 12% of what?

20. Three hundred fifty is what percent of 1110?

21. Pam gets a 20% discount on a $242 chair. How much does she pay for the chair?

22. There are 56 men and 47 women in a certain lounge. What is the percent of women?

23. Roberta gets a $60-per-month raise. This is a $5\frac{1}{2}$% raise. What was Roberta's monthly income before the raise?

24. If Rachel can drive 212 miles on 10.1 gallons of gas, how many gallons will she need to drive 3000 miles cross country?

25. If 159.6 out of 100,000 people die of cancer each year, how many people out of 220,000,000 Americans die of cancer each year?

26. Tom, Dick, and Harriet buy a store together. Tom puts up $6000, Dick puts up $7000, and Harriet puts up $9000. Four years later they sell the store for $40,000. What is Harriet's share?

27. How many meters tall is a 6-foot 2-inch man?

28. How many kilograms is a 118-pound woman?

29. How many grams are in a 0.75-kilogram box?

30. How many gallons are in 21 liters?

31. Fifty miles per hour is how many meters per second?

32. A 1500-meter race is how many kilometers?

*East wing, National Gallery of Art, Washington, D.C.
(Dennis Brack/Black Star)*

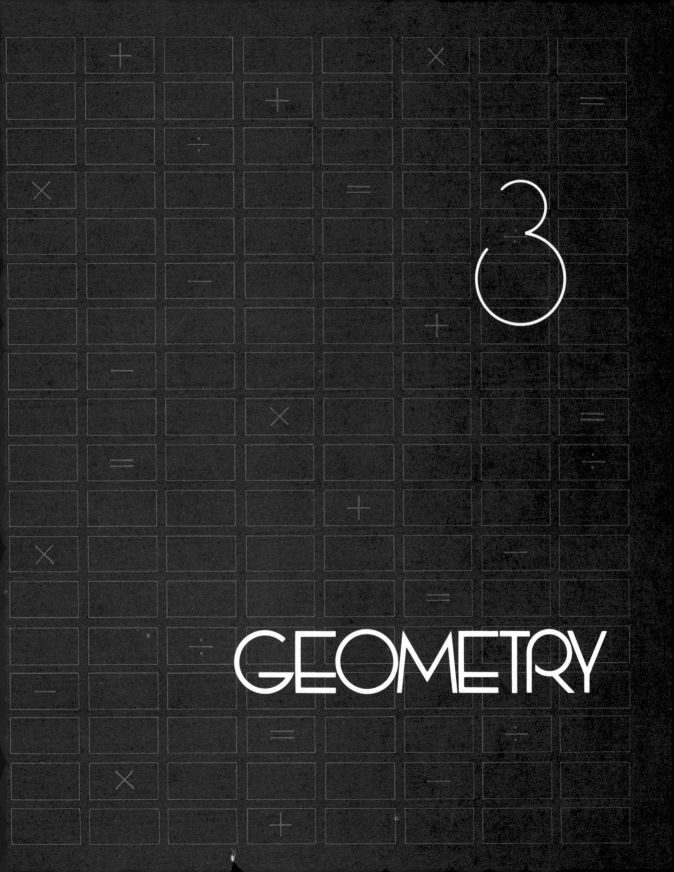

3

GEOMETRY

Question   What is geometry?

ANSWER   **Geometry** comes from two Greek words meaning *measuring the earth*. Although there are other aspects to geometry, we will concentrate in this chapter on measurement.

We will study the four basic measurements of geometry: length, area, volume, and angles.

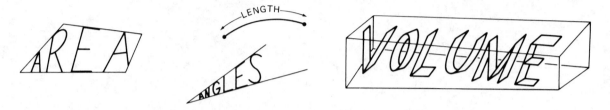

Question   What are the basic terms of geometry?

ANSWER   Rather than attempting to give technical definitions of these terms, we will instead give simple sketches of them. This is one of the nice aspects of geometry: Just about everything can be drawn. All these terms will be discussed further throughout the chapter.

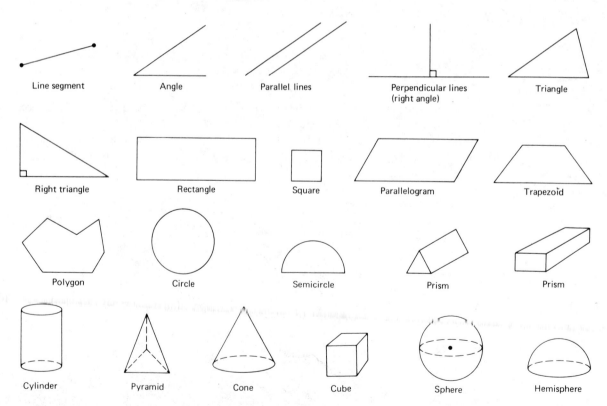

3-1 ANGLES AND LENGTH

Question How do we measure angles?

ANSWER The most common way to measure **angles** is in degrees. An entire circle is 360°, and all other angles are some fraction of 360°. Some examples of angles are given.

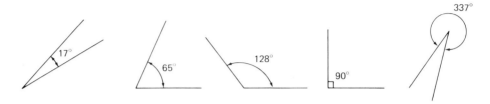

Sometimes, we can make a good approximation just by looking at the angle. Usually, we can come within about 5° either way. To measure an angle accurately, we use a protractor.

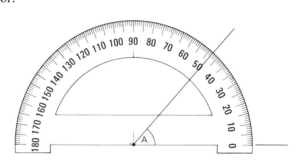

The figure shows how the protractor is used. The center of the semicircle of the protractor is laid on the **vertex** (common point) of the angle. One edge of the protractor is laid on one side of the angle. We can read the number of degrees by seeing where the other side of the angle passes through the protractor. In this case, the particular angle is about 48°.

(© Joel Gordon)

121

Problem A 48° angle is what percent of an entire circle.

ANSWER Since an entire circle is 360°, we can translate this to a percentage problem.

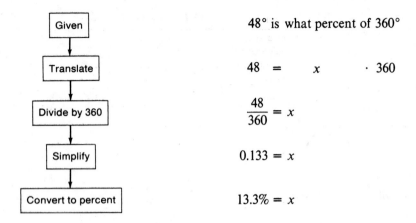

48° is what percent of 360°

| Given |

Translate $48 \ = \ x \ \cdot \ 360$

Divide by 360 $\dfrac{48}{360} = x$

Simplify $0.133 = x$

Convert to percent $13.3\% = x$

Pie diagrams are often used to visualize percents. In this case we must convert all quantities to degrees.

Problem Louise is the treasurer of a townhouse group. This year's expenses are as follows. How does she make a pie diagram to display this information?

Insurance	$ 5,600
Administration	11,000
Maintenance	13,000
Reserves	7,500
Total	$37,100

ANSWER A **pie diagram** is a full circle that is sliced into appropriate proportions. We will do "Insurance" in detail to demonstrate this. What we are looking for is the number of degrees that make the slice for "Insurance."

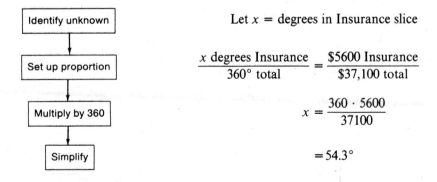

Identify unknown   Let x = degrees in Insurance slice

Set up proportion $\dfrac{x \text{ degrees Insurance}}{360° \text{ total}} = \dfrac{\$5600 \text{ Insurance}}{\$37,100 \text{ total}}$

Multiply by 360 $x = \dfrac{360 \cdot 5600}{37100}$

Simplify $= 54.3°$

We will need a 54.3° slice of the pie for insurance. If we solve for the Administration, Maintenance, and Reserves slices in the same way, we get the following table of results. The accompanying figure shows the final pie diagram.

Item	Amount ($)	Angle
Insurance	$5,600	54°
Administration	11,000	107°
Maintenance	13,000	126°
Reserves	7,500	73°
Total	$37,100	360°

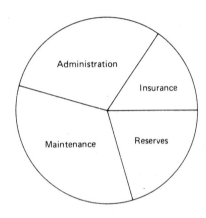

Problem Leon is designing a clock face and wants to know at what angles to place the numerals.

ANSWER There are 12 divisions on a clock face: 12, 1, 2, . . . , 10, 11. There are 360° in the whole circle of a clock face. So each division gets 360/12 = 30°. Thus we get the following table.

Numeral	12	1	2	3	4	5	6	7	8	9	10	11
Angle from top	0°	30°	60°	90°	120°	150°	180°	210°	240°	270°	300°	330°

PROBLEM SET 3-1-1
1. Define or discuss:
   (a) Angle.
   (b) Pie diagram.

Use a protractor to measure angles in problems 2 through 7.

2.

3.

4.

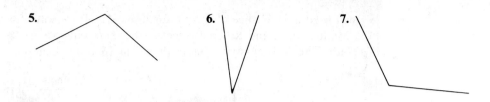

5.   **6.**   **7.**

Complete the following table.

	Angle	Percent of whole circle
8.	90°	?
9.	52°	?
10.	?	16
11.	?	45
12.	225°	?
13.	?	95

14. The class sizes at Millard Fillmore College are given below. Draw a pie diagram to illustrate this.

Class	Freshman	Sophomore	Junior	Senior	Total
Size	520	470	430	410	1,830

15. Mary does a study of how she spends her time in a typical week ($7 \times 24 = 168$ total hours). Draw a pie diagram to illustrate her findings.

Activity	In class	Studying	Sleeping	Eating	Dressing, washing	Socializing	TV	Other	Total
Time (hr.)	15	25	56	14	10	25	10	13	168

16. Al and Nancy set up a budget for themselves. Draw a pie diagram to show how they budget their combined $1500-per-month income.

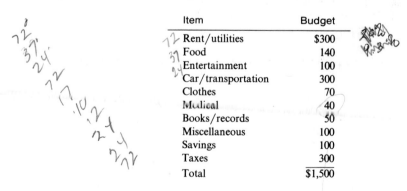

Item	Budget
Rent/utilities	$300
Food	140
Entertainment	100
Car/transportation	300
Clothes	70
Medical	40
Books/records	50
Miscellaneous	100
Savings	100
Taxes	300
Total	$1,500

Setting up a budget. (Mimi Forsyth/Monkmeyer Press Photo Service)

17. Marilyn is designing a nine-sided flower pot. What should each angle be? Draw the figure yourself. $9 \div 360° =$

18. The distance around the world is about 25,000 miles. The distance from New York to Los Angeles is about 2800 miles. How many degrees is this?

19. There are 24 time zones in the world. How many degrees are in each time zone?

20. Chicago is about 42° latitude. If the world is about 25,000 miles around, how many miles is Chicago from the equator?

Length

Question What is length?

ANSWER **Length** is the measurement of any curve or line segment. It is the distance from one end to the other, measured along the curve or line segment. Some examples are shown.

Question How do we measure length?

ANSWER To measure length, we have standard units. In one system the standard units are 1 inch, 1 foot, 1 yard, and 1 mile. In another system, the standard units are 1 centimeter, 1 meter, and 1 kilometer.

These standards are supposed to be defined in a way to be uniform for all people. Originally, the inch was the width of a man's thumb. Then an inch was defined to be three barleycorns laid end to end. Needless to say, these are not very exact definitions.

Originally, the meter was 1/10,000,000th the distance from the North Pole to the equator. Now the meter is defined to be 1,650,763.73 wavelengths of orange-red light of excited krypton of mass number 86. That isn't very useful, is it? For our purposes, length is the measurement we get from a 15-cent plastic ruler sold at every corner store.

Question   What is perimeter?

ANSWER   **Perimeter** is the distance around a closed geometric figure, such as a triangle, rectangle, or circle.

Question   How do we find the perimeter of a polygon?

ANSWER   A **polygon** is a closed figure made up of line segments attached together, such as triangles, rectangles, and octagons. The formula for perimeter is as follows:

Formula 3–1–1

$$P_{polygon} = \text{sum of lengths of sides}$$

This is one of the simplest formulas in geometry. We simply add up all the lengths of the sides.

Examples   For the figures shown, we use Formula 3–1–1.

1. $P = 8 + 10 + 12 = 30$

2. $P = 7 + 20 + 7 + 20 = 54$

Notice that we filled in the lengths of the missing sides since they must also be 7 and 20.

Problem   Paul is putting floor trim around the walls of the room with the floor plan shown.

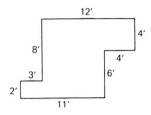

(a) How much wood does he need?
(b) At 21 cents per foot, how much will this cost?

ANSWER   (a) By Formula 3–1–1, we get

$$P = 12 + 4 + 4 + 6 + 11 + 2 + 3 + 8 = 50 \text{ ft}$$

So Paul needs 50 feet of wood.
(b) To find the cost, we multiply:

$$50 \text{ ft} \times \frac{21¢}{1 \text{ ft}} = 1050¢ = \$10.50$$

Question   What is the perimeter of a circle?

ANSWER   A **circle** is a shape in which all the points are the same distance from a center. This distance (from center to circle) is called the **radius**. The distance across the circle, through the center, is the **diameter**. Notice that the diameter is twice the radius. Generally, the distance around a circle is called the **circumference**.

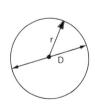

pie fixed
constant π

3.14

The formulas are

Formula 3–1–2

$$\boxed{C = 2\pi r}$$

or

Formula 3–1–3

$$\boxed{C = \pi D}$$

where C = circumference, π = 3.14 (approximately), r = radius, and D = diameter. Since the diameter is twice the radius, these formulas are really the same.

Examples   We will use Formulas 3–1–2 and 3–1–3 to find the following circumferences:

1. $C = 2\pi r = 2(3.14)(10) = 62.8$

2. $C = \pi D = (3.14)(25) = 78.5$

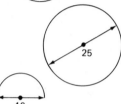

3. $C = \frac{1}{2} \cdot \pi D = (0.5)(3.14)(16) = 25.12$

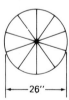

Notice that if we are given the radius we use Formula 3–1–2; if we are given the diameter we use Formula 3–1–3; if we have half a circle, we use the formula and then divide by 2.

Question   A bike has a 26-inch tire. How many times will the tire go around in 1 mile of travel?

ANSWER   We first compute the circumference, which is one turn.

$$C = \pi D = (3.14)(26) = 81.64 \text{ in. per turn}$$

Now we convert this to turns per mile.

$$1 \text{ mile} \times \frac{1 \text{ turn}}{81.64 \text{ in.}} \times \frac{12 \text{ in.}}{1 \text{ ft}} \times \frac{5280 \text{ ft}}{1 \text{ mi}} = 776.09 \text{ turns}$$

The wheel turns 776 times per mile.

Hand Calculator Instant Replay

PUNCH	DISPLAY	MEANING
C	0.	Clear
3 · 1 4	3.14	π
× 2 6	26.	Times D
=	81.64	Circumference
C	0.	Clear
1 2	12.	Conversion
× 5 2 8 0	5280.	factors
÷ 8 1 · 6 4	81.64	
=	776.09015	Answer

Problem Stu is planting a circular garden. He has 20 feet of fence to use to enclose the garden. What diameter will the circle have?

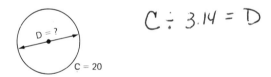

$$C \div 3.14 = D$$

ANSWER The 20 feet of fence is the circumference. We now use Formula 3–1–3 to solve for the diameter.

Formula 3–1–3

$$C = \pi D$$

Substitute

$$20 = (3.14)D$$

Divide by 3.14

$$\frac{20}{3.14} = D$$

Simplify

$$6.37 \text{ ft} = D$$

Problem   The indoor track at the Bean Gymnasium is shown in the sketch.

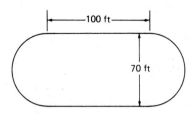

(a) How far around is one lap?
(b) How many laps to a mile?

ANSWER

(a) To find the perimeter, we break the track up into four pieces: two straightaways (100 feet each) and two semicircles (diameter 70 feet). Then we add all the lengths together.

$$2 \text{ straightaways} = 2 \times 100 = \qquad\qquad 200.00$$

$$2 \text{ semicircles} \quad = 2(\tfrac{1}{2}\pi D) = (3.14)(70) \quad = 219.80$$
$$\overline{\phantom{2 \text{ semicircles} \quad = 2(\tfrac{1}{2}\pi D) = (3.14)(70) \quad}P = 419.80}$$

So one lap is 419.80 feet.

(b) To find out how many laps to a mile, we treat this as a conversion problem.

$$1 \text{ mi} \times \boxed{?} = \text{laps}$$

$$1 \text{ mi} \times \frac{5280 \text{ ft}}{1 \text{ mi}} \times \frac{1 \text{ lap}}{419.80 \text{ ft}} = 12.6 \text{ laps}$$

Notice that we had to use feet as an extra unit to determine that 12.6 laps is a mile at this track.

**PROBLEM
SET 3–1–2**

1. Define or discuss:
   (a) Length.   (b) Perimeter.   (c) Radius.   (d) Circle. *distance out from center*
   (e) Polygon. *lil seg*   (f) Diameter.   (g) Circumference. *distance around center*

Find the following perimeters.

2.

3.

4.   *11.6*

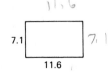

5.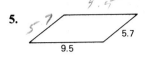

6.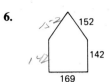

7.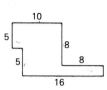

Find the following circumferences.

8.

9.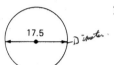

10.

11.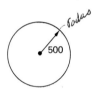

12.

13.

Find the following perimeters.

14.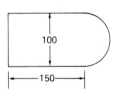

15.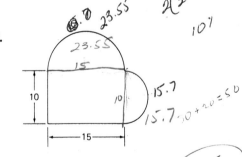

16.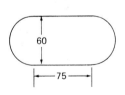

17.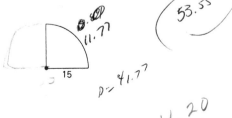

A sketch of Ron's family room is shown. He wants to put molding along the floor boards.

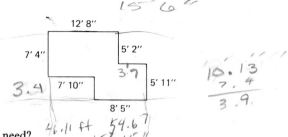

18. How many feet will he need?

19. At 17 cents per foot, what will it cost?

20. In the same room, Ron is putting in a suspended ceiling. This requires that wall angles be put around the perimeter of the wall. If these come in 8-foot lengths, how many will Ron need?

21. Jenny is planting a garden next to her townhouse. The rules are that she can only plant up to 3 feet from the wall. If she has 20 feet of fence, how long can she make the garden?

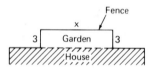

22. Alice wants to plant a garden in a corner. She has 4 meters of fence to make a quarter-circle. What will the radius be?

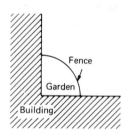

The indoor track in the Joque Gymnasium is shown.

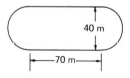

23. How far around is one lap?

24. How many laps are there to 1 kilometer?

25. How many laps are there to 1 mile?

A sketch is given of the track surrounding the football field at Beaverville College. They want the track to be a quarter-mile around.

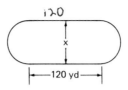

26. How many yards are in a quarter-mile?

27. How wide must the field be to be exactly a quarter-mile?

3-2 AREA

Question   What is area, and why is it important?

ANSWER   Basically, **area** is the measurement of a flat surface. We use area when we paint or paper walls, carpet or tile floors, plant grass or gardens, plan office or home space, calculate the light coming through a window or camera lens, and so on.

Question   How do we measure area?

ANSWER   Just as length had certain standard units, such as 1 inch, 1 centimeter, 1 foot, 1 meter, and so on, area has its own standard units. These are squares whose sides are the standard lengths. Two examples, 1 square inch and 1 square centimeter, are shown. The standard notation for these units is in.^{2} and cm^2. Most floor tiles are now 1 foot by 1 foot. This is 1 square foot. The area of a floor is then the exact number of tiles needed to cover the floor.

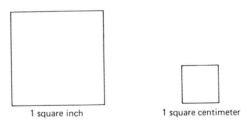

1 square inch   1 square centimeter

Simple Figures

Question   What is the formula for the area of a polygon?

ANSWER   There is no single formula. The simplest formula is for the area of a **rectangle**.

Formula 3–2–1

$$A_{\text{rectangle}} = l \cdot w$$

where l = length and w = width.

Examples   The areas shown are computed with Formula 3–2–1.

1. $A = l \cdot w = 3 \cdot 5 = 15$ cm^2

   Notice that here we can even count the 15 cm^2.

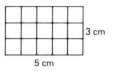

2. $A = l \cdot w = (12.7)(8.2) = 104.14$ m^2

3. $A = l \cdot w = (7.2)(7.2) = 51.84$ ft^2

Notice that we can get a special formula for a **square**:

Formula 3–2–2

$$A_{\text{square}} = s^2$$

where s = length of a side.

If we look carefully at a triangle, we see that it is really only half of a rectangle. So we get the formula for a **triangle**:

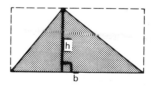

Formula 3–2–3

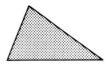

$$A_{\text{triangle}} = \left(\tfrac{1}{2}\right)b \cdot h$$

where b = base and h = height.

Examples   The figures shown use Formula 3–2–3.

1. $A = \dfrac{1}{2}b \cdot h = \dfrac{1}{2}(65)(44) = 1430$

2. $A = \dfrac{1}{2}b \cdot h = \dfrac{1}{2}(17.2)(10.1) = 86.86 \text{ cm}^2$

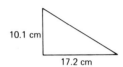

3. $A = \dfrac{1}{2}b \cdot h = \dfrac{1}{2}(8.5)(7.8) = 33.15$

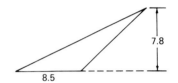

The other two simple polygons are the **parallelogram** and the **trapezoid**. The formulas for their areas are

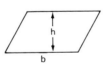
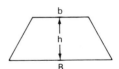

Formula 3–2–4

$$A_{\text{parallelogram}} = b \cdot h$$

where b = base and h = height, and

Formula 3–2–5

$$A_{\text{trapezoid}} = \dfrac{1}{2}(B + b) \cdot h$$

where B = big base, b = little base, and h = height.

Examples The figures shown use Formula 3–2–4.

1. $A = b \cdot h = 11 \cdot 8 = 88$

2. $A = b \cdot h = (12.7)(3.2) = 40.64$ m^2

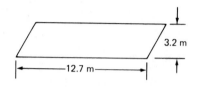

Examples The figures shown use Formula 3–2–5.

1. $A = \frac{1}{2}(B + b)h = \frac{1}{2}(8 + 4)3$

$= \frac{1}{2}(12)3$

$= 18$ cm^2

2. $A = \frac{1}{2}(B + b)h = \frac{1}{2}(82 + 61)47$

$= \frac{1}{2}(143)47$

$= 3360.5$

The circle is not a polygon, of course, but it is the only other figure for which we can give a simple formula.

Formula 3–2–6

$$\boxed{A_{\text{circle}} = \pi r^2}$$

where $\pi = 3.14$ (approximately) and r = radius.

Examples The figures shown use Formula 3–2–6.

1. $A = \pi r^2 = (3.14)(5)^2$
   $= (3.14)(25) = 78.5 \text{ mi}^2$

2. $A = \pi r^2 = (3.14)(1.41)^2$
   $= (3.14)(1.9881) = 6.24 \text{ m}^2$

(handwritten: $R = 1.41$ $2.82 \div 2 = R$)

Notice that we first found the radius, 1.41, which is half the diameter, 2.82.

3. $A = \dfrac{1}{2}\pi r^2 = \dfrac{1}{2}(3.14)(1.7)^2$
   $= \dfrac{1}{2}(3.14)(2.89)$
   $= 4.54 \text{ km}^2$

Notice that here we only had half a circle (a **semicircle**), so we multiplied the full area by $\dfrac{1}{2}$.

Sometimes, we are given the area, and we need to find a missing dimension. The next problem gives examples.

Problem Find the missing dimensions for the given shapes and areas.

(a)

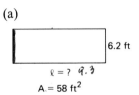

$\ell = ?$ *9.3*
A. = 58 ft^2

(b)

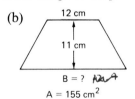

$B = ?$
A = 155 cm^2

(c)

$r = ?$
A = 90

ANSWER To solve these, we substitute what we are given into the appropriate equation and solve for the unknown.

Formula 3–2–1

↓

Substitute

↓

Divide by 6.2

↓

(a) $A = \ell \cdot w$

$58 = \ell \cdot (6.2)$

$\dfrac{58}{6.2} = \ell$

| Simplify |

$9.4 \text{ ft} = l$

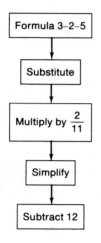

(b) $A = \frac{1}{2}(B + b)h$

| Formula 3-2-5 |

| Substitute |

$155 = \frac{1}{2}(B + 12) \cdot 11$

| Multiply by $\frac{2}{11}$ |

$\frac{2}{11} \cdot 155 = B + 12$

| Simplify |

$28.2 = B + 12$

| Subtract 12 |

$16.2 \text{ cm} = B$

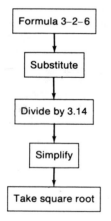

(c) $A = \pi r^2$

| Formula 3-2-6 |

| Substitute |

$90 = (3.14)r^2$

| Divide by 3.14 |

$\frac{90}{3.14} = r^2$

| Simplify |

$28.66 = r^2$

| Take square root |

$5.35 \approx r$

**PROBLEM
SET 3-2-1**

1. Define or discuss:

    (a) Area.   (b) Rectangle.   (c) Triangle.   (d) Parallelogram.

    (e) Trapezoid.   (f) Semicircle.   (g) Square.

Find the area of the following shapes.

mult iples of 4

2.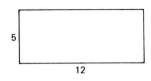
5
12

3.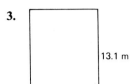
13.1 m
8.9 m

4.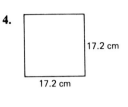
17.2 cm
17.2 cm

5.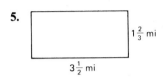
$1\frac{2}{3}$ mi
$3\frac{1}{2}$ mi

6.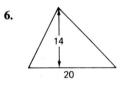
14
20

7.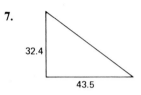
32.4
43.5

8.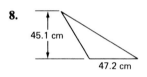
45.1 cm
47.2 cm

9.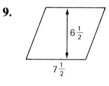
$6\frac{1}{2}$
$7\frac{1}{2}$

10.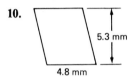
5.3 mm
4.8 mm

11.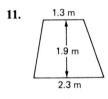
1.3 m
1.9 m
2.3 m

12.
2.3
2.5
0.7

13.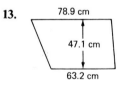
78.9 cm
47.1 cm
63.2 cm

14.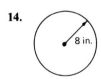
8 in.

15.
9.3 km

16.
29.6

17.
18 cm

18.
14.8

19.
15 m

Find the dimension missing in the following figures.

20.

14

A = 150

21.

s = ?

A = 500

22.

h = ?

25

A = 130

23.

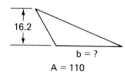

16.2

b = ?

A = 110

24.

h = ?

75 m

A = 11,000 m^2

25.

31 cm

h = ?

53 cm

A = 1300 cm^2

26.

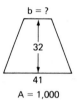

b = ?

32

41

A = 1,000

27.

r = ?

A = 200 ft^2

28.

D = ?

A = 1,000

29.

r = ?

A = 100

Less Simple Figures

Question How do we find the area of a figure that isn't as simple as a triangle, rectangle, or circle?

ANSWER Sometimes the figure is a combination of simple figures.

Problem Find the area of each figure.

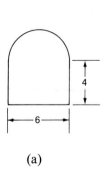

(a)

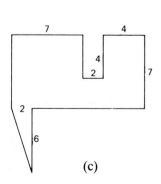

(b)

(c)

ANSWER   Each figure is a combination of simple figures: rectangles, triangles, and circles. To solve the figure, we will cut it into simple pieces.

(a) We cut this figure into a rectangle plus a semicircle. Since the diameter of the semicircle is 6, the radius must be 3.

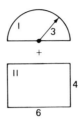

$$A_{\mathrm{I}} = \tfrac{1}{2}\pi r^2 = \tfrac{1}{2}(3.14)(3)^2 = 14.13$$

$$+ A_{\mathrm{II}} = l \cdot w = 6 \cdot 4 \qquad = 24.00$$

$$\overline{A_{total}} \qquad\qquad\qquad = \overline{38.13}$$

(b) Here we have the big circle minus the little circle. Notice that the radii are half the given diameters.

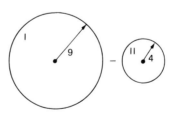

$$A_{\mathrm{I}} = \pi r^2 = (3.14)(9)^2 = \quad 254.34$$

$$- A_{\mathrm{II}} = \pi r^2 = (3.14)(4)^2 = - \quad 50.24$$

$$\overline{A_{total}} \qquad\qquad\qquad = \quad \overline{204.10}$$

MATH FLASH

1 2 3 4 5 6 7 8

√	%	+/−	÷	MC
7	8	9	X	MR
4	5	6	−	M−
1	2	3	+	M+
CE	0	.	=	C

Hand Calculator Instant Replay

PUNCH	DISPLAY	MEANING
C MC	0.	Clear everything
3 . 1 4	3.14	π
× 9 × 9	9.	Times r^2
=	254.34	A_{I}
M+	254.34	Store in memory
3 . 1 4	3.14	π
× 4 × 4	4.	Times r^2
=	50.24	A_{II}
M−	50.24	*Subtract* from memory
MR	204.10	Answer: $A_{\mathrm{I}} - A_{\mathrm{II}}$

(c) We have cut this figure into three rectangles and one triangle. We compute all the areas and add.

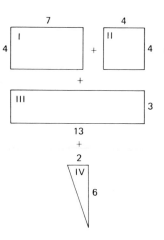

$$A_I = l \cdot w = 7 \cdot 4 \quad = \quad 28$$

$$A_{II} = l \cdot w = 4 \cdot 4 \quad = \quad 16$$

$$A_{III} = l \cdot w = 13 \cdot 3 \quad = \quad 39$$

$$A_{IV} = \frac{1}{2} bh = \frac{1}{2} \cdot 2 \cdot 6 = \; + \; 6$$

$$\overline{A_{total}} \qquad\qquad\qquad = \quad \overline{89}$$

PROBLEM SET 3–2–2 Find the area of the following figures (the dark regions are cut out).

1.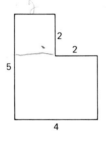

2.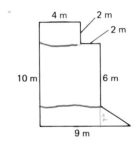

3.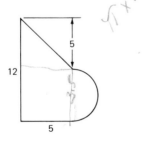

4.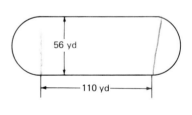

5.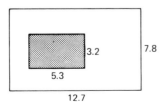

6.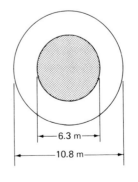

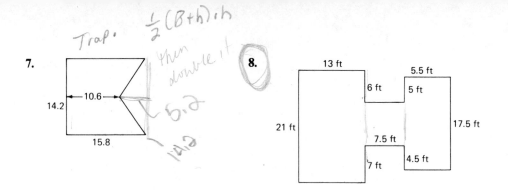

7.

8.

3-3 APPLICATIONS OF AREA

Question   How do we convert from one unit of area to another?

ANSWER   We use the following rule.

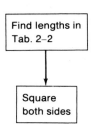

> To convert area units:
> 1. Find the corresponding length conversions in Table 2–2 , page 105.
> 2. *Square* both sides of the length conversion.
> 3. Use these numbers in the area conversion.

Examples

Find lengths in Tab. 2–2

↓

Square both sides

1. 1 in. = 2.54 cm
    1 in.^{2} = $(2.54)^2$ = 6.45 cm^2

2. 1 yd = 3 ft
    1 yd^2 = 3^2 = 9 ft^2

3. 1 mi = 1.61 km
    1 mi^2 = $(1.61)^2$ = 2.59 km^2

Problem   A gallon of paint usually covers 450 square feet. How many square meters is this?

ANSWER   From Table 2–2, we see that 1 foot = 0.305 meter. So 1 square foot = $(0.305)^2$ square meter. Therefore,

$$450\ \text{ft}^2 \times \frac{(0.305)(0.305)\ \text{m}^2}{1\ \text{ft}^2} = 41.86\ \text{m}^2$$

Problem The Wilkins' lot is 525 square meters. How many square kilometers is this?

ANSWER We see in Table 2–2 that 1 kilometer = 1000 meters. So 1 square kilometer = $1000^2 = 1,000,000$ square meters. Therefore,

| Set up units to cancel |

$$525 \cancel{m^2} \times \frac{1 \text{ km}^2}{1,000,000 \cancel{m^2}} = \frac{525}{1,000,000} \text{km}^2$$

| Simplify |

$$= 0.000525 \text{ km}^2$$

Just as in Section 2–5, we see that this conversion within the metric system is merely a decimal shift six places to the left: $0\,\underset{\wedge}{.}000525$.

Problem A sketch is shown of the two bedrooms and hall that Pam and Bob want to carpet.
  (a) Find the area in square feet.
  (b) Convert the area to square yards.
  (c) If carpeting is $11 per square yard, find the cost.

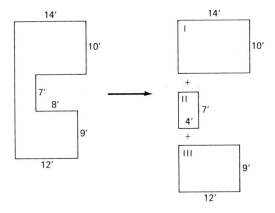

ANSWER In the figure we show the rooms cut into three rectangles.

(a) $A_I = l \cdot w = (14)(10) = 140$

$A_{II} = l \cdot w = (7)(4) \quad = 28$

$A_{III} = l \cdot w = (12)(9) \quad = 108$

$\overline{A_{total}} \qquad\qquad = \overline{276} \text{ ft}^2$

The total area is 276 ft².
(b) To convert to square yards, we square the foot-to-yard conversion.

$$276 \cancel{ft^2} \times \frac{1 \text{ yd}^2}{3 \cdot 3 \cancel{ft^2}} = 30.7 \quad \text{round} \quad 31$$

Probably, we should round this up to 31 square yards, since carpeting is sold by the square yard.

(c) To compute the cost, we multiply by $11.

$$31 \ \cancel{yd^2} \times \frac{\$11}{1 \ \cancel{yd^2}} = \$341$$

Thus the job will cost about $341.

Problem Mary wants to wallpaper the wall shown in the sketch. The wallpaper comes in long rolls, 15 inches wide.

    (a) How many square feet is the wall?

    (b) How long a roll does Mary need to cover the wall exactly?

    (c) How long a roll should Mary get if she wants an extra 10% to allow for waste or mistakes?

    (d) If the wallpaper is 79 cents per yard of length, how much will the job cost?

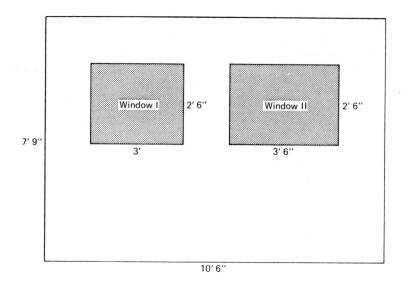

ANSWER (a) We first compute the total area and convert all the units to feet. We will find the area of the whole wall, and then subtract the area of the windows.

$$A_{wall} = l \cdot w = (10.5)(7.75) = 81.375$$
$$-A_{window \ I} = l \cdot w = 3(2.5) = -7.500$$
$$-A_{window \ II} = l \cdot w = (3.5)(2.5) = -8.750$$
$$\overline{A_{total} = 65.125 \ ft^2}$$

Notice that we changed 10 feet, 6 inches to 10.5 feet, since 6 inches is $\dfrac{6}{12}$ foot or 0.5 foot. Also, we changed 7 feet, 9 inches to $7\frac{9}{12}$ feet to 7.75. It is usually easier to work with decimals.

(b) Now we must find out how long a roll Mary needs. The rolls are 15 inches wide. This is $1\frac{3}{12}$ feet = 1.25 feet. To solve this, imagine a very long rectangle that is 1.25 feet wide and l (unknown) feet long, with a total area of 65.125 square feet.

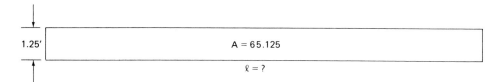

We now use Formula 3–2–1 to solve for l:

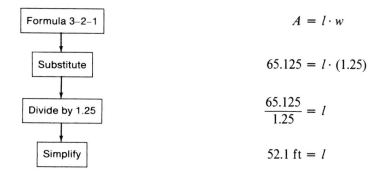

$$A = l \cdot w$$

$$65.125 = l \cdot (1.25)$$

$$\frac{65.125}{1.25} = l$$

$$52.1 \text{ ft} = l$$

Mary needs a piece of wallpaper exactly 52.1 feet long.

(c) Many decorating books suggest that people add an extra 10% for waste, matching patterns, and mistakes. Let

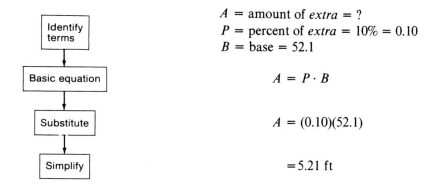

A = amount of *extra* = ?
P = percent of *extra* = 10% = 0.10
B = base = 52.1

$$A = P \cdot B$$

$$A = (0.10)(52.1)$$

$$= 5.21 \text{ ft}$$

If Mary adds the extra 5.2 feet, she will need 52.1 + 5.2 = 57.3 feet.
(d) To find the cost, we convert feet to yards, and then yards to dollars.

$$57.3 \, \cancel{ft} \times \frac{1 \, \cancel{yd}}{3 \, \cancel{ft}} \times \frac{0.79 \, \$}{1 \, \cancel{yd}} = \$15.09$$

Thus this project will cost Mary about $15.

Problem Dan is trying to determine the size of his house lot. On his city block $\left(\frac{1}{8} \text{ mile by } \frac{1}{16} \text{ mile}\right)$, there are a total of 18 houses (9 on each side). If each lot is about the same size, how big is each lot?

ANSWER First we find the area of the whole block.

$$A = l \cdot w = \left(\frac{1}{8}\right)\left(\frac{1}{16}\right) = \frac{1}{128} \text{ mi}^2$$

Since 1 square mile = 640 acres, we have

$$\frac{1}{128} \, \cancel{mi^2} \times \frac{640 \text{ acres}}{1 \, \cancel{mi^2}} = 5 \text{ acres}$$

So this city block is 5 acres. Since there are 18 lots, each lot is then about $\frac{5}{18} = 0.28$ acre. Thus, Dan's lot is about 0.28 acre, or a little over a $\frac{1}{4}$ acre.

Problem Peter is planting a quarter-circular garden where two walls meet. Its radius is 20 feet.
(a) How much fence does he need to enclose it?
(b) How much area does he have?

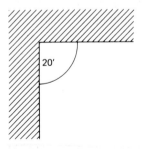

ANSWER (a) This is a perimeter problem. Since it is a quarter-circle, we multiply Formula 3–1–2 by $\frac{1}{4}$.

$$P = \frac{1}{4} 2\pi r = \frac{1}{4} \cdot 2 \cdot (3.14) \cdot 20 = 31.4 \text{ ft}$$

(b) For the area, we multiply Formula 3–2–6 by $\frac{1}{4}$.

$$A = \frac{1}{4}\pi r^2 = \frac{1}{4}(3.14)(20)^2 = 314 \text{ ft}^2$$

Problem Tom is trying to compute how much air conditioning he needs for a room with a window facing west. The window is shown in the sketch.
(a) What is the area of the window?
(b) If he needs 150 BTU's per square foot, how many BTU's does the window let in?

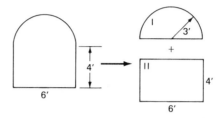

ANSWER (a) In the figure we show the window cut into a semicircle and a rectangle. Notice that the length of the rectangle, 6 feet, is also the diameter of the semicircle. Therefore, the radius is 3 feet, half the diameter.

$$A_I = \frac{1}{2}\pi r^2 = \frac{1}{2}(3.14)(3)^2 = 14.13$$

$$A_{II} = l \cdot w = 6 \cdot 4 \qquad = 24.00$$

$$\overline{A_{total}} \qquad\qquad\qquad = \overline{38.13} \text{ ft}^2$$

(b) The area is 38.13 square feet. We now convert this to BTU's.

$$38.13 \text{ ft}^2 \times \frac{150 \text{ BTU}}{1 \text{ ft}^2} = 5719.5 \text{ BTU}$$

Unfortunately for Tom, this is more BTU's than most room air conditioners provide (and we haven't even considered the heat that the ceiling, floors, and walls let in). With a shade on the window, only 65 BTU's per square foot is needed:

$$38.13 \text{ ft}^2 \times \frac{65 \text{ BTU}}{1 \text{ ft}^2} = 2478.45 \text{ BTU}$$

By putting a shade over the window, Tom can save over 3000 BTU's of cooling.

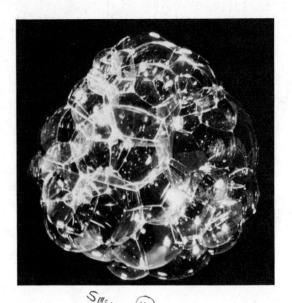

We cannot begin to count the number of possible geometric configurations formed by a soap film spanning a wire frame. (Photo by Fritz Goro)

SUSAN S. ☺

PROBLEM SET 3–3–1
Use Table 2–2 to help fill in the following table of conversions (1 square mile = 640 acres). Do not do a conversion with an ×.

* Multiples of 64
start
by 8

	in.²	ft²	yd²	mi²	acre	cm²	m²	km²
1.		200		×	×			×
2.	6000			×	×			×
3.	×	×			500	×		
4.	×		120	×	×	×		×
5.				×	×	450		×
6.	×	×				×		1,100,000
7.				×	×		600	×

A company has a floor that is 180 feet by 120 feet. *21,600 sq ft*

8. What is the area of the floor?

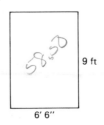

58.50 9 ft

6′ 6″

9. What is the area of the office space shown in the sketch? *59.4*

10. How many such offices can be put on this floor?

150 Ch. 3 / Geometry

A wallet-sized photo is $2\frac{1}{2}$ in. by $3\frac{1}{2}$ in. *~3½ inches*

11. What is the area of this photo? *8.75*

12. What is the area of a 14- by 20-in. sheet of developing paper. *280 in*

13. How many wallet-sized photos can be cut from one 14- by 20-in. sheet of developing *32* paper?

Mike and Ann live on a lot that is 60 feet by 115 feet.

14. How many square feet is their lot? *6900*

15. If 1 acre = 43,560 square feet, how many acres is this lot? *0.16*

Ron and Judy's yard is shaped like the figure shown.

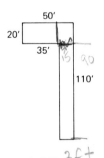

16. What is the area in square feet? *2350 2ft*

17. If grass seed covers 200 square feet per pound, how much seed will they need for the whole yard? *Know area of yard in sq ft.*

18. If seed is $4.50 per pound, how much will this cost?

19. What is the area in square yards? *261.1 yd²*

20. If sod cost 79 cents per square yard, how much will it cost to sod the yard? *sq yard = 9*

The sketch shows the Beaver City football field.

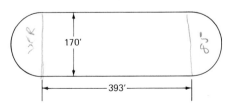

21. What is the perimeter in feet?

22. What is the perimeter in miles?

~~84,496.6~~

89,496.5 ft²

23. What is the area in square feet?

24. What is the area in square yards?

25. At 79 cents per square yard, what will it cost to sod the football field?

26. If artificial turf is $4.50 per square yard, what will it cost to cover the field?

Pat is going to fix up the two walls shown in the sketches.

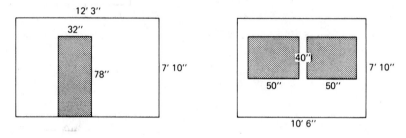

133.1 ft²

27. What is the area of the walls?

28. If paint covers 110 square feet per quart, how many quarts will she need for both walls?

29. If paint costs $3.50 per quart, how much will it cost to paint the walls?

Suppose that Pat (Problems 27–29) decides to wallpaper the walls instead. Her wallpaper comes in rolls 18 inches wide.

30. How long a roll would she need to cover the walls exactly?

31. If an extra 5% is added for waste, how long must the roll be?

32. At 95 cents per yard of length, what will it cost to paper the wall?

Dan and Dianne are planning to carpet the living–dining room shown in the sketch.

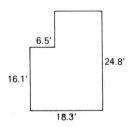

33. What is the area in square feet?

34. What is the area in square yards?

35. If the carpeting is $12.50 per square yard, how much will it cost?

The ceiling in Renee's basement is shown in the sketch.

5/12

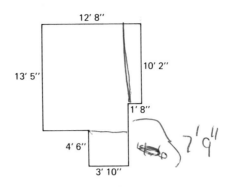

36. Find the missing dimensions.

37. Find the area of the ceiling.

38. If ceiling tiles are 2 feet by 4 feet, how many tiles will Renee need to cover the ceiling exactly?

39. Allowing an extra 10% for waste, how many tiles should Renee buy?

40. At $1.19 per tile, how much will the tiles cost?

Joanne puts together a jigsaw puzzle chessboard that is 20 inches by 20 inches.

400 in²

41. How many square inches is this?

42. How many square feet is this?

43. She wants to cover it with a new glass-hard plastic lacquer. One package covers 3 square feet. How many packages will she need?

What is the area of a

44. Ten-inch pizza pie?

45. Twelve-inch pizza pie?

46. Fourteen-inch pizza pie?

47. Sixteen-inch pizza pie?

$A = \pi r^2$

3-4 VOLUME

Question What is volume?

ANSWER **Volume** is a measure of the inside capacity of a solid.

Question How do we measure volume?

ANSWER Just as length and area are based on standard units, so is volume. Here the basic unit is the cubic inch, the cubic centimeter, the cubic foot, and so on. These are usually written in.^{3}, cm^3, ft^3, and so on.

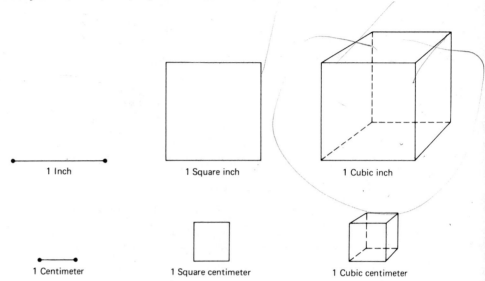

1 Inch   1 Square inch   1 Cubic inch

1 Centimeter   1 Square centimeter   1 Cubic centimeter

The accompanying figure gives a comparison of the three measurements: *length* is measured by standard line segments; *area* is measured by standard squares; *volume* is measured by standard cubes.

Question What are the formulas for the volumes of the basic figures?

ANSWER The simplest formula is for the volume of a **rectangular box**.

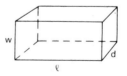

Formula 3–4–1

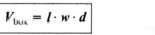

$$V_{box} = l \cdot w \cdot d$$

where l = length, w = width, and d = depth.

Example For the box shown, we have

$$V = l \cdot w \cdot d = 8 \cdot 5 \cdot 6 = 240 \text{ cm}^3$$

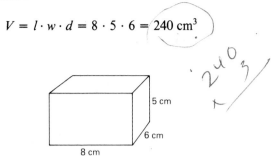

Notice that Formula 3–4–1 has an area term ($l \cdot w$) times a third dimension, depth. This is true for any **prism**. We take the area of the face and multiply it by the depth.

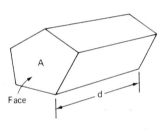

Prism

Formula 3–4–2

$$\boxed{V_{\text{prism}} = A \cdot d}$$

where A = area of the face and d = depth.

Examples The following use Formula 3–4–2. In each case, we first find the area of the face.
  1.

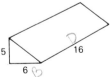

We first find the area of the triangle.

$$A_{\text{face}} = \frac{1}{2}(b \cdot h) = \frac{1}{2}(5 \cdot 6) = 15$$

$$V = A \cdot d = 15 \cdot 16 = 240$$

2.

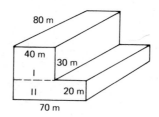

The face can be broken into two parts.

$$A_{\mathrm{I}} = l \cdot w = 40 \cdot 30 \quad = 1200$$

$$+ A_{\mathrm{II}} = l \cdot w = 70 \cdot 20 \quad = 1400$$

$$\overline{A_{\mathrm{face}}} \qquad\qquad\qquad = \overline{2600} \ \mathrm{m}^2$$

$$V = A \cdot d = 2600 \cdot 80 = 208{,}000 \ \mathrm{m}^3$$

3.

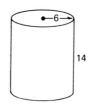

The figure shown is generally called a **cylinder**, but it can be solved using Formula 3–4–2.

$$A_{\mathrm{face}} = \pi r^2 \quad = (3.14)(6)^2$$

$$= 113.04$$

$$V = A \cdot d = (113.04)(14) = 1582.56$$

We will also give two special cases of Formula 3–4–2:

Formula 3–4–3

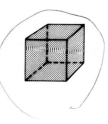

where s = any side.

Formula 3–4–4

$$V_{\text{cylinder}} = \pi r^2 h$$

where $\pi \approx 3.14$, r = radius, and h = height.

The volume of a sphere (or ball) is given by the next formula, which we will not prove or explain here:

Formula 3–4–5

$$V_{\text{sphere}} = \frac{4\pi r^3}{3}$$

where $\pi \approx 3.14$ and r = radius.

Examples   The following use Formula 3–4–5.

1.

Here we have the radius.

$$V = \frac{4\pi r^3}{3} = \frac{4(3.14)(7^3)}{3} = 1436.03$$

2.

First, we find the radius, which is half the diameter, to be 15 cm.

$$V = \frac{4\pi r^3}{3} = \frac{4(3.14)(15^3)}{3}$$

$$= 14,130 \text{ cm}^3$$

3.

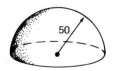

Since this is a hemisphere, we take half the volume in Formula 3–4–5.

$$V = \frac{1}{2}\left(\frac{4\pi r^3}{3}\right) \quad = \frac{1}{2}\frac{(4)(3.14)(50^3)}{3}$$
$$= 261{,}666.7$$

Finally, we have the volume of a **pyramid** or a **cone**. The pyramid has a polygon for a base, and the cone has a circle for a base. But they are both "pinched" at the top.

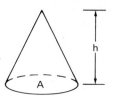
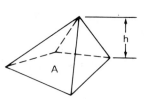

Formula 3–4–6

$$V_{\text{pyramid}} = \frac{Ah}{3}$$

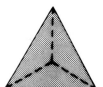

where A = area of base and h = height.

Formula 3–4–7

$$V_{\text{cone}} = \frac{\pi r^2 h}{3}$$

where $\pi \approx 3.14$, r = radius, and h = height. Since $A = \pi r^2$ for a circle, these two are really the same formula.

Examples The following use Formulas 3–4–6 and 3–4–7. In all cases we first find the area of the base.

1. $A = \frac{1}{2}b \cdot h = \frac{1}{2}(35)(20) = 350$

    $V = \frac{Ah}{3} = \frac{(350)(55)}{3} = 6416.7$

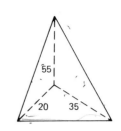

2. $A = l \cdot w = (10)(8) = 80 \text{ m}^2$

    $V = \frac{Ah}{3} = \frac{1}{3}(80)(7) = 186.7 \text{ m}^3$

$\frac{l \times w \times H}{3}$

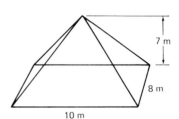

3. $A = \pi r^2 = (3.14)(4.2)^2 = 55.39$

    $V = \frac{\pi r^2 h}{3} = \frac{Ah}{3} = \frac{1}{3}(55.39)(5.1) = 94.163$

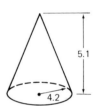

Just as with area, we may often know what the volume is and have to solve for some missing dimension. In this case, we find the proper formula, substitute what we know, and solve.

7.06

Problem Find the missing dimensions for the given volumes.

$l \times w \times D$ $V = l \times H$ $\frac{\pi R^2 \times H}{3} = V$ γR^2

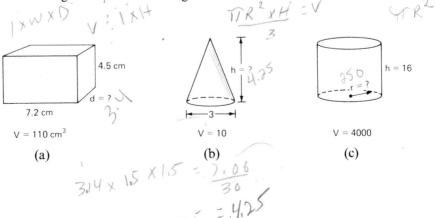

V = 110 cm³   V = 10   V = 4000

(a)   (b)   (c)

$3.14 \times 1.5 \times 1.5 = \frac{7.06}{30}$

$30 \div 7.05 = 4.25$

$30 \div 7.05$

ANSWER   In each case, we will first write down the proper volume formula. Then we substitute everything we have and solve for the unknown.

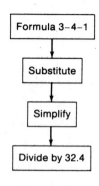

(a) $V = l \cdot w \cdot d$

$110 = (7.2)(4.5)d$

$110 = (32.4)d$

$3.4 \text{ cm} = d$

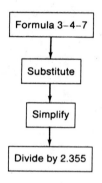

(b) $V = \frac{1}{3}\pi r^2 h$

$10 = \frac{1}{3}(3.14)(1.5)^2 h$

$10 = 2.355h$

$4.25 = h$

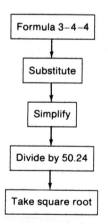

(c) $V = \pi r^2 h$

$4000 = (3.14)r^2(16)$

$4000 = 50.24r^2$

$79.6 = r^2$

$8.9 = r$

PROBLEM SET 3-4-1   1. Define or discuss:

    (a)   Volume.   (b)   Prism.   (c)   Cylinder.
    (d)   Cube.   (e)   Sphere.   (f)   Hemisphere.
    (g)   Pyramid.   (h)   Cone   (i)   Rectangular box.

Find the volume of the following shapes.

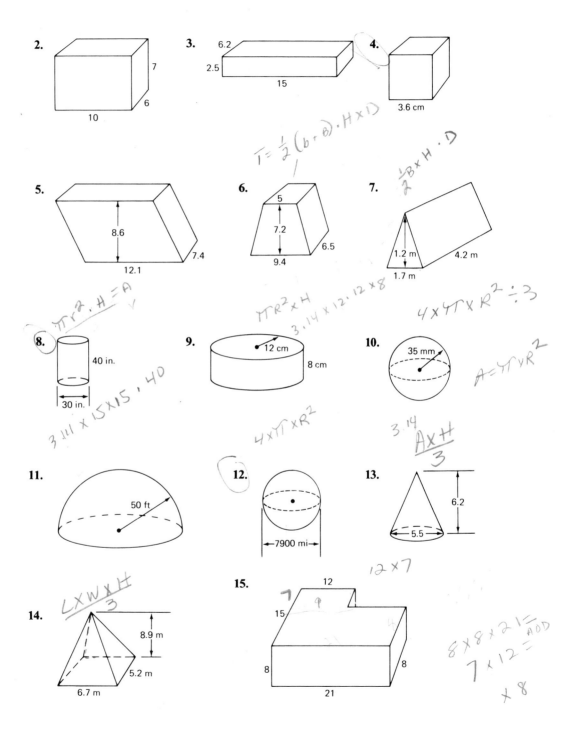

2.

7

6

10

3. 6.2

2.5

15

4.

3.6 cm

5.

8.6

12.1

7.4

6.

5

7.2

9.4

6.5

7.

$T = \frac{1}{2}(b+8)\cdot H \times D$

$\frac{1}{2}B \times H \cdot D$

1.2 m

4.2 m

1.7 m

8.

40 in.

30 in.

$\pi r^2 . H = A$

$3.14 \times 15 \times 15 \cdot 40$

9.

12 cm

8 cm

$\pi R^2 \times H$

$3.14 \times 12 \cdot 12 \times 8$

$4 \times \pi \times R^2$

10.

35 mm

$4 \times 4 \pi \times R^2 \div 3$

$A = 4 \pi v R^2$

11.

50 ft

12.

←7900 mi→

$4 \times \pi \times R^2$

13.

$\frac{A \times H}{3}$

3.14

6.2

5.5

12×7

14.

$\frac{L \times W \times H}{3}$

8.9 m

5.2 m

6.7 m

15.

12

7

15

9

8

21

8

8

$8 \times 8 \times 21 =$ AOD

$7 \times 12 =$

$\times 8$

In the following cases, the volume is given, but one of the dimensions is missing. Find the missing dimension.

16.

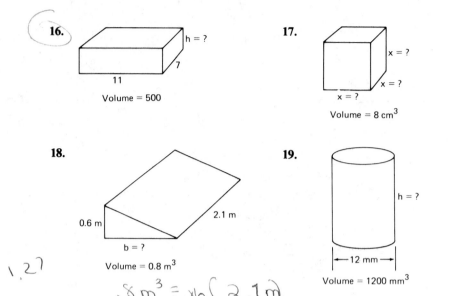

h = ?

11

7

Volume = 500

17.

x = ?

x = ?

x = ?

Volume = 8 cm^3

18.

0.6 m

2.1 m

b = ?

Volume = 0.8 m^3

1.2?

$.8\ m^3 = .6 \times h(2.1m)$

19.

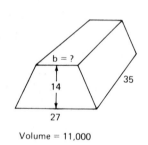

h = ?

|← 12 mm →|

Volume = 1200 mm^3

20.

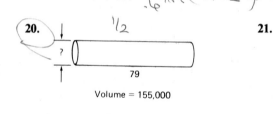

$^1/_2$

?

79

Volume = 155,000

21.

b = ?

14

35

27

Volume = 11,000

22.

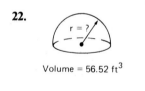

r = ?

Volume = 56.52 ft^3

23.

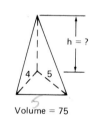

h = ?

4   5

5

Volume = 75

Combinations and Applications

Question   What happens if we have a combination figure?

ANSWER   Just as with area, we break it into a sum or difference of simple figures.

Problem   Find the volumes of the figures given.

(a) 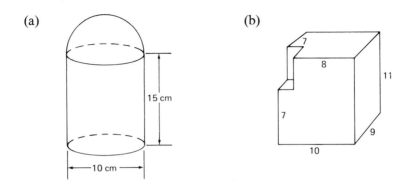   (b)

ANSWER   (a) In both shapes (hemisphere and cylinder), we must first find the radius, which is half the diameter; so $r = 5$. Also, I is a hemisphere; so we take half the volume of a full sphere.

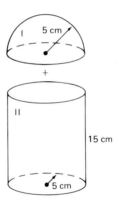

$$V_1 = \frac{1}{2} \cdot \frac{4\pi r^3}{3} = \frac{1}{2} \cdot \frac{4(3.14)5^3}{3} = \quad 261.7$$

$$+ V_{II} = \quad \pi r^2 h \quad = (3.14)(5^2)(15) = +1177.5$$

$$\overline{V_{total}} = \qquad\qquad\qquad = \overline{\quad 1439.2 \text{ cm}^3}$$

MATH FLASH

| 1 2 3 4 5 6 7 8 |

√	%	+/-	÷	MC
7	8	9	×	MR
4	5	6	−	M-
1	2	3	+	M+
CE	0	·	=	C

Hand Calculator Instant Replay

PUNCH	DISPLAY	MEANING
C MC	0.	Clear everything
· 5	0.5	$\frac{1}{2}$
× 4	4.	Times 4
× 3 · 1 4	3.14	Times π
× 5 × 5 × 5	5.	Times r^3
÷ 3	3.	Divided by 3
=	261.66666	V_{I}
M+	261.66666	Store in memory
3 · 1 4	3.14	π
× 5 × 5	5.	Times r^2
× 1 5	15.	Times h
=	1177.5	V_{II}
M+	1177.5	Add to memory
MR	1439.1666	Answer: $V_{\text{I}} + V_{\text{II}}$

(b) For this shape, we subtract from the volume of the larger box (I) the volume of the smaller box (II) that was cut away. We find the dimensions of II by subtracting.

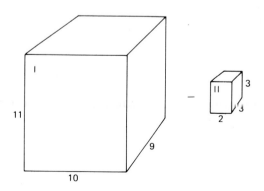

$$V_1 = l \cdot w \cdot d = 10 \cdot 9 \cdot 11 = \quad 990$$

$$-V_{II} = l \cdot w \cdot d = 2 \cdot 3 \cdot 3 \quad = - \quad 18$$

$$\overline{V_{total}} \qquad\qquad\qquad = \overline{\quad 972}$$

Question   How do we convert from one unit of volume to another?

ANSWER   We use the following rule.

> To convert volume units:
>
> 1. Find the corresponding length conversions in Table 2-2.
> 2. *Cube* both sides of the length conversion.
> 3. Use these numbers in the volume conversions.

Notice that this is almost the same rule as for converting area units, except here we cube the units.

Examples

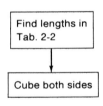

1. 1 yd = 3 ft
   $$1 \text{ yd}^3 = 3^3 \text{ ft}^3 = 27 \text{ ft}^3$$

2. 1 in. = 2.54 cm
   $$1 \text{ in.}^3 = 2.54^3 \text{ cm}^3 = 16.39 \text{ cm}^3$$

3. 1 ft = 12 in.
   $$1 \text{ ft}^3 = 12^3 \text{ in.}^3 = 1728 \text{ in.}^3$$

Problem   A refrigerator has 15 cubic feet of storage. How many cubic meters is this?

ANSWER

Find lengths in Tab. 2-2

Cube both sides

Set up units to cancel; multiply

1 ft = 0.305 m

$$1 \text{ ft}^3 = 0.305^3 \text{ m}^3 = 0.028 \text{ m}^3$$

$$15 \text{ ft}^3 \times \frac{0.028 \text{ m}^3}{1 \text{ ft}^3} = 0.42 \text{ m}^3$$

The refrigerator has 0.42 cubic meters of storage space.

Problem A desk drawer has a volume of 0.03 cubic meters. How many cubic centimeters is this?

ANSWER We see in Table 2-2 that 1 meter = 100 centimeters. So 1 cubic meter = 100^3 cubic centimeters = 1,000,000 cubic centimeters. Therefore,

| Set up units to cancel; simplify |

$$0.03 \; m^3 \times \frac{1,000,000 \; cm^3}{1 \; m^3} = 30,000 \; cm^3$$

Like all the other metric system conversions, we just move the decimal point; in this case, six places right: 0.030000.

Problem Bob is putting down a small walkway as shown in the sketch.

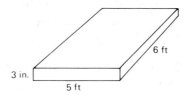

(a) What is the volume?
(b) If blacktop comes in $1\frac{1}{2}$-cubic-foot bags, how many bags will he need?
(c) At $3.30 per bag, what will the job cost?

ANSWER

(a) To find volume, we first convert 3 in. to $\frac{3}{12} = \frac{1}{4} = 0.25$ feet. Now

$$V = l \cdot w \cdot d = (6)(5)(0.25) = 7.5 \; ft^3$$

(b) To find the number of bags, we can treat this as a conversion problem.

$$7.5 \; ft^3 \times \frac{1 \; bag}{1.5 \; ft^3} = \frac{7.5}{1.5} = 5 \; bags$$

Therefore, Bob needs 5 bags.
(c) At $3.30 per bag, this will cost 5(3.30) = $16.50.

Problem Pam and Rudy are moving themselves by renting the truck shown in the sketch. How many cubic feet will this truck hold?

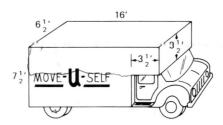

ANSWER The truck is basically a prism, so we can use Formula 3–4–2, $V = A \cdot d$. First we find A, area, by cutting the face into two rectangles. (We get the missing dimensions by subtracting $7\frac{1}{2} - 3\frac{1}{2} = 4$.)

$$A_{\mathrm{I}} = l \cdot w = 16 \cdot 3\tfrac{1}{2} = 56$$

$$A_{\mathrm{II}} = l \cdot w = 12\tfrac{1}{2} \cdot 4 = 50$$

$$\overline{A_{total} \phantom{=l \cdot w = 12\tfrac{1}{2} \cdot 4}} = \overline{106 \text{ ft}^2}$$

Now we compute the volume.

$$V = A \cdot d = (106)(6.5) = 689 \text{ ft}^3$$

Problem How can Pam and Rudy tell if the truck is big enough for their move?

ANSWER The best way is to try to estimate the volume of the possessions that they are moving. For example, let's look at their kitchen.

Items	Dimensions (ft)	Total volume (ft³)
1 refrigerator	$6 \times 2\frac{1}{2} \times 2$	30
1 stove	$3\frac{1}{2} \times 4 \times 2$	28
1 table	$2\frac{1}{2} \times 3 \times 3$	22.5
4 chairs	$2\frac{1}{2} \times 1\frac{1}{2} \times 1\frac{1}{2}$ (each)	22.5
8 filled boxes	$2\frac{1}{2} \times 1\frac{1}{2} \times 1$ (each)	30.0
V_{kitchen}		$= 133.0$

The answer, 133 cubic feet, is probably a high estimate. Since the chairs and table are not solid, some clever packing on the truck could get extra space (under the table, for example). But it is safest to get the highest estimate, just in case.

Problem Kent is planning to add insulation to his attic (shown in the sketch). It is recommended that he add insulation $3\frac{1}{2}$ inches deep.

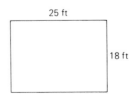

25 ft

18 ft

(a) What volume does he need?
(b) If a bag of pouring wool holds $4\frac{1}{2}$ cubic feet, how many bags does he need?

(c) If each bag is $2.60, how much will the job cost?

ANSWER (a) To find volume, we first must convert all units to feet. To do this, we convert $3\frac{1}{2}$ inches $= \dfrac{3.5}{12}$ feet $= 0.29$ feet.

$$V = l \cdot w \cdot d = (25)(18)(0.29) = 130.5 \text{ ft}^3$$

(b) To find the number of bags needed, we convert cubic feet to bags.

$$130.5 \text{ ft}^3 \times \frac{1 \text{ bag}}{4.5 \text{ ft}^3} = \frac{130.5}{4.5} = 29 \text{ bags}$$

(c) Since each bag is $2.60, this will cost $29 \times \$2.60 = \$75.40$.

Problem Kent, in the preceding problem, is also considering using insulation rolls, which are 15 inches wide, $3\frac{1}{2}$ inches deep, and 56 feet long.

56 ft

$1\frac{1}{4}$ ft

(a) How many rolls will he need?
(b) At $8.40 per roll, what will this cost?

ANSWER   Since these rolls are the proper depth ($3\frac{1}{2}$ inches), we can forget the depth and just look at the area of the attic and the area of roll.

$$A_{\text{attic}} = l \cdot w = (25)(18) = 450 \text{ ft}^2$$

$$A_{\text{roll}} = l \cdot w = (56)\left(\frac{15}{12}\right) = 70 \text{ ft}^3$$

Thus the attic is 450 ft^3 and each roll covers 70 ft^3. So Kent needs $\dfrac{450}{70} = 6.4$ rolls. Since rolls are not split, he will have to buy 7 rolls. This will cost $7 \cdot \$8.40 = \$58.80$.

Proper insulation can cut home heating costs. (Johns-Manville.)

Problem   A local radio station is having a contest to guess the number of beans in the drum shown in the sketch. What can Jerry and Donna do to get a good guess?

30″

18″

ANSWER   First, they can estimate the dimensions of the drum. (The radio station won't let people measure exactly.) From this, they can compute the volume of the drum. We use Formula 3–4–4. Notice that $r = 9$ inches.

$$V_{\text{drum}} = \pi r^2 h = (3.14)(9^2)(30) = 7630.2 \text{ in.}^3$$

Now Jerry and Donna count the number of beans in a small box that they can measure. For this, they pick the box sketched. They count 842 beans in the box. We need the volume of the box.

3″
3″
3″

$$V_{\text{box}} = l \cdot w \cdot d = 3 \cdot 3 \cdot 3 = 27 \text{ in.}^3$$

Now we set a proportion, since we assume that the beans-to-volume ratio is the same.

Identify unknown

↓

Set up proportion

↓

Multiply by 7630.2

↓

Simplify

Let x = beans in drum

$$\frac{x \text{ beans}}{7630.2 \text{ in.}^3} = \frac{842 \text{ beans}}{27 \text{ in.}^3}$$

$$x = \frac{(7630.2)(842)}{27}$$

$$= 237,949$$

Jerry and Donna can submit a guess of about 237,949. If they are allowed several guesses, they can make slightly higher and lower guesses since this answer is just an approximation.

**PROBLEM
SET 3–4–2**

Find the volume of the following figures.

v/c

1.

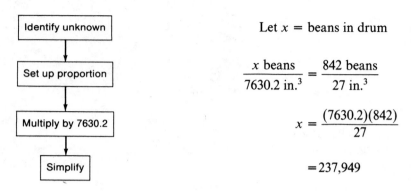

46

38

cylinder

2.

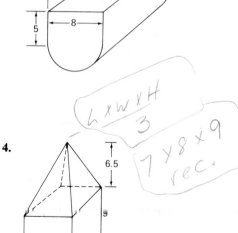

18

5   8

$\dfrac{L \times W \times H}{3}$

$7 \times 8 \times 9$

rec.

3.   4.2 mm

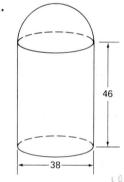

6.6 mm

3.7 mm

cone

$\dfrac{\pi R 2 H}{3}$

4.

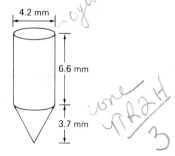

6.5

9

8

7

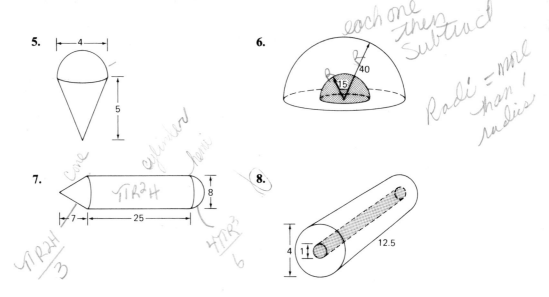

5. |← 4 →|

5

6. each one then subtract

40

15

Radii = more than 1 radius

7. cone   cylinder   hemi

$\pi R^2 H$

|←7→|← 25 →|

$\frac{\pi R^2 H}{3}$ $\frac{4\pi R^3}{6}$

8.

4 1

12.5

Use Table 2-2 to help fill in the following table of conversions.

	ft³	in.³	yd³	cm³	m³
9.	35				
10.		580			
11.			14		
12.				10	
13.					1600

2.54 cm = 1 in

14. Find (or approximate) the volumes (total space taken up) by the following everyday objects. Give answers in both the English and metric systems.

(a) Refrigerator   (b) Stove
(c) Desk   (d) Car
(e) Couch   (f) Waste basket
(g) Bookshelf (and books)   (h) Beer (or pop) can
(i) Basketball   (j) Typewriter
(k) Textbook   (l) Pencil (with and without point)
(m) Your room   (n) Mattress
(o) Hostess snowball   (p) Record album

Sue is filling in a section of walkway as shown.

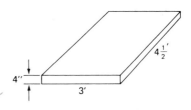

$4\frac{1}{2}'$

4″

3′

15. What is the volume?

16. If blacktop comes in $1\frac{1}{2}$-cubic-foot bags, how many bags will this require?

17. At $3.50 per bag, what will this cost?

18. Jack and Marsha are building the A-frame house in the given figure. Find the volume of the house.

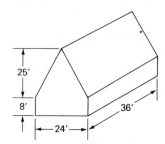

19. Mike and Amy are going to rent a truck. Find the volume of the truck shown.

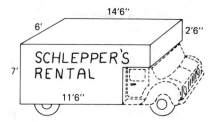

20. Jan is filling a flower pot like the one shown. What is the volume?

21. Bruce is filling a flower pot like the one shown. Assume that it is a sphere and find its volume.

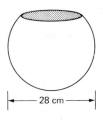

Ron is going to insulate his attic, which is shown in the sketch. He wants to put in a 6-inch depth of pouring wool insulation.

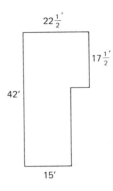

22. What is the volume of insulation needed?

23. If one bag is $4\frac{1}{2}$ cubic feet, how many bags does he need?

24. If a bag is $2.80, how much will the job cost?

25. Harvey orders a huge load of black topsoil for his lawn. It is dumped in a heap approximating a hemisphere with a diameter of 16 feet. How many cubic yards is the mound of dirt?

A paper drinking cup used at many cafeterias is shown.

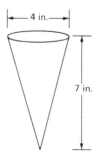

26. Find the volume of the cup.

27. If 1 ounce = 1.8 cubic inches of liquid, how many ounces are in the cup?

The Wheatville High School swimming pool is sketched here.

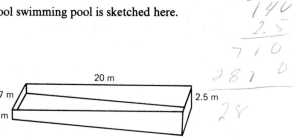

28. Find the volume in cubic meters.

29. If 1 cubic meter = 1000 liters of water, how many liters are needed to fill the pool?

An ice cream cone is sketched here. Assume that the cone is filled and include the scoop on top.

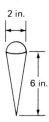

30. What is the volume of ice cream?

31. If 1 ounce = 1.8 cubic inches of ice cream, how many ounces of ice cream are in the whole cone?

The largest of the Egyptian pyramids is shown. (The base is a square.)

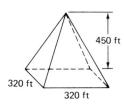

32. Find the volume in cubic feet.

33. Find the volume in cubic meters.

The earth has a diameter of about 7900 miles.

34. Find the volume of the earth in cubic miles.

35. Find the volume of the earth in cubic kilometers.

Find the volume of the following pizza pies.

	Diameter (in.)	Thickness (in.)
36.	8	$\frac{1}{2}$
37.	8	1
38.	10	$\frac{1}{2}$
39.	10	1
40.	10	$1\frac{1}{2}$
41.	12	$\frac{1}{2}$
42.	12	1
43.	12	$1\frac{1}{2}$
44.	14	$\frac{1}{2}$
45.	14	1

3-5 PROPORTIONALITY

Question   How do we use proportionality in geometry?

ANSWER   Remember that a proportion is two equal ratios. With proportion, we can take something we know and get information about something else.

We have the following theorem from geometry. Two figures are called **similar** if they have the same shape. (They probably won't have the same size.)

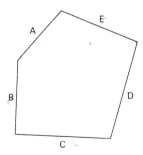

Theorem 3-5-1 The corresponding sides of similar figures are proportional. In other words, the ratios of corresponding sides are equal.

The figures shown are similar. They have the same shape (but obviously different sizes). Put in terms of an equation, Theorem 3–5–1 tells us the following:

$$\frac{a}{A} = \frac{b}{B} = \frac{c}{C} = \frac{d}{D} = \frac{e}{E}$$

Notice that we were careful to keep the terms of the ratios the same: little polygon to big polygon.

Page from a book dated 1690. (The New York Public Library)

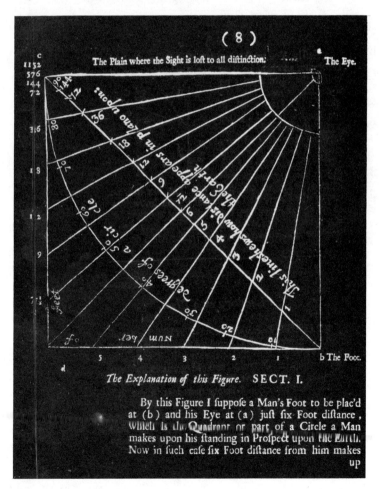

Question   How do we use Theorem 3–5–1?

ANSWER   Suppose that we have two similar triangles. And suppose that we know the size of one of them completely and one side of the other. Then we can find out all the dimensions of the second.

Problem   Find the missing dimensions of the given triangle.

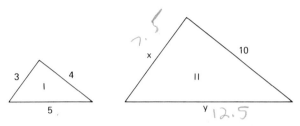

ANSWER   The two triangles are similar. We know triangle I completely, but we only know a little about triangle II. We can use Theorem 3–5–1 to solve for the missing dimensions. We will solve for x first.

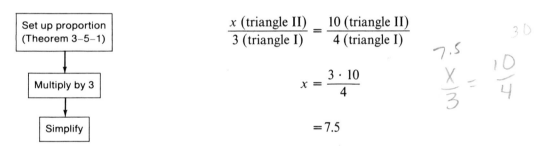

$$\frac{x \text{ (triangle II)}}{3 \text{ (triangle I)}} = \frac{10 \text{ (triangle II)}}{4 \text{ (triangle I)}}$$

$$x = \frac{3 \cdot 10}{4}$$

$$= 7.5$$

Similarly, we have

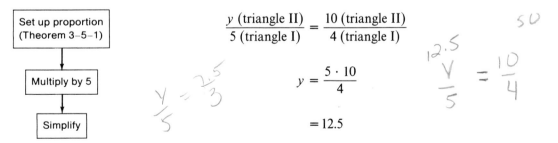

$$\frac{y \text{ (triangle II)}}{5 \text{ (triangle I)}} = \frac{10 \text{ (triangle II)}}{4 \text{ (triangle I)}}$$

$$y = \frac{5 \cdot 10}{4}$$

$$= 12.5$$

Notice that we compared corresponding sides of the triangles.

Problem   Jennifer would like to measure the height of a building. She measures that it casts an 11-meter shadow. She also measures that a meter stick casts a 0.4-meter shadow. How tall is the building?

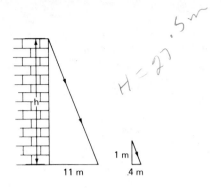

H = 27.5 m

1 m

11 m   .4 m

ANSWER   We can see that the two imaginary triangles that are formed by the sun's rays are similar. So we can use Theorem 3–5–1.

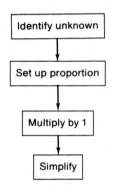

Let h = height of building

$$\frac{h \text{ (building)}}{1 \text{ (stick)}} = \frac{11 \text{ (building)}}{0.4 \text{ (stick)}}$$

$$h = \frac{1 \cdot 11}{0.4}$$

$$= 27.5 \text{ m}$$

Question   How else can we use proportionality?

ANSWER   Suppose that we wish to enlarge (or shrink) a drawing. Since we want the drawing to have the same shape, it must have proportional dimensions (by Theorem 3–5–1).

Problem   A caricature of a distinguished author is shown on the next page. How can we enlarge this to four times its size (twice as tall, twice as wide)?

ANSWER   First we put a grid of squares on it. We get a 7- by 6-square pattern. Now we make another 7- by 6-square pattern with squares twice as big in both dimensions.

We take the new larger grid, and we duplicate each little square's design in the corresponding big square. If we had tried to copy the entire picture exactly, it would have been very hard. But by enlarging it in little pieces, it is much simpler.

PROBLEM SET 3–5–1

1. Define or discuss similar shapes.

Find the missing dimensions in the following similar shapes.

2.

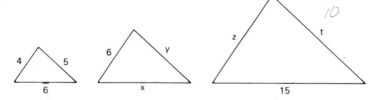

3.

4.

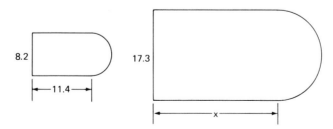

5.

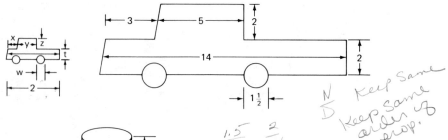

$\dfrac{N}{D}$ Keep Same

Keep Same

order if

prop.

$\dfrac{1.5}{w}$ $\dfrac{2}{14}$

6.

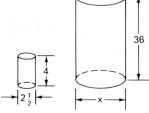

7.

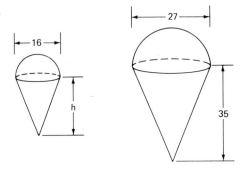

8. The shadow of a building is 25 meters. A meter stick casts a 0.62-m shadow. How tall is the building?

9. Sue is enlarging a $2\frac{1}{2}$- by $3\frac{1}{2}$-inch photograph. The large dimension must be 20 inches. What is the smaller dimension of the enlarged photo?

10. Sally is sewing mother–daughter look-alike outfits. The patterns are shown with the dimensions for the larger. Find the missing dimensions x, y, z, and t for the smaller pattern.

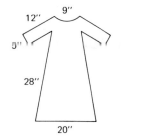

By using a grid of squares, reproduce each of the following sketches to be
  (a) twice their size (in both dimensions);
  (b) half their size (in both dimensions).

11.    **12.**

Another Application

Another useful theorem about proportionality is the following:

Theorem 3–5–2   If two lines are cut by a set of parallel lines, the corresponding pieces of the two lines are proportional.

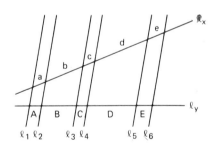

The figure helps display this theorem. Lines l_x and l_y are any two lines. They are cut by six parallel lines: l_1, l_2, l_3, l_4, l_5, and l_6. Theorem 3–5–2 tells us that

$$\frac{a}{A} = \frac{b}{B} = \frac{c}{C} = \frac{d}{D} = \frac{e}{E}$$

Again, we are careful always to keep the correspondence the same: l_x to l_y. We also have the proportions

$$\frac{a}{b} = \frac{A}{B}, \qquad \frac{a}{c} = \frac{A}{C}, \qquad \frac{b}{c} = \frac{B}{C}, \qquad \text{and so on}$$

Question How do we use Theorem 3–5–2?

ANSWER Theorem 3–5–2 is usually used more for constructions than for solving for unknowns. Generally, we have a line we want to divide into a certain ratio. What we do is construct a line that we know has the needed ratio. Then we pass parallel lines through the two lines.

Problem Divide the following line segment into a 5-to-7 ratio.

●————————————————●

ANSWER We first construct another line segment off the given line that we know has a 5-to-7 ratio. Then, as in the figure, we connect the endpoints with l_2, and draw l_1 parallel to l_2. Theorem 3–5–2 tells us that $\dfrac{a}{b} = \dfrac{5}{7}$, which is what we wanted.

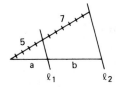

Problem Connie is making a poster and must divide it into seven equal columns. What is the easiest way to do this?

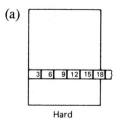

Hard

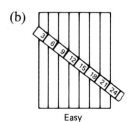

Easy

ANSWER Connie measures the width and gets $17\frac{3}{8}$ inches. Dividing this by 7, she would get

$$17\tfrac{3}{8} \div 7 = \frac{139}{8} \times \frac{1}{7} = \frac{139}{56} = 2\tfrac{27}{56} \text{ in.}$$

Needless to say, it would be no fun to try to measure $2\frac{27}{56}$ inches. This is the "hard" way shown in the sketch. Instead we use Theorem 3–5–2.

First we take a convenient multiple of 7 just over $17\frac{3}{8}$. In this case, it is 21 inches. Now we turn the ruler on an angle until it measures exactly 21 inches across, as in part (b) of the sketch. This distance is easy to divide by 7. Now we mark off the multiples of 3: 3, 6, 9, 12, 15, 18, 21. Then we draw lines parallel to edges through these marks. This is the easy way.

Notice that by turning the ruler on an angle we can make a distance easy to divide. Theorem 3–5–2 tells us that the parallels cut it the same as they cut the top.

Problem   Divide the box in the figure given into 10 equal rows.

(a)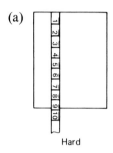

Hard

(b)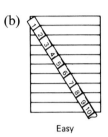

Easy

ANSWER   If we measure it, we get $8\frac{7}{8}$ inches. Again, it would be very difficult to divide this by 10 and then measure it. So we turn the ruler on an angle until we can measure 10 inches from top to bottom.

Now we make a mark at 1, 2, 3, 4, 5, 6, 7, 8, 9, 10. Then we draw lines parallel to the top and bottom through these marks. This is the "easy" way to divide the box into 10 equal rows.

PROBLEM SET 3–5–2   Divide the following line segments as directed,

$\frac{1}{5} = \frac{x}{\text{Length of Line}}$

1. •————————•   Into 5 equal parts.

2. •————————•   Into 3 equal parts.

3. •————————————•   Into 10 equal parts.

4. •————————•   Into 2 parts, in a 3-to-7 ratio.

5. •————•   Into 3 parts, in a 2-to-3-to-4 ratio.

Divide a standard $8\frac{1}{2}$- by 11-inch sheet of paper in the following ways. (Don't forget, you can use either the inch or the centimeter side of your ruler.)

6. Into 6 equal columns   7. Into 10 equal columns

8. Into 5 equal rows   9. Into 12 equal rows

10. Into an 8-row by 6-column grid   11. Into a 3-row by 10-column grid

Proportionality of Areas

Question   What is the relationship between the areas of similar figures?

ANSWER   The relationship is a little different than for the sides.

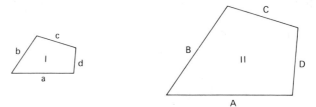

Theorem 3–5–3 The areas of similar shapes are proportional to the squares of any corresponding dimensions. In the figure shown, we have

$$\frac{\text{Area}_{\text{II}}}{\text{Area}_{\text{I}}} = \frac{A^2}{a^2} = \frac{B^2}{b^2} = \cdots$$

For example, suppose that A is twice as much as a, or $A = 2a$. Then $A^2 = 4a^2$. So

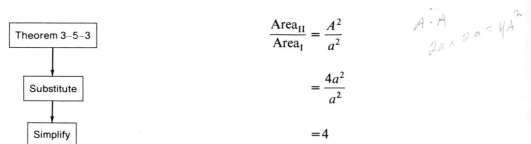

Thus Area$_{\text{II}}$ is four times Area$_{\text{I}}$. This is true for all shapes, both polygons and circles.

1. Double the sides, the area is four times bigger.
2. Triple the sides, the area is nine times bigger.
3. Quadruple the sides, the area is 16 times bigger.

Problem   Suppose that we want to double the area of a circle. How much bigger should the radius (or diameter) be?

ANSWER   Obviously, all circles have the same shape, so they are similar. Thus we can use Theorem 3–5–3 to relate their areas.

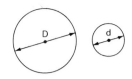

Let d = diameter of little circle

D = diameter of big circle

area = area of little circle

AREA = area of big circle

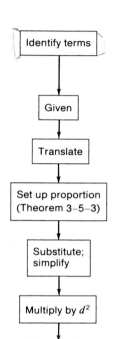

The area of the big circle is twice that of the little circle:

Identify terms

Given

Translate

$$\text{AREA} = 2 \text{ area}$$

Set up proportion
(Theorem 3–5–3)

$$\frac{D^2}{d^2} = \frac{\text{AREA}}{\text{area}}$$

Substitute;
simplify

$$\frac{D^2}{d^2} = \frac{2 \text{ area}}{\text{area}} = 2$$

Multiply by d^2

$$D^2 = 2d^2$$

Take square root

$$D = \sqrt{2}\, d \approx 1.41d$$

Thus, to double the area, we must multiply the little diameter (or radius) by $\sqrt{2}$, or about 1.41.

Question Where is Theorem 3–5–3 used?

ANSWER (a) In photography, the light coming through the lens is proportional to the area of the lens. Thus, if we double the lens area, we double the light; if we quadruple the lens area, we quadruple the light.

But in the last problem we saw that the diameter has to go up by 1.4 to double the area. Camera people refer to the lens opening sizes as $f/stops$. The numbers follow this pattern: f/1.4, f/2, f/2.8, f/4, f/5.6, f/8, f/11, f/16, and so on.

Notice that each number is about 1.4 times the number before it. Each stop up *decreases* the light by half. Each stop down doubles the light. (The order is reversed: increase the f/stop, lower the light.) For example, going from f/8 to f/11 cuts the light in half; going from f/5.6 to f/2 increases the light 8 times (2 × 2 × 2).

(b) Another example is pizza size. This $\sqrt{2}$ result tells us that a 14-inch pizza is just about twice the size of a 10-inch pizza.

(c) Another example is actually an abuse of Theorem 3–5–3. Suppose that a country wanted to tell its people that its income had doubled in the last 20 years. It would seem reasonable to draw a dollar bill twice as big.

In the figure, the 1980 dollar bill is drawn twice as big in both dimensions. The fallacy is that the 1980 dollar bill is really drawn with *four* times the area. Thus the increase in income appears much bigger than it really is.

1960

1980

The triangles shown are similar. Complete the following table.

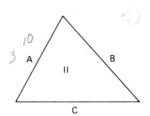

	Side$_I$	Side$_{II}$	Area$_I$	Area$_{II}$
1.	$a = 5$	$A = 10$	40	?
2.	$b = 2$	$B = 3$	?	54
3.	$c = ?$	$C = 16$	50	200
4.	$a = 12$	$A = ?$	180	300
5.	$b = 4$	$B = ?$	120	?

6. Stan is opening up a pizza place. He determines he should charge $2.50 for a 10-inch pizza. Assuming that price is proportional to area, complete the table below to help determine his prices.

Pizza size (in.)	10	12	14	16	18
Price	$2.50	?	?	?	?

7. By what factor should the Freedonia dollar, shown previously, have been increased so that the area was doubled?

A certain dance floor is a square 40 by 40 meters.

8. What is the area? 1600 sq m

9. If each side is increased by 30%, what is each new side? 52² =

10. What is the new area? 2704 sq m

11. What is the percent of increase in area?

12. Suppose that we want to triple the area of a circle. How much bigger should the diameter (or radius) be?

13. Ellen wants to plant a garden 30% bigger (more area) this year. How many times bigger must the sides of the garden be so the area is 30% larger? (The garden is square.)

For each of the following changes in f/stop, tell how much the light increases or decreases.

	Change in f/stop	Change in light
14.	f/2.8 to f/5.6	
15.	f/11 to f/4	
16.	f/2 to f/8	
17.	f/16 to f/8	
18.	f/1.4 to f/16	
19.	f/8 to f/1.4	
20.	f/5.6 to f/1.4	
21.	f/4 to f/2.8	

3-6 PYTHAGOREAN THEOREM

Question What is the Pythagorean theorem?

ANSWER This is a formula known to the Greeks, Chinese, and Egyptians at least 2500 years ago. It is true only for right triangles. The **hypotenuse** is the side opposite the right angle. The other two sides are called **legs**. The theorem relates the hypotenuse to the two legs.

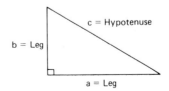

Theorem 3–6–1   If c is the hypotenuse of a right triangle, and a and b are the legs (see the sketch), then

Formula 3–6–1

$$c^2 = a^2 + b^2$$

Another way to write this is by taking square roots:

Formula 3–6–2

$$c = \sqrt{a^2 + b^2}$$

Examples   In the following figures, we can find the hypotenuse using Formula 3–6–2.

1. $c = \sqrt{a^2 + b^2} = \sqrt{3^2 + 4^2}$
   $\phantom{c = \sqrt{a^2 + b^2}} = \sqrt{25}$
   $\phantom{c = \sqrt{a^2 + b^2}} = 5$

2. $c = \sqrt{a^2 + b^2} = \sqrt{7^2 + 9^2}$
   $\phantom{c = \sqrt{a^2 + b^2}} = \sqrt{49 + 81}$
   $\phantom{c = \sqrt{a^2 + b^2}} = \sqrt{130}$
   $\phantom{c = \sqrt{a^2 + b^2}} \approx 11.4$

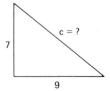

Notice that we substitute the legs into Formula 3–6–2 and simplify. Usually we will have to take a square root at the end.

Problem   Find the missing dimensions of the given right triangle.

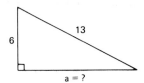

ANSWER   In this case, we will substitute the given information into Formula 3–6–1 and solve.

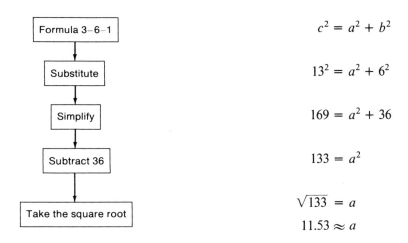

$$c^2 = a^2 + b^2$$

$$13^2 = a^2 + 6^2$$

$$169 = a^2 + 36$$

$$133 = a^2$$

$$\sqrt{133} = a$$

$$11.53 \approx a$$

Problem   Three streets in Jeff's town are shown. He wants to go from home to the movies.
(a) How far is it if he takes Raccoon Rd and Squirrel St?
(b) How far is it if he takes Beaver Blvd directly?
(c) How much shorter is it to take Beaver Blvd?
(d) What percent saving is this?

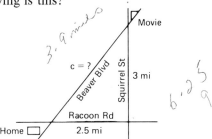

ANSWER   (a) For Raccoon and Squirrel, the distance is simply 2.5 + 3.0 = 5.5 miles.
(b) To find the distance taking Beaver, we use the Pythagorean theorem.

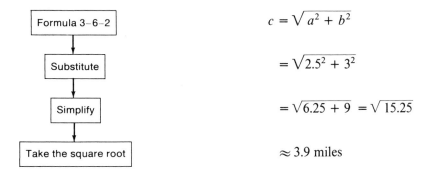

$$c = \sqrt{a^2 + b^2}$$

$$= \sqrt{2.5^2 + 3^2}$$

$$= \sqrt{6.25 + 9} = \sqrt{15.25}$$

$$\approx 3.9 \text{ miles}$$

(c) This is a saving of $5.5 - 3.9 = 1.6$ miles.

(d) As a percentage problem, we have

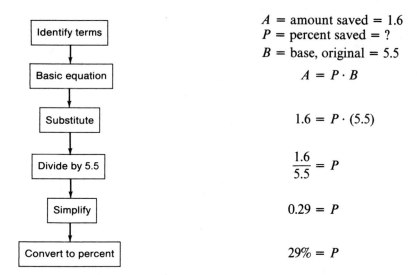

A = amount saved = 1.6
P = percent saved = ?
B = base, original = 5.5

$$A = P \cdot B$$

$$1.6 = P \cdot (5.5)$$

$$\frac{1.6}{5.5} = P$$

$$0.29 = P$$

$$29\% = P$$

Thus Jeff saves 29% of the distance (and time and gasoline) by taking the "angle street."

Problem   Jackie buys a 19-inch color TV. What is the area of the screen?

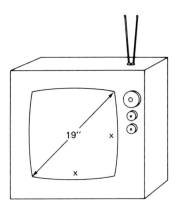

ANSWER   If we didn't know better, we might say $19 \times 19 = 361$ square inches. This is wrong! Advertisers use the diagonal to measure the screen so that the set will sound larger.

To find the area, we must first find the actual dimensions of the sides of the screen. We assume that the screen is a square (to simplify the problem), and we call the unknown sides x. We now use Formula 3–6–1 on this right triangle.

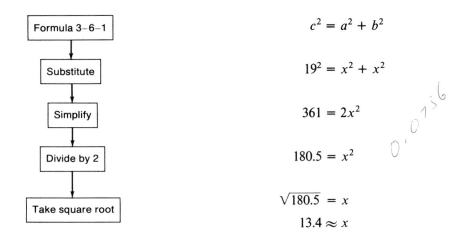

Formula 3–6–1	$c^2 = a^2 + b^2$
Substitute	$19^2 = x^2 + x^2$
Simplify	$361 = 2x^2$
Divide by 2	$180.5 = x^2$
Take square root	$\sqrt{180.5} = x$
	$13.4 \approx x$

The sides of the screen are really only about 13.4 inches. Obviously, an advertiser would rather try to sell a 19-inch TV than a 13.4-inch TV. The area is easy to find, using Formula 3–2–2.

$$\text{Area} = s^2 = (13.4)^2 = 180.5 \text{ in.}^2$$

Problem The Sears Tower in Chicago is 1454 feet tall. How far can one see from the top of this building?

ANSWER Consider the sketch.

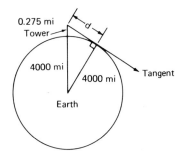

The radius of the earth is about 4000 miles. In miles, the height of the Sears Tower is

$$1454 \text{ ft} \times \frac{1 \text{ mi}}{5280 \text{ ft}} = 0.275 \text{ mi}$$

The distance one can see from the top of the Sears Tower is given by d. This is on a line that is tangent to (barely touching) the earth. It is proved in geometry that this line is perpendicular to the radius. Thus we can use Pythagorean's theorem to find d. Notice that the hypotenuse $c = 4000 + 0.275$.

The Sears Tower, Chicago. (Harr, Hedrich Blessing)

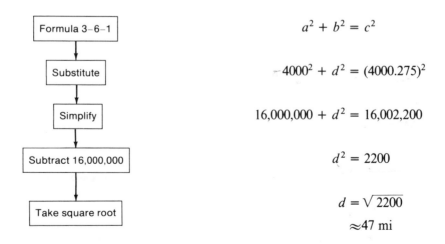

Formula 3–6–1	$a^2 + b^2 = c^2$
Substitute	$4000^2 + d^2 = (4000.275)^2$
Simplify	$16{,}000{,}000 + d^2 = 16{,}002{,}200$
Subtract 16,000,000	$d^2 = 2200$
Take square root	$d = \sqrt{2200}$ $\approx 47 \text{ mi}$

Thus, one can see about 47 miles from the Sears Tower.

PROBLEM SET 3–6–1

1. Define or discuss:
   (a) Hypotenuse.
   (b) Legs.

Find the missing dimensions in the following right triangles and rectangles.

2.

3.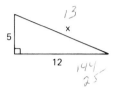

4.

5.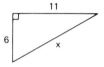

6.

7.

8.

9.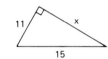

10.

11.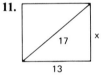

12.

13.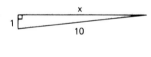

Bob is driving from home to school. He can use routes 1 and 2 or route 3.

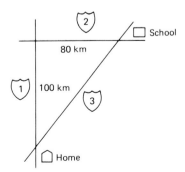

14. How far is the trip by routes 1 and 2?

15. How far is the trip by route 3?

16. What is the percent of saving in distance by taking route 3?

17. How fast is 55 miles per hour in kilometers per hour?

18. If Bob can average 55 miles per hour on routes 1 and 2, how much time will the trip take by these roads?

19. How fast is 40 miles per hour in kilometers per hour?

20. If Bob can average 40 miles per hour on route 3, how much time will the trip take by this route?

Joanne is a traveling sales representative. She must drive from town *A* to town *B*. The speed limits are given. (Recall that distance is rate times time.)

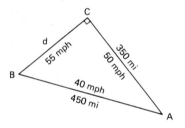

21. What is the distance from *C* to *B*?

22. What is the time from *A* to *C*?

23. What is the time from *C* to *B*?

24. What is the time from *A* to *B*?

25. Which takes the shortest time, *A* to *B* directly, or *A* to *B* through *C*?

Using the technique shown in the text, find the distances one can see from the following places. (Be sure to convert height to miles.)

	Place	Height
26.	Empire State Building (New York)	1250 ft
27.	Prudential Tower (Boston)	745 ft
28.	Bank of America (San Francisco)	700 ft
29.	Washington Monument (Washington, D.C.)	555 ft
30.	Eiffel Tower (Paris)	0.298 km
31.	Great Wall (China)	30 ft
32.	Mt. Everest (Tibet)	8.8 km

Linda and Al buy a 12-inch (diagonal) TV set.

$12^2 = 144 \div 2$

$\sqrt{72}$

8.48

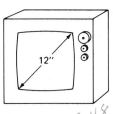

33. How long is each side of the square screen? 8.48

— **34.** What is the area of the screen?

Barb and Tom buy a larger set, a 25-inch-diagonal set.

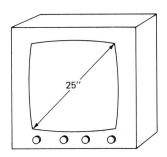

35. How long is each side of the square screen?

36. What is the area of the screen?

37. Reggie wants to put his stereo speakers on adjoining walls 14 feet apart. How far from the corner must the speakers be placed?

38. On a baseball diamond, the distance between bases is 90 feet. What is the distance from home to second base?

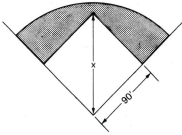

(Courtesy of Star Publications. Williams Press Newspaper Group)

39. An equilateral triangle has sides equal to 10. What is the area? (*Hint*: Find *h* first.)

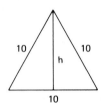

1. What percent of a whole circle is 112°?

2. The class sizes at Byrd College are as given. Draw a pie diagram to show these data.

Class	Size
Freshman	650
Sophomore	520
Junior	440
Senior	390

Find the perimeters (circumferences) of the following figures.

3.
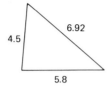

4.
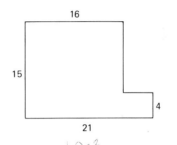

5.

6.
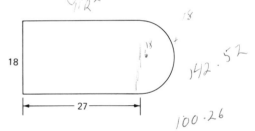

Find the areas of the following figures.

7.

8.
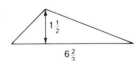

9.
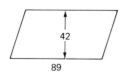

10.
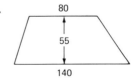

11.

12.

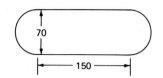

Find the missing dimensions.

13.

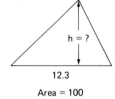

12.3

Area = 100

14.

Area = 78.5

15. How many square meters is 100 square yards?

16. How many square centimeters is 2.4 square meters?

Bev and John's living-dining room is sketched at the right.

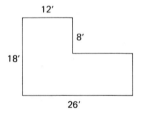

17. What is the area in square feet?

18. What is the area in square yards?

19. If carpeting is $12 per square yard, what will it cost to carpet these rooms?

20. Find the area of the window shown.

1.2 m

1.4 m

Find the volumes of the following figures.

21.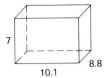

22.

23.

24.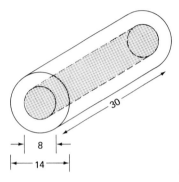

25.

26.

27. How many cubic meters is 50,000 cubic centimeters?

28. How many cubic feet is 50 cubic meters?

The Jacksons buy a backyard swimming pool of the size shown.

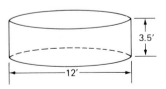

29. What is the volume in cubic feet?

30. If 1 cubic foot is about 7.5 gallons, how many gallons does it take to fill the pool?

31. Find the missing dimensions of the similar figures shown.

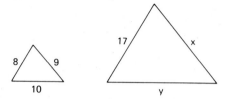

32. Use proportions to double (in both dimensions) the given sketch.

33. Divide a 3-inch line segment into seven equal parts.

34. Divide an $8\frac{1}{2}$- by 11-inch piece of paper into six equal columns and four equal rows.

35. Find the missing area (the triangles are similar).

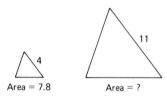

Find the missing dimensions in the following right triangles.

36.

37.

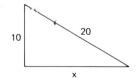

Assume that the TV screen shown is a square with a 15-inch diagonal.

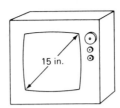

38. What is the length of each side?

39. What is the area of the screen?

Network Operations Center of AT&T's Long Lines Department in Bedminister, New Jersey. (Reproduced with permission of AT&TCo)

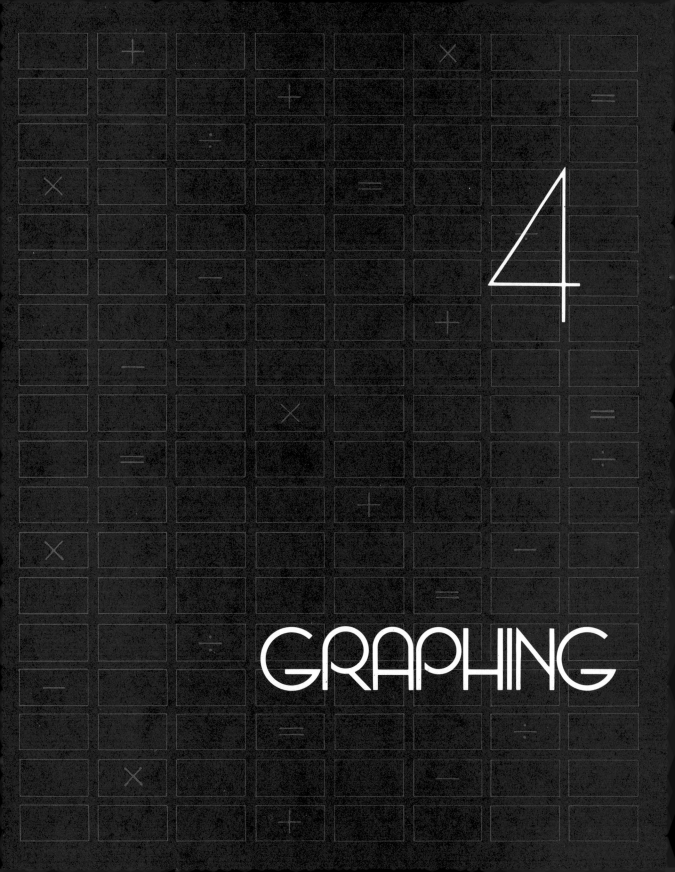

4

GRAPHING

Question Why are graphs important?

ANSWER Pictures are worth a thousand words. Often, a well-drawn graph can tell a better story than a thousand numbers.

4–1 CARTESIAN COORDINATE SYSTEM

Question What is the simplest graph?

ANSWER A line. We take a line and mark off numbers on it. It looks like a thermometer turned on its side. Three number lines are shown in the figure.

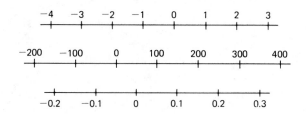

Question Why do these lines look different?

ANSWER We are allowed to put the 0 wherever we want to. Also, we are allowed to make the space between numbers any size we want (or need) to.

Problem What is the location of the points A, B, and C on the line shown?

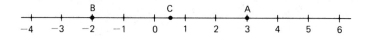

ANSWER We call the number location of a point its **coordinate**. Here we see $A = 3$, $B = -2$, and $C = \frac{1}{2}$.

Problem Suppose that we have the number line shown. Locate the points $D = 20$, $E = 47$, and $F = -15$.

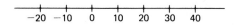

ANSWER $D = 20$ is easy to find. For $E = 47$ and $F = -15$ we will have to approximate their locations. Notice that we simply had to guess the location of the numbers in between the markings.

The Plane (Two dimensions)

Question   Why do we need to study two-dimensional graphs?

ANSWER   In science and in everyday life, we are usually comparing two quantities: for example, height and weight, population and time, caloric need and age. We use a two-dimensional coordinate system to help us locate the points on the plane. We call this a **Cartesian coordinate system**, named for the mathematician René Descartes.

Problem   What are the coordinates of the points *A*, *B*, *C*, and *D* in the given figure?

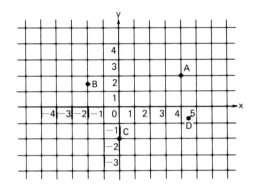

ANSWER   The figure shows a typical coordinate system. It has two perpendicular **axes**, *x* (horizontal) and *y* (vertical). Each axis is marked off as a number line. Where the two axes meet, *O*, is called the **origin**. We always start at *O* and "walk" to the point: first horizontally, then vertically.

1. To get from *O* to *A*, we walk 4 to the *right*, and 2 *up*. Thus we say that $A = (4, 2)$.
2. To get from *O* to *B*, we walk 2 to the *left*, and $1\frac{1}{2}$ *up*. Thus we say that $B = (-2, 1\frac{1}{2})$.
3. To get from *O* to *C*, we walk no units *right* or *left*, and 2 *down*. Thus we say that $C = (0, -2)$.
4. To get from *O* to *D*, we walk $4\frac{1}{2}$ units to the *right* and $\frac{2}{3}$ *down*. Thus we say that $D = (4\frac{1}{2}, -\frac{2}{3})$.

> To determine the coordinates of a point in the plane,
>
> 1. Start at the origin *O*.
> 2. Walk horizontally toward the point (right is positive, left is negative).
> 3. Walk vertically to the point (up is positive, down is negative).
> 4. Write down this pair of numbers as the coordinates.

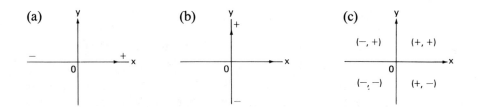

(a)   (b)   (c)

Parts (a) and (b) of the figure show how right and up are positive, and left and down are negative. Part (c) shows the possible combinations of positives and negatives.

Problem   Locate the points $A = (6, 3)$, $B = (-5, 2\frac{1}{2})$, $C = (3, 0)$, and $D = (-2\frac{1}{2}, -1)$ on a graph.

ANSWER   We show these on the accompanying graph.

(a) $A = (6, 3)$ means 6 to the right, and 3 up.
(b) $B = (-5, 2\frac{1}{2})$ means 5 to the left, and $2\frac{1}{2}$ up.
(c) $C = (3, 0)$ means 3 to the right and no units up or down.
(d) $D = (-2\frac{1}{2}, -1)$ means $2\frac{1}{2}$ to the left, and 1 down.

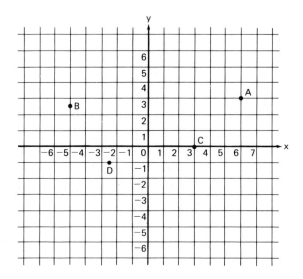

Graphs are used to display information that comes in pairs. We show an example of this in the next problem.

Problem   Bob and Ann open a record store. They keep careful records of the number of customers each day.

Day	0	1	2	3	4	5	6	7	8	9	10	11	12
Customer	0	17	29	40	49	57	64	68	73	77	79	81	82

Can a graph help them study this data?

ANSWER   Yes. We put the day number on the *x*-axis (horizontal), and the number of customers on the *y*-axis (vertical). We then locate all the (day, customers) pairs. For example, $A = (0, 0)$, $B = (1, 17)$, $C = (2, 29)$, and $D = (3, 40)$. This is done in the accompanying figure, and we can see more clearly what is happening. The number of customers went up quickly at first; but then leveled off at about 80.

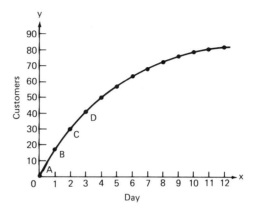

Notice that the scaling on the two axes is different. On the *x*-axis, the numbers go up by 1's since the range of days is 0 to 12. But the *y*-axis goes up by 10's since the range of customers is 0 to 80.

We can always adjust the scale of each axis to include as many points as possible.

An oscilloscope displaying a man whistling. (Courtesy of Carol Olson)

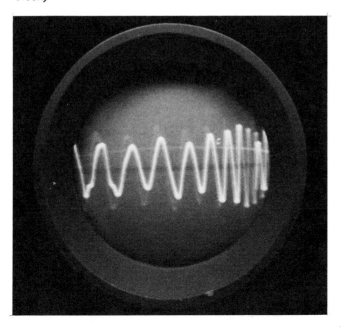

1. Define or discuss:
   (a) Graph.
   (b) Coordinate.
   (c) Axis.
   (d) Origin.
   (e) Cartesian coordinate system.

2. Give the coordinates of the points on the following number line.

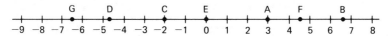

3. Locate the following points on a number line.

$$A = 4 \qquad B = -2 \qquad C = 5\tfrac{1}{2}$$
$$D = -3\tfrac{1}{3} \qquad E = 4.2 \qquad F = -0.3$$
$$G = 0 \qquad H = 4\tfrac{4}{7} \qquad I = -5.7$$

Give (or approximate) the coordinates of the points on the following graphs.

4.

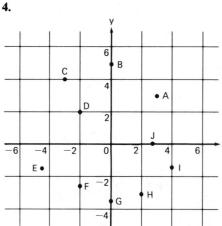

5.

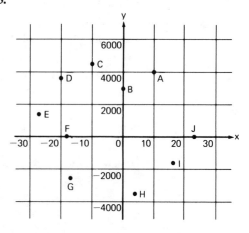

Use graphs like the ones shown to graph the following data. Then try to draw a smooth line through the points.

6. The following data are the average height of a growing boy. Use the graph of part (a).

Age (years)	0	1	2	4	6	8	10	12	14	16	18	20
Height (in.)	22	30	34	40	44	46	49	55	60	65	68	69

7. In a certain survey, 1157 people were asked their incomes. The following are the percents having certain incomes (rounded off). Use the graph of part (b).

Income ($1000)	6	8	10	12	14	16	18	20	22	24
Percent	3	5	11	19	21	18	11	6	3	2

(a)   (b)

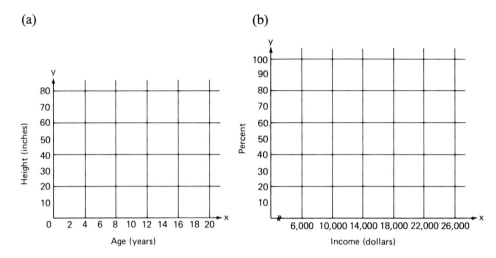

8. According to the U.S. census, the population has been growing as follows. Graph the data on the graphs of parts (a) and (b). Notice that the y-axis in part (b) is not even, but approximate the points anyway.

Year	1760	1780	1800	1820	1840	1860	1880	1900	1920	1940	1960
U.S. population (in millions)	1.8	2.9	5.3	9.8	17.1	31.4	50.1	76.0	109.7	131.7	179.3

Can you extend the graph of parts (a) and (b) to guess the U.S. population in 1980? in 2000?

(a)   (b)

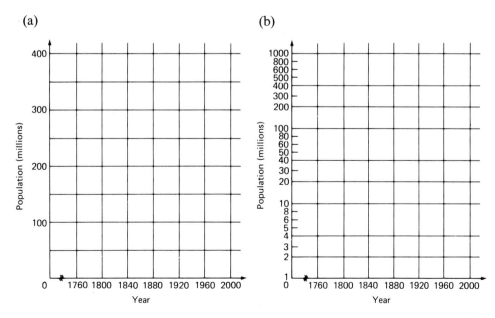

Scaling

Question How can we make sure that all the data numbers fit on the graph?

ANSWER The trick is to scale each axis properly. We can *stretch*, *shrink*, or even *cut* the axis to fit in all the numbers.

Problem Suppose that we are dealing with people's weights. What sort of scale would we use?

ANSWER Most people's weights vary between 0 and 350 pounds. So we shrink the scale to fit the numbers 0 to 350 on it.

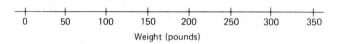

Weight (pounds)

Problem Suppose that we are dealing with the national debt. What scale would we use?

ANSWER The national debt might run as high as $100,000,000,000, or $100 billion. We can scale this as follows.

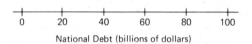

National Debt (billions of dollars)

Notice that each number really stands for $1 billion. For example, 40 means $40 billion.

Problem Suppose that we are dealing with the progress of a moving snail. What scale would we use?

ANSWER Here the numbers are very small; we might stretch the axis to run between 0 and 2 centimeters.

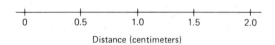

Distance (centimeters)

Problem Suppose that we wanted to look at the Dow Jones stock market average in a year where it only varied between 800 and 1050. How would the scale look?

ANSWER We could make a graph to run from 0 to 1050. But then all the points would be cramped at the right between 800 and 1050. We can do better by *breaking* the axis as follows.

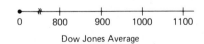

Dow Jones Average

Notice that the wiggle mark means that an unneeded part of the axis was cut away. This is called a **broken scale**.

Problem What would the scale be for the heights of a group of men?

ANSWER Most men's heights are between 5 and 7 feet. So we break the axis and only show this range.

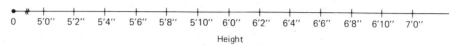

Height

Problem Suppose that we are dealing with numbers that have a very wide range: very small to very big. For example, how would we scale the following?
(a) The populations of cities and towns in the United States.
(b) The distance to various planets and stars.

ANSWER In both situations, the numbers might have a very wide range.
(a) With U.S. cities and towns, we can have Oakland Mills, Iowa, with 23 people on the same graph with New York City with about 8 million people. To scale this we use the powers of 10 as follows.

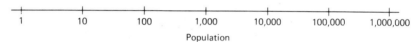

Population

(b) The distances to planets and stars can vary from 220,000 miles (the moon) to 4×10^{15} miles (the star Canopus) and beyond, of course. Here again we use the powers of 10.

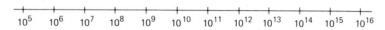

This power of 10 scale can be called an **exponential** or **log scale**. Although this scale is handy to use with numbers of all sizes, it is also tricky to use since the numbers do not come up evenly between the markings. Rather, they come up as follows.

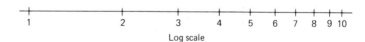

Log scale

PROBLEM SET 4–1–2

1. Define or discuss:
   (a) Broken scale.
   (b) Log (exponential) scale.

For each of the following situations, give a reasonable range for the numbers involved. Then draw the appropriate axis for this range.

2. The body temperature of a human body.

3. The number of push-ups a college freshman can do.

4. The IQ of a college freshman.

5. The number of dates per month of a college freshman.

6. The number of records sold by a singing star.

7. The annual income of a college professor.

8. The life span of a human being.

9. The distance a snail can travel in 1 minute.

10. The number of U.S. congressmen for a certain issue.

11. The temperature of your hometown (in Fahrenheit and in Celsius).

12. The price of a car (new or used).

13. The number of children a family will have.

14. The net worth of an American family. (Remember, some families are in debt.)

15. The distance from your school to various cities of the world.

The following data come in pairs. Determine the proper ranges for each. Then scale the *x*- and *y*-axes to show all the values. Then plot the points and draw a smooth line through the points. The top quantity is the *x*-axis; the bottom is the *y*-axis.

16.

Annual income	8000	9000	10,000	11,000	12,000	13,000	14,000	15,000
Number in a group	25	125	260	300	200	70	20	10

17.

Year	1953	1954	1955	1956	1957	1958	1959	1960	1961	1962	1963	1964	1965
Most home runs in National League	47	49	51	43	44	47	46	41	46	49	44	47	52

18.

Age (years)	5	10	15	20	25	30	35	40	45	50	55
Hours of TV per week	25	35	20	15	20	23	26	30	30	30	30

19.

Age (years)	5	10	15	20	25	30	35	40	45	50	55	60	65	70	75
Expected remaining years to live	67	62	57	53	48	44	39	34	30	26	22	18	15	12	9

20.

Car weight (lb)	2000	2500	3000	3500	4000	4500	5000
Miles per gallon	28	22.5	19	16	14	12	11

21.	Age of car (yr)	0	1	2	3	4	5	6	7	8	9	10
	Value ($)	6000	4500	3375	2531	1898	1423	1068	800	600	450	338

22.	Years invested at 5%	0	5	10	15	20	25	30	35	40
	Growth of $100	100	128	163	208	265	339	432	552	704

4–2 SCIENTIFIC GRAPHS

Question   How are graphs used?

ANSWER   Graphs are used to picture either scientific data or an algebraic equation.

Question   What is a scientific graph?

ANSWER   A **scientific graph** is any graph that pictures numbers or data taken from real-life situations. This might mean the price of a can of peas or the energy level of an atomic particle. The important thing is that the numbers were observed by some person or persons who can then use a graph to look for information.

Question   How do we make scientific graphs?

ANSWER   This is the rule we use.

> To make a scientific graph:
> 1. Gather the data (either in the library or by experiment)
> 2. Determine the range of the data for both quantities.
> 3. Choose a proper scale to display all the data.
> 4. Scale the x- and y-axes accordingly.
> 5. Plot the data points on the graph.
> 6. Try to fit a curve or line through the points.

Frequency Graphs

Question   What is a frequency graph?

ANSWER   A **frequency graph** pictures how often or frequently the different numbers from a situation occur. With this type of graph, we will always have one quantity (age, height, income) on the *x*-axis. On the *y*-axis, we will have the number of occurrences or percent of occurrences.

Problem One hundred and two students are asked to rate a new record from 0 to 10 (0 is horrible; 10 is great). The results are

Rating	0	1	2	3	4	5	6	7	8	9	10
Number	1	2	3	5	7	9	13	17	21	15	9

Can we use a frequency graph for these data?

ANSWER Yes. We have a quantity (rating of record) and the number of times it occurs. We put the rating on the *x*-axis and the number on the *y*-axis. This is shown in the accompanying graph. We see that the graph vividly shows that most of the ratings are lumped near 8. We can predict that the record will probably be a hit.

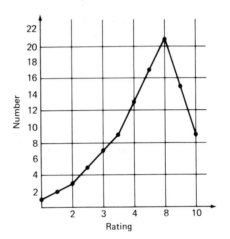

Problem Can we use a frequency graph to study the following weights of the members of a college football squad? *Weight* (lb): 182, 186, 191, 193, 196, 197, 197, 200, 201, 203, 204, 206, 206, 207, 207, 208, 210, 211, 211, 211, 213, 214, 215, 215, 215, 217, 217, 218, 219, 220, 221, 221, 221, 223, 224, 224, 225, 226, 226, 228, 229, 230, 231, 233, 237, 239, 240, 244, 248, 254.

ANSWER Yes; but if we try to graph them just as they are, we won't get too far. There are too many weights and not enough players at each weight.

Instead, we lump the weights in 5-pound groups, for example, 180–184, 185–189, and so on. Now we count the number in each group.

Weight (lb)	180 –184	185 –189	190 –194	195 –199	200 –204	205 –209	210 –214	215 –219	220 –224	225 –229	230 –234	235 –239	240 –244	245 –249	250 –254
Number	1	1	2	3	4	5	6	7	7	5	3	2	2	1	1

But how do we graph a range of numbers like 190–194? We have two choices. We can graph the range with a **bar graph**, as shown in part (a).

Another possibility is to use the middle point of the range to represent the whole range. For example, let 192 represent the range 190–194, as shown in part (b). Notice that the graphs are very similar.

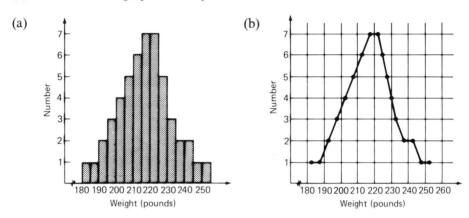

(a)

(b)

Weight (pounds)

Weight (pounds)

Question   Can we graph percents instead of numbers in a frequency graph?

ANSWER   Yes. The graph will have exactly the same shape. Consider the following problem.

Problem   The table shows the age distribution for the House and Senate of the 94th Congress. Put the data on a single graph to compare them.

Age	House (%)	Senate (%)
25–34	5.6	1
35–44	20.9	11
45–54	36.4	33
55–64	25.5	31
65–74	10.0	18
75–84	1.6	6

(American Heart Association)

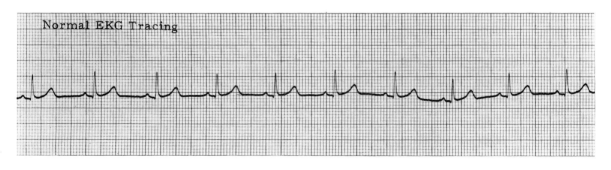

Normal EKG Tracing

ANSWER   Here we really have two sets of data: the House data and the Senate data. Normally, we would make two frequency graphs. But here we wish to put them on the same graph to compare them.

To do this, we cannot use bar graphs. That would get messy. Instead we will use the middle point to represent the range; for instance, 40 will stand for 35–44.

Also, to keep the two graphs separate, we will use an × to locate a House point, and we will use a ○ to locate a Senate point. The accompanying figure shows these data. We can conclude pretty clearly that Senators tend to be older than Representatives (by about 5 years on the average).

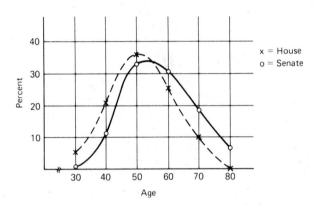

**PROBLEM
SET 4-2-1**

1. Define or discuss:
   (a) Scientific graph.
   (b) Frequency graph.
   (c) Bar graph.

Plot frequency graphs for the following samples. Remember to put the top quantity on the x-axis and the number or percent on the y-axis.

2.

Height (in.)	63	64	65	66	67	68	69	70	71	72	73	74	75
Number	135	210	317	421	531	603	644	621	519	389	260	177	93

3.

Household size	1	2	3	4	5	6	7	8	9	10
Number	29	42	52	39	25	11	5	2	1	1

4.

Age of arrested person	14	15	16	17	18	19	20	21	22	23	24	25
Number	251	276	314	302	283	241	201	200	161	146	138	125

5.

Hours of nightly TV	0	$\frac{1}{2}$	1	$1\frac{1}{2}$	2	$2\frac{1}{2}$	3	$3\frac{1}{2}$	4	$4\frac{1}{2}$	5	$5\frac{1}{2}$	6
Percent	5	7	8	10	11	13	12	9	8	7	6	3	1

6.

Hour of traffic accident (P.M.)	12	1	2	3	4	5	6	7	8	9
Number in week	8	15	28	38	45	51	47	41	32	32

7.

Grade in school	1	2	3	4	5	6	7	8	9	10	11	12
Number (millions)	3.5	3.6	3.7	3.8	3.8	3.8	3.9	3.8	4.0	3.8	3.4	3.0

The following data give a range of values for the measured quantity. In each case, draw a bar graph and a line graph (using the midpoint of the range).

8.

Test score	0 –9	10 –19	20 –29	30 –39	40 –49	50 –59	60 –69	70 –79	80 –89	90 –100
Number	2	5	9	17	32	49	66	84	52	25

9.

Age	0 –4	5 –9	10 –14	15 –19	20 –24	25 –29	30 –34	35 –39	40 –44	45 –49
Number in U.S. (millions)	7.9	9.0	8.8	8.0	7.0	5.7	4.9	5.0	5.4	5.3

	50 –54	55 –59	60 –64	65 –69	70 –74	75 –79	80 –84
	4.8	4.2	3.5	2.8	2.0	1.5	0.8

10.

Annual income	6000 –7999	8000 –9999	10,000 –11,999	12,000 –13,999	14,000 –15,999	16,000 –17,999	18,000 –19,999
Number	179	257	496	816	611	337	218

11.

Monthly rainfall (in.)	0 –1	1 –2	2 –3	3 –4	4 –5	5 –6	6 –7	7 –8	8 –9	9 –10
Number of months	45	83	103	92	75	51	33	18	9	2

12.

IQ	40–55	55–70	70–85	85–100	100–115	115–130	130–145	145–160
Percent	0.5	2	13.5	34	34	13.5	2	0.5

13.

Doctor's age	20–30	30–40	40–50	50–60	60–70	70–80
Percent	11.3	27.3	24.6	18.1	11.3	7.3

14.

Age of suicide victim	5–14	15–24	25–34	35–44	45–54	55–64	65–74
White male	0.5	13.9	19.9	23.3	29.5	35.0	41.1
White female	0.1	4.2	9.0	13.0	13.5	12.3	8.5
Black male	0.2	11.3	19.8	12.6	14.1	10.5	10.8
Black female	0.2	4.1	5.8	4.3	4.5	2.2	3.6

All suicide rates are per 100,000 population.

15.

Number of children (born or expected)	0	1	2	3	4	5	6	7
Percent of women, 18–24 years	4.1	8.6	57.5	20.8	6.5	1.5	0.7	0.2
Percent of women, 35–39 years	4.3	7.9	25.0	27.1	17.5	9.2	4.4	3.2

Put both curves on the same graph to visualize differences.

16.

Number of rooms in house/apartment	1	2	3	4	5	6	7 or more
Percent of total	1	3	10	19	25	21	21

The following data are given just as received. Place the data into groups or ranges, and then make a bar or line graph.

17. Annual incomes of a group: $11,235; $12,738; $15,719; $19,811; $20,903; $15,114; $17,739; $13,006; $14,192; $12,128; $19,156; $24,862; $19,529; $16,084; $15,172; $14,807; $13,342; $17,775; $23,157; $20,041; $19,814.

18. Scores on a math test: 81, 45, 75, 73, 91, 79, 62, 59, 66, 82, 74, 69, 72, 53, 81, 74, 72, 85, 66, 76, 95, 87, 73, 72, 65, 61, 51, 41, 79, 66, 71, 59, 78.

19. Sales in a day: 44, 28, 37, 52, 41, 39, 35, 31, 41, 45, 36, 32, 28, 39, 47, 41, 58, 42, 44, 35, 31, 38, 27, 30, 35, 44, 51, 42, 38, 35, 39, 29, 31, 35, 38.

20. Most home runs in National League (1945–1974): 28, 23, 51, 40, 54, 47, 42, 37, 47, 49, 51, 43, 44, 47, 46, 41, 46, 49, 44, 47, 52, 44, 39, 36.

Time Graphs

Question   What is a time graph?

ANSWER   A **time graph** shows how some quantity, such as population, use of sugar, height, or the Dow Jones average, changes with time. Time always goes on the x-axis, and the other quantity goes on the y-axis. With a time graph, we can visualize whether there is a trend upward, downward, or oscillating up and down.

Problem   Use a time graph to find a trend in men's clothing style.

ANSWER   As an example, we can look up in the *Statistical Abstracts for the United States* the following data about men's suits.

Year	1950	1955	1960	1965	1970	1973
Sales (in millions)	23.7	20.3	21.3	21.9	17.7	16.6

We graph this as shown at the top of page 219. The graph gives a strong visual impression of what is happening with men's suits. The sales go up and down, but overall the trend for men's suits is down.

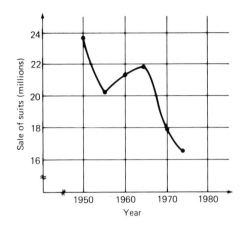

Problem   How has the number of bachelor's degrees given each year grown over the last 100 years?

ANSWER   Whenever we see the word *grown*, we know that we are dealing with a time graph. From the *Historical Statistics of the United States*, we get the following table.

Year	1870	1880	1890	1900	1910	1920	1930	1940	1950	1960	1965	1970
Bachelor's or first professional degree (in thousands)	9.3	12.9	15.5	27.4	37.2	48.6	122.5	186.5	432.1	396	539	833

The figure shows two different graphs of the data: (a) with a regular scale, and (b) with a log scale. In both cases, the year 1950 seems to be a peculiar point. (This is probably because of the huge number of World War II veterans returning to school on the G.I. Bill.) In part (a) we had to cramp the early years in at the bottom; the later years seem to fly off the graph paper. In part (b), however, the data points are all displayed fairly evenly. We also have a much better idea of where the graph is going.

(a)

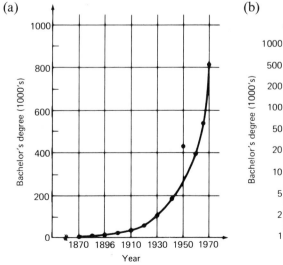

(b)

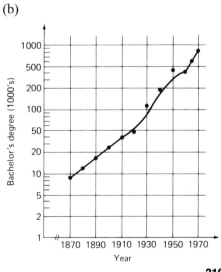

Problem Can we compare two quantities on a time graph, such as egg consumption and heart disease?

ANSWER For years there has been a debate about whether the cholesterol in eggs leads to heart disease. We can look up statistics on the per person egg consumption and the heart disease rate. We get the following table.

Year	1900	1910	1920	1930	1940	1950	1960	1970	1974
Egg consumption (per person)	—	306	299	331	319	389	335	311	287
Cardiovascular disease (per 100,000)	345	372	365	414	486	495	515	496	495

To get the full impact of the data, we will put them on the same graph. We will use × to mark the egg points, and ○ to mark the cardiovascular disease points.

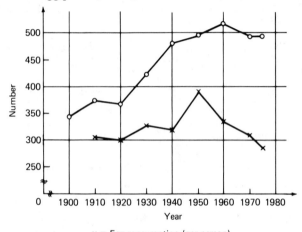

x = Egg consumption (per person)
o = Cardiovascular disease (per 100,000)

Like many things in science, the conclusion here is not clear. In the last 25 years, it appears that egg consumption is going down, while cardiovascular disease is staying about the same. Certainly, a poultry farmer would say eggs have nothing to do with heart disease.

PROBLEM SET 4–2–2

1. Define or discuss a time graph.

Make time graphs for the following data. Remember that time always goes on the x-axis.

2.

Year	1959	1960	1962	1963	1964	1965	1966	1967	1968
Consumer price index	100	101.6	103.8	105	106.4	108.3	111.3	114.3	119.3

Year	1969	1970	1971	1972	1973	1974
Consumer price index	125.7	133.2	139	143.5	152.4	169.5

3.

Year	1920	1925	1930	1935	1940	1945	1950	1955	1960	1965	1970	1975
Energy consumption per person (million BTU)	186	180	181	150	181	238	226	243	249	278	330	345

4.

Year	1950	1955	1960	1965	1970	1975
Percent of one-person households	10.9	10.9	13.1	15.0	17.0	19.1

5.

Time (seconds)	0	1	2	3	4	5	6	7
Distance an object falls (feet)	0	16	64	144	256	400	576	784

6. A 70°F spoon is thrown into a tub of 40°F water.

Time (sec)	0	1	2	3	6	9	12	15	18	21	24	27
Temperature of spoon (°F)	70	64	59	55	47.5	43.6	41.9	40.9	40.4	40.2	40.1	40.05

7.

Month (Jan. is 1)	1	2	3	4	5	6	7	8	9	10	11	12
Average temp., town A	27.9	31.1	40.3	51.5	61.9	71.6	76.1	73.8	67.1	56.0	41.5	30.7
Average temp., town B	42.0	43.2	44.7	47.5	50.9	54.0	55.5	55.7	54.5	51.9	47.4	44.1

Put both curves on the same graph.

8.

Time (min)	0	1	2	3	4	5	6	7	8
Heart beats per min after exercise	104	84	76	72	69	66	64	64	64

9.

Year	1950	1955	1960	1965	1970	1975
Cost of a gall bladder operation ($)	361	487	660	839	1397	2208

10.

Year	1920	1930	1940	1950	1960	1965	1970	1973
Percent of women who are single	29.4	28.4	27.6	19.6	19.0	12.4	13.7	13.9

11.

Year	1940	1950	1960	1965	1970	1973
Fresh fruit per person (lb)	139	109	93	81	81	78
Processed fruit per person (lb)	34	44	48	46	50	51

Put both curves on the same graph.

12.

Month (Jan. is 1)	1	2	3	4	5	6	7	8	9	10	11	12
Index of industrial output	115	112	111	111	113	116	117	119	121	122	123	124

13.

Month	0	1	2	3	4	5	6	7	8
Time to run 1 mile (min)	9.5	8.1	7.2	6.8	6.5	6.3	6.2	6.1	6.0

14. Keep track of your moods for a 30- or 60-day period. Use the following scale for your mood: -3 = depressed; -2 = blue; -1 = blah; 0 = neutral; 1 = good; 2 = happy; 3 = ecstatic. Keep the following chart and then graph it. See if there is any pattern.

Day	1	2	3	4	5	etc.
Mood						

15. Keep track of your weight for 10 weeks (or more) by weighing yourself the same time once or twice a week. Then graph the table.

Weeks	0	$\frac{1}{2}$	1	$1\frac{1}{2}$	2	$2\frac{1}{2}$	etc.
Weight							

General Relation Graphs

Question What is a general relation graph?

ANSWER All graphs picture a relation between two quantities. One goes on the x-axis, the other on the y-axis. We have already looked at two special graphs, the frequency and time graphs. Since we have no better name, we will call everything else **general relation graphs**.

Problem The Klik Pen Co. is trying to decide what price to charge for its new pen. Naturally, they want to make as large a profit as possible. Can they use a relation graph?

ANSWER Yes. Here we want to see if there is a relation between price and profit. We suspect there is. If they charge too little, they won't make much profit. If they charge too much, very few people will buy the pens, and they won't make much profit. Somewhere in the middle is the best price. Suppose that they test different prices and get the following data.

Price (¢)	10	20	30	40	50	60	70	80	90	100
Profit ($)	-1000	0	800	1300	1750	1800	1500	1000	500	200

We have graphed the data in the accompanying figure. As suspected, if the price is too high or low, the profit is very small (or negative). We can also see that the best price to charge is about 57 cents, which will give a profit of a little over $1850.

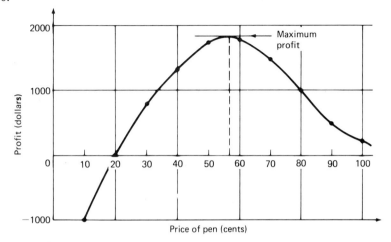

Notice that we had to use the negative part of the y-axis, since charging only 10 cents meant a $1000 loss.

Problem Is there a relation between gas mileage (miles per gallon) and the speed at which the car is driven?

ANSWER Auto testers have come up with the following data for an average car.

Speed (mph)	20	30	40	50	60	70	80
Miles per gallon	17.8	18.2	18.3	17.7	15.4	13.2	11.4

We put speed on the x-axis, and miles per gallon on the y-axis, and obtain the graph shown.

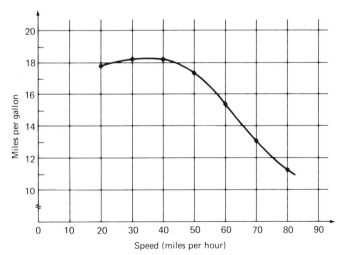

We can tell from this graph that a car's best gas mileage is between 30 and 40 miles per hour. After 45 or 50 miles per hour, the car's gas mileage goes down very quickly.

Problem Is there a relationship between a family's income and its likelihood to own one car or to own two or more?

ANSWER To answer this question, we can look up the data in sources such as the *Statistical Abstracts of the United States*. It is available in almost every library. In this case we get the following data:

Annual income ($)	2000	4000	6250	8750	12,500	20,000	32,000
Percent with at least one car	40.6	68.0	84.2	91.3	94.9	96.5	93.0
Percent with at least two cars	5.3	12.6	23.2	32.2	45.6	58.4	66.6

We want to put both of these on the same graph. We use an × for the one-car data and a ○ for the two-car data. We can see from the graph shown that past $10,000 just about everyone has at least one car. The second graph tells us that the likelihood of owning two cars or more is roughly proportional to one's income (between $2,000 and $14,000), and then after $14,000 the graph flattens out slowly.

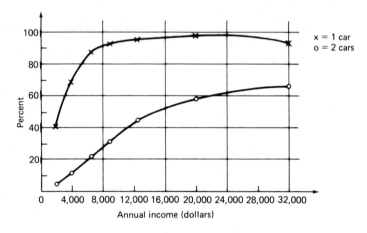

Problem Suppose that we wish to see how the scores of one test relate to the scores on a second test. Can we do this?

ANSWER Yes, but the graph will not be as simple. For each student, we plot a pair of points (first score, second score). The data appear on the next page. There are 26 students, so we will get 26 points.

Notice on the resulting graph that the points just seem to be scattered on the graph. For that reason, this is sometimes called a **scattergram**. Of course, it is practically impossible to draw a line or curve through these points. What can be done is to sketch an oval that encloses all or most of the points.

Student	Test 1	Test 2	Student	Test 1	Test 2
1	75	80	14	86	82
2	83	91	15	81	87
3	62	67	16	65	67
4	91	85	17	53	57
5	84	85	18	95	91
6	72	77	19	84	91
7	85	82	20	59	68
8	53	59	21	66	62
9	71	69	22	75	79
10	88	91	23	78	81
11	68	73	24	71	68
12	41	56	25	84	79
13	72	61	26	69	73

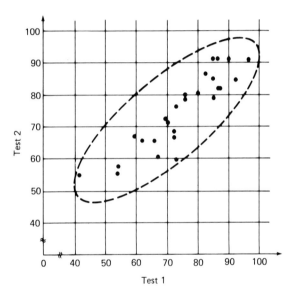

When the oval goes up to the right, as this one does, we can conclude that there is some **correlation** between the two quantities. This means that people who scored high on test 1 *tended* to score high on test 2, and people who scored low on test 1 *tended* to score low on test 2. If the oval had been down to the right, we would conclude the opposite: people who did well on test 1 tended to do poorly on test 2, and vice versa. If the oval is more like a circle, with no direction up or down, this probably means that there is little or no relation between the variables.

PROBLEM SET 4-2-3

1. Define or discuss:
   (a) General relation graph.
   (b) Correlation.
   (c) Scattergram.

Prepare a relation graph to display the following data. Put the top quantity on the *x*-axis, and the bottom quantity on the *y*-axis.

2.

Cost of tire ($)	11.43	13.85	14.86	16.82	21.45	22.75	37.25
Total miles wear	12,500	15,000	17,500	20,000	25,000	30,000	40,000

3. Dan buys a $4400 car. His monthly payments depend on how much his down payment is.

Down payment ($)	400	800	1200	1600	2000	2400	2800
Monthly payment ($)	138	124	110	96.50	83	69	55

4. Bob is a stereo nut. He rates stereos from 0 to 100. This rating will depend on price.

Cost of stereo ($)	50	100	200	400	800	1600	3200
Bob's rating	15	30	45	60	75	90	97

5. The liability coverage of auto insurance depends on the premium.

Premium ($)	100	109	112	119	128	134
Liability limits ($)	25,000	40,000	50,000	100,000	300,000	500,000

6. The price one can charge for a radio depends on the supply on the market.

Supply (millions)	1	2	3	4	5	6
Price ($)	50	44	39	35	32	30

7. When plucking a string on a guitar, the pitch depends on the fret at which the string is held down.

Fret	0 = open	1	2	3	4	5	6
Pitch (frequency)	E = 326	F = 345	F♯ = 366	G = 388	G♯ = 411	A = 435	A♯ = 461

	7	8	9	10	11	12
	B = 488	C = 517	C♯ = 548	D = 581	D♯ = 615	E = 652

8. With 35-millimeter camera lenses, the viewing angle depends on the focal length of the lens.

Focal length (mm)	21	28	35	50	75	105	135	150	200	300	400	500
Angle of view	91	76	63	46	32	23	18	16	12	8	6	5

9. According to the theory of relativity, the mass of an object depends on its velocity, as follows.

Velocity (mi/sec)	0	18,600	62,000	93,000	124,000	167,400	184,140
Mass (g)	1	1.005	1.06	1.15	1.34	2.29	7.08

10. On a certain farm, the output of wheat depends on the number of laborers working the farm.

Number of laborers	1	2	3	4	5	6	7	8
Bushels of wheat	60	150	234	312	384	450	510	564

11. The income tax a family of four pays depends on its gross income.

Gross income ($)	5000	10,000	15,000	20,000	25,000	30,000	35,000	40,000
Tax ($)	0	450	1,385	2,536	3,871	5,424	7,242	9,226

Make a scattergram for the following data. Then make a conclusion (if you can) about the data.

12.

Student	A	B	C	D	E	F	G	H	I	J	K	L	M
Math score	79	63	77	91	86	65	62	65	70	81	44	83	77
English score	83	85	66	72	52	71	85	96	80	77	63	72	81

Student	N	O	P	Q	R	S	T	U	V	W
Math score	63	97	85	35	56	75	62	91	84	75
English score	89	73	71	49	84	69	88	87	64	74

13.

Baseball team	Det	Bos	Bal	NY	Cle	Mil	Oak	Chi
Total games won in 1972	86	85	80	79	72	65	93	87
Total runs scored in 1972	558	640	519	557	472	493	604	566

Team	Min	KC	Cal	Tex	Pit	Chi	NY	StL	Mon	Phi	Cin
Games won	77	76	75	54	96	85	83	75	70	59	95
Runs scored	537	580	454	461	691	685	528	568	513	503	707

Team	Hou	LA	Atl	SF	SD
Games won	84	85	70	69	58
Runs scored	708	584	628	662	488

14.

State	Ala.	Ariz.	Calif.	Colo.	Fla.	Ga.	Ha.	Ill.
Income from crops ($ millions)	285	310	3041	285	1011	557	178	1624
Income from livestock ($ millions)	526	402	1883	1064	411	701	43	1174

State	Iowa	Kan.	Md.	Mich.	Minn.	Mo.	Neb.	N.J.	N.C.
Crops	1295	764	130	457	830	616	719	151	925
Livestock	2730	1492	267	514	1454	1087	1524	89	601

State	N.D.	Ohio	Okla.	Pa.	S.D.	Tex.	Wash.	Wisc.	Wyo.
Crops	554	648	269	276	278	1132	558	240	41
Livestock	300	768	858	293	867	2122	291	1425	223

15.

Student	A	B	C	D	E	F	G	H	I	J	K	L	M	N
Dates per month	6	4	10	2	0	5	6	7	3	8	1	4	5	6
Grade point average	2.4	3.1	2.1	3.3	3.8	2.7	2.8	2.5	3.0	3.1	3.2	2.6	3.1	2.5

4–3 SLOPES AND LINES

Slopes

Question What is the slope of a line segment?

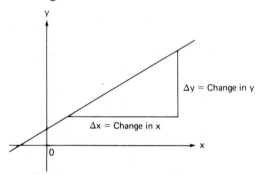

ANSWER The **slope** of a line segment is the change in the y-coordinate divided by the change in the x-coordinate. We use the following notations:

$$\Delta y = \text{change in the } y\text{-coordinate}$$
$$\Delta x = \text{change in the } x\text{-coordinate}$$
$$m = \text{slope}$$

Using these symbols, we have the following formula.

Formula 4–3–1

$$m = \frac{\Delta y}{\Delta x} = \frac{\text{change in } y}{\text{change in } x}$$

Formula for finding a Slope

As we will see, the slope measures whether the line is going up or down and how fast. There is one exception to Formula 4–3–1. We will soon see that a vertical line segment has no slope.

Question Is there a simple way to find the slope?

ANSWER Yes. The trick is to remember that the word *change* usually means *subtraction*. Suppose that we have two points $A = (a, b)$ and $B = (c, d)$, and we want to find the slope of the line segment between them, as shown in the accompanying figure.

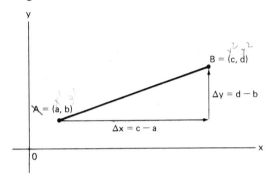

The change in y is $d - b$. The change in x is $c - a$. If we put this into Formula 4–3–1, we get

2 pt

Formula 4–3–2

$$m = \frac{d - b}{c - a}$$

$slope = \dfrac{y_2 - y_1}{x_2 - x_1} = slope$

To use Formula 4–3–2 correctly, it is important to be very careful with plus and minus signs.

Problem Find the slopes of the line segments shown.

(a)

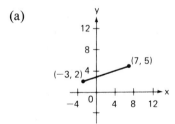

(b)

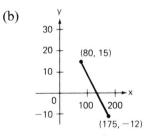

(c)

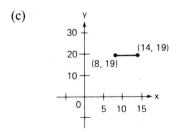

(d)

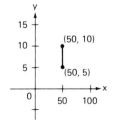

A fun slope. (Suzanne Szasz/Photo Researchers)

ANSWER   For all these, we use Formula 4–3–2. Notice the care with which we handle the plus and minus signs.

(a) $m = \dfrac{d-b}{c-a} = \dfrac{5-2}{7-(-3)} = \dfrac{3}{10} = 0.3$

(b) $m = \dfrac{d-b}{c-a} = \dfrac{-12-15}{175-80} = \dfrac{-27}{95} = -0.28$

(c) $m = \dfrac{d-b}{c-a} = \dfrac{19-19}{14-8} = \dfrac{0}{6} = 0$

(d) $m = \dfrac{d-b}{c-a} = \dfrac{10-5}{50-50} = \dfrac{5}{0} = \text{undefined}$

Since we are not allowed to divide by 0, we say that the slope is *undefined* in case (d) and in all other vertical line segments.

Question   What is the relation between the direction of a line and its slope?

ANSWER   The four previous examples show us clearly what the four cases are. The following table summarizes the results.

Direction of line	Slope
(a) Upward	Positive
(b) Downward	Negative
(c) Horizontal	0
(d) Vertical	Undefined

Question   What is the easiest way to measure the slope of a line?

ANSWER   Here is the rule we use.

> To measure the slope of a line:
>
> 1. Choose two points with convenient x-coordinates; then subtract them to get Δx.
>
> 2. Determine the y-coordinates of these points; subtract them (in the same order as the x-coordinates) to get Δy.
>
> 3. Divide $\dfrac{\Delta y}{\Delta x}$ to get m.

Question   Why do we choose the x-coordinates to be the simpler numbers?

ANSWER   With fractions, it is always easier to divide if the denominator is simple. For example, $\dfrac{6.928}{3}$ is much simpler to compute than $\dfrac{4}{1.732}$, even though both give the same answer, 2.309.

Since Δx is in the denominator of $m = \dfrac{\Delta y}{\Delta x}$, we pick the x terms to be as simple as possible. Then Δx will be a simple number.

Problem   Find the slope of the line shown in part (a) of the figure.

(a)

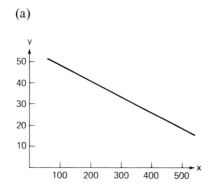

(b)

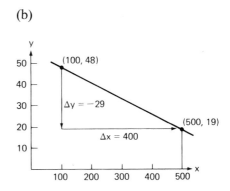

ANSWER　In part (b) we have chosen two convenient x points, $x = 100$ and $x = 500$; so $\Delta x = 400$. Now we measure the corresponding y-coordinates as $y = 48$ and $y = 19$. We must be careful here to subtract $19 - 48 = -29$ for Δy since this was the order in subtracting the x values.

Using Formula 4–3–1, we get

$$m = \frac{\Delta y}{\Delta x} = \frac{-29}{400} = -0.0725$$

Notice that the slope is negative because the line is headed downward. If we had reversed the order in subtraction for Δy, we would have gotten a positive slope. This would have been a clue that we made a mistake.

Question　Why are slopes important?

ANSWER　Slopes measure how fast or slow a line is rising or falling. This can tell us how fast prices are rising, how fast a population is decreasing, and the like. Slopes tell us how things are changing.

PROBLEM SET 4–3–1

1. Define or discuss slope.

Complete the following table (using Formula 4–3–1).

	m	Δy	Δx
2.	?	12	4
3.	?	−20	100
4.	14	?	8
5.	250	3000	?
6.	−15	?	−4
7.	−0.2	?	3
8.	1.3	7.5	?
9.	?	−25	600

Locate each of the following pairs of points on graph paper. Then compute the slope of the line segment between them. (Remember to be careful with the + and − signs.)

10. $A = (2, 7)$, $B = (4, 12)$　　11. $C = (6, 3)$, $D = (9, 15)$

12. $E = (4, -5)$, $F = (10, 14)$　　13. $G = (-4, 4)$, $H = (3, -6)$

14. $I = (-3, -7)$, $J = (11, -2)$　　15. $K = (6, 150)$, $L = (11, 212)$

16. $M = (1960, 180 \text{ million})$,　　17. $P = (15, -23)$, $Q = (18, -9)$
　　$N = (1970, 210 \text{ million})$

18. $R = (15, 1200)$, $S = (25, 1750)$　　19. $T = (30, 5600)$, $U = (40, 3100)$

Find the approximate slopes of all the following lines. Also, find the coordinates of the points where they cross.

20.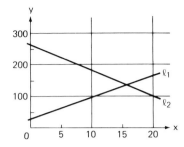

21.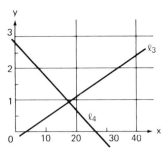

22.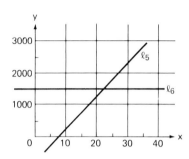

23.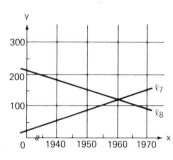

24.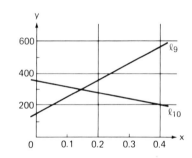

25.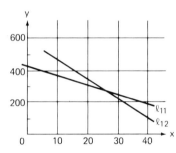

The Straight Line

Problem A car-rental agency charges $15 plus 14 cents per mile for 1 day. What is the formula for the charge in terms of miles driven? Make a graph of the charges and miles.

ANSWER To find the formula, we give each of the quantities a letter and then translate as we did in Chapter 2.

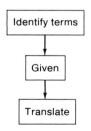

Let y = charge to rent car
x = miles driven

The charge is $15 plus 14 cents per mile:

$$y = 15 + (0.14) \cdot x$$

Notice that 14 cents per mile translates as $0.14x$. To graph this we need data points. Since we are not given any, we will have to use the formula to get a few. For example, if $x = 50$ miles, then the charge $y = 15 + (0.14)(50) = 15 + 7 = 22$. We take x points 50 miles apart for convenience.

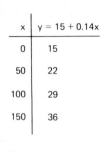

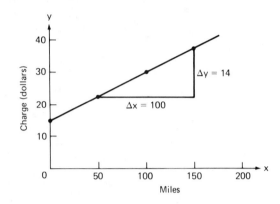

The accompanying figure shows a table of points and the graph. Notice that the graph is a straight line.

Question What is the slope of the line shown?

ANSWER We can take any two points. We use points $(50, 22)$ and $(150, 36)$ in Formula 4–3–2.

$$m = \frac{36 - 22}{150 - 50} = \frac{14}{100} = 0.14$$

Notice that this is the same number that multiplies the x in the formula $y = 0.14x + 15$.

Question What does the 15 mean in the formula?

ANSWER Look at the graph. Notice that the line hits the y-axis at 15. So the 15 in the formula tells us where the line hits the y-axis. This is called the **y-intercept**. It also tells us the charge before we even drive 1 mile.

In general, the straight line graph has the following form:

Formula 4–3–3

$$\boxed{y = mx + p}$$

Here the terms m and p each have their own meanings. The m term is always the slope, and p is always the y-intercept (see the given figure). The vertical line cannot use Formula 4–3–3 since it has no slope m. But we will not be dealing with vertical lines in this section.

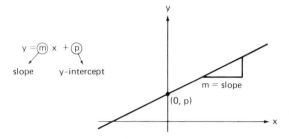

Question What is the general rule for finding the equation of a line?

ANSWER To find the equation of a line, we will need two things: (1) the *slope*, and (2) a *point* on the line. We must have both.

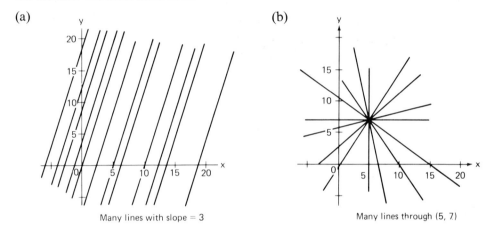

(a) Many lines with slope = 3

(b) Many lines through (5, 7)

The illustration shows what happens if we only have the slope or the point, but not both. In part (a), we get a whole set of parallel lines, all with slope 3. In part (b), we get an entire set of lines all through (5, 7). But there is *only one* line with slope 3 *and* through (5, 7).

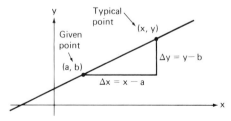

Suppose that we are given that the slope of a line is m, and that it passes through (a, b). Let us take (x, y) to be a typical point on the line (see the accompanying figure). Now we use Formula 4–3–1:

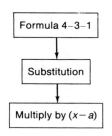

$$\text{slope} = \frac{\Delta y}{\Delta x}$$

$$m = \frac{y - b}{x - a}$$

$$m(x - a) = y - b$$

We now have a valuable formula:

Formula 4–3–4

$$\boxed{y - b = m(x - a)}$$

where m = slope and (a, b) is on the line.

Problem Find the equation of the line with slope 12 and passing through (1, 9).

ANSWER We are given the slope and a point, so we can use Formula 4–3–4. Here we have $m = 12$, $a = 1$, and $b = 9$.

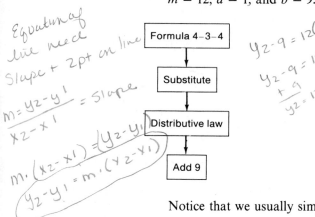

$$y - b = m(x - a)$$

$$y - 9 = 12(x - 1)$$

$$y - 9 = 12x - 12$$

$$y = 12x - 3$$

Notice that we usually simplify the equation so that y is alone on the left side.

Problem A drug store charges \$3 to develop and print a 12-exposure roll of color film. They charge \$4.44 to develop and print a 20-exposure roll of color film.
- (a) Graph these data with the number of exposures on the x-axis and the price on the y-axis.
- (b) Find the equation of the line connecting these points.
- (c) Explain what the terms of this equation mean.

ANSWER
- (a) First we graph the data as shown in the figure.
- (b) To find the equation, we need the slope and a point. We have a point; we will have to find the slope.

$$m = \frac{\Delta y}{\Delta x} = \frac{1.44}{8} = 0.18$$

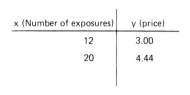

x (Number of exposures)	y (price)
12	3.00
20	4.44

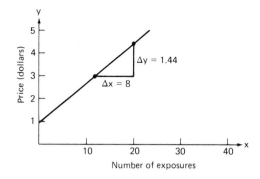

Now we can use Formula 4–3–4 with (12, 3) as the point (a, b).

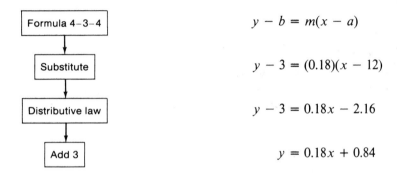

Formula 4–3–4 $\qquad y - b = m(x - a)$

Substitute $\qquad y - 3 = (0.18)(x - 12)$

Distributive law $\qquad y - 3 = 0.18x - 2.16$

Add 3 $\qquad y = 0.18x + 0.84$

This is the equation of the line. We can use this to find the price of other sized rolls. For example, a 36-exposure roll would cost $y = (0.18)(36) + 0.84 = 6.48 + 0.84 = \$7.32$.

(c) The number 0.84 is the y-intercept of the line. This also tells us the cost to develop the roll (before any prints are made). The number 0.18 in the equation is the slope of the line. It also gives the per print cost. In other words, the prints cost 18 cents each.

Problem When Pam started her diet, she weighed 147 pounds. Twenty days later she weighed 144.5 pounds.

(a) Graph these data with weight on the y-axis and days on the x-axis.

(b) Find the equation of the line connecting these points.

(c) If she loses at this steady rate, how long will it take her to get down to 125 pounds?

ANSWER

(a) Since the weights are close together, we cut the y-axis so that the data will fit well, as illustrated on page 238.

(b) To find the equation we must first find the slope. This is

$$m = \frac{\Delta y}{\Delta x} = \frac{-2.5}{20} = -0.125$$

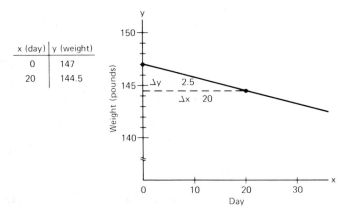

x (day)	y (weight)
0	147
20	144.5

We can use this and (0, 147) as (*a*, *b*) in Formula 4–3–4.

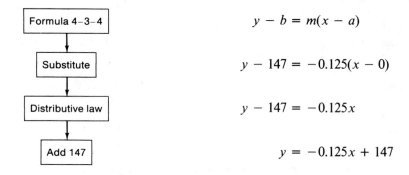

Formula 4–3–4

$$y - b = m(x - a)$$

Substitute

$$y - 147 = -0.125(x - 0)$$

Distributive law

$$y - 147 = -0.125x$$

Add 147

$$y = -0.125x + 147$$

This equation tells us that Pam starts her diet at 147 pounds. The -0.125 term tells us that Pam is losing weight at a rate of about 0.125 (or $\frac{1}{8}$) pound per day. (This is probably an average rate, since weight rarely comes off at a steady rate.)

(c) Now we want to know how long it will take to get down to 125 pounds. Here we know the final weight, $y = 125$, and the unknown is x, the number of days.

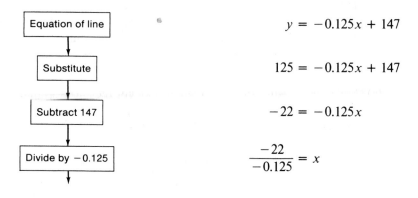

Equation of line

$$y = -0.125x + 147$$

Substitute

$$125 = -0.125x + 147$$

Subtract 147

$$-22 = -0.125x$$

Divide by -0.125

$$\frac{-22}{-0.125} = x$$

Simplify $176 = x$

Therefore, it will take Pam about 176 days (almost $\frac{1}{2}$ year) to lose the weight.

Problem   The boiling point of water is 100°C or 212°F. The freezing point of water is 0°C or 32°F.
(a) Put these points on a graph and draw the line through them.
(b) Determine the equation of this line.

ANSWER   Just as with length, mass, and capacity, there are two different units for temperature: Fahrenheit (English system) and Celsius (SI or metric system).
(a) We can make a graph by putting °C on the y-axis and °F on the x-axis, as illustrated.

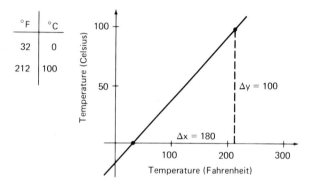

(b) To determine the equation of the line, we must first find the slope. We use the two given points:

$$m = \frac{\Delta y}{\Delta x} = \frac{100 - 0}{212 - 32} = \frac{100}{180} = \frac{5}{9}$$

Now we use this slope $m = \frac{5}{9}$ with the point $(a, b) = (32, 0)$.

Formula 4–3–4 $y - b = m(x - a)$

Substitute $y - 0 = \frac{5}{9}(x - 32)$

Replace y by °C
and x by °F $°C = \frac{5}{9}(°F - 32)$

This is the equation to change °F into °C. We can also solve this for °F.

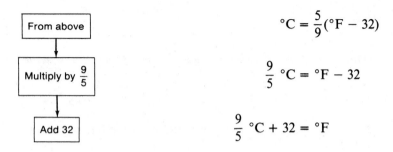

$$°C = \frac{5}{9}(°F - 32)$$

$$\frac{9}{5}°C = °F - 32$$

$$\frac{9}{5}°C + 32 = °F$$

This is the standard conversion from °C into °F.

PROBLEM SET 4-3-2 Find the equations of the lines with the following slopes m and points Q. Leave the answer as $y = mx + p$.

1. $m = 10$, $Q = (6, 4)$ 2. $m = 5$, $Q = (-1, 12)$

3. $m = -4$, $Q = (15, 9)$ 4. $m = -12$, $Q = (-16, -7)$

5. $m = 1.2$, $Q = (1.9, -6.3)$ 6. $m = -4.7$, $Q = (-12.5, 20.9)$

Find the equations of the lines with the following two points, M and N. Leave the answer as $y = mx + p$.

7. $M = (3, 3)$, $N = (5, 7)$ 8. $M = (4, 9)$, $N = (10, 21)$

9. $M = (-6, 10)$, $N = (-1, -5)$ 10. $M = (12, -25)$, $N = (25, 110)$

11. $M = (-100, 7000)$, $N = (200, -5500)$ 12. $M = (1700, -8300)$, $N = (2500, -5700)$

Approximate the equations of the lines drawn below. Leave the answer as $y = mx + p$. Also, find the coordinates of the points where the lines cross.

13.
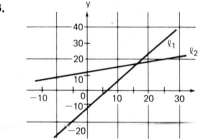

14.
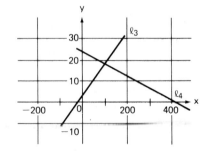

A rule of thumb for telling what is normal blood pressure is 100 plus your age. (This is the highest it should be.)

15. Find the equation for this rule, with y = blood pressure and x = age.

16. Graph this equation.

One doctor has determined the following rules for a person's ideal weight.
Men: The ideal weight is 4 times the height (in inches) minus 130.
Women: The ideal weight is 3.5 times height (in inches) minus 110.

17. Find the equations for these rules, with y = weight and x = height in inches.

18. Graph these equations for heights between 58 and 72 inches for women and between 64 and 78 inches for men.

Pat and Ken figure that it costs them $770 plus 6 cents a mile to run their car every year.

19. Compute the cost of running the car 0 miles, 5000 miles, 10,000 miles, and 20,000 miles.

20. Graph the data in Problem 19, with miles on the x-axis and cost on the y-axis.

21. Find the equation of the line in the graph.

Mary has just started a new job. She starts at $11,000 per year with an $800 raise every year.

22. What is her salary after 0 years, 1 year, and 2 years?

23. Graph the data in Problem 22, with years on the x-axis and salary on the y-axis.

24. Find the equation of the line in the graph.

25. How many years will it take her to reach $20,000?

Steven goes on a diet. After 10 days he weighs 221 pounds. After 35 days, he weighs 216 pounds.

26. Graph these data, with days on the x-axis and weight on the y-axis.

27. Find the equation of the line in the graph.

28. If he loses at this rate, how long will it take him to get down to 175 pounds?

4–4 INFORMATION FROM A GRAPH

Problem   What is the graph of the following data?

Year	1960	1965	1970	1975	1980
Population of Little Beaver, Ill.	5590	6970	8280	9610	10,935

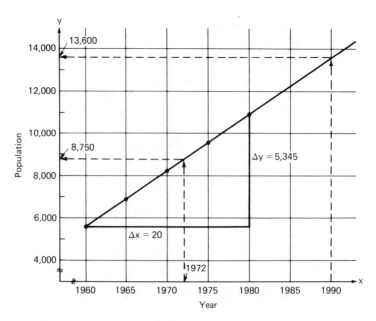

Problem What was the population of Little Beaver in 1972?

ANSWER For this, we must go *between the data points*, (1970, 8280) and (1975, 9610) on the graph. If we start from 1972 on the *x*-axis and read up to the line, we see this has a *y*-coordinate of about 8750. Even though we have no data from 1972, we can conclude from the graph that the population was about 8750.

Problem About what will the population of Little Beaver be in 1990?

ANSWER For this, we must extend the curve and go *beyond the data points* for the information. If we start from 1990 on the *x*-axis and read up the extended line, we see this has a *y*-coordinate of about 13,600. So, using the graph, we can project that the population in 1990 will be about 13,600.

Problem How fast is the population increasing each year?

ANSWER Here is where the slope becomes important. From the slope, we can get the rate of population increase. From the figure we use the years 1960 and 1980 to get our points for the slope.

$$ m = \frac{\Delta y}{\Delta x} = \frac{5345 \text{ people}}{20 \text{ years}} = 267.25 \, \frac{\text{people}}{\text{year}} $$

We can conclude that the population is going up about 267 people per year.

Problem Can we use a graph to find square roots; for example, $\sqrt{57}$?

Milestones in Pete Rose's Career

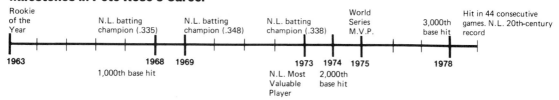

Rookie of the Year | N.L. batting champion (.335) | N.L. batting champion (.348) | N.L. batting champion (.338) | World Series M.V.P. | 3,000th base hit | Hit in 44 consecutive games. N.L. 20th-century record

1963   1968   1969   1973   1974   1975   1978

1,000th base hit   N.L. Most Valuable Player   2,000th base hit

(Photograph by Robert Walker and art © 1978 by The New York Times Company. Reprinted by permission.)

ANSWER   Yes. Since we want square roots, we will graph $y = \sqrt{x}$. For the table, we will only use numbers where we know the exact square root.

x	0	1	4	9	16	25	36	49	64	81	100
$y = \sqrt{x}$	0	1	2	3	4	5	6	7	8	9	10

We first graph these points as shown. Now we are ready to approximate $\sqrt{57}$. We find 57 on the x-axis. We then read up to the curve and over to the y-axis to find the y-coordinate to be about 7.55. We conclude that $\sqrt{57}$ is about 7.55.

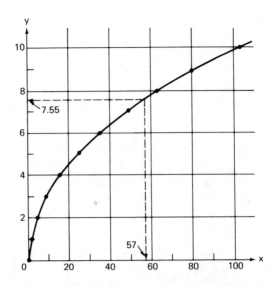

In this problem, we again went in between data points; in this case, we were between (49, 7) and (64, 8). Whenever we go *between* points to get information, we call this **interpolation**.

Problem Use the following data about college presidents' salaries to project into the future, 1984 for example.

Year	1960	1962	1964	1966	1968	1970	1972	1974
Average salary ($)	13,827	15,375	17,330	19,638	22,303	25,979	29,750	31,342

ANSWER First we make the graph as shown. Beyond 1974, we use a dashed line to extend the graph. This reminds us that we are just guessing this portion of the curve. We see that, for x = 1984, the y-coordinate of the extended curve is about $47,000. We

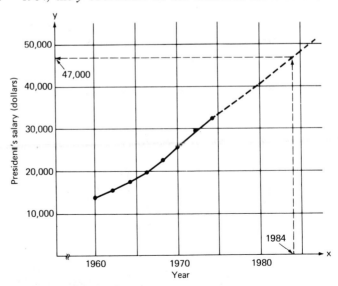

might predict that the average college or university president will earn about $47,000 in 1984.

In this problem, we had to go beyond the data points. Whenever we go *beyond* the data points to get information, see trends, or make predictions, we call this **extrapolation**. Since extrapolation is beyond the data, it must be done very carefully.

Problem A builder notices a relation between the size of a house he builds and the cost to build it. What is the relation?

Floor space (ft²)	480	920	1350	1960
Cost ($)	10,000	19,000	28,000	40,000

ANSWER Let us first graph these data as shown. We see that the curve forms a straight line. Therefore, we can compute its slope. This will tell us the rate at which building costs go up in terms of dollars per square foot.

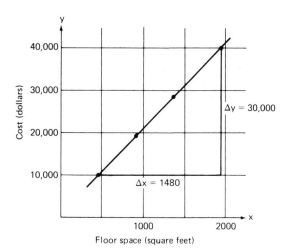

$$m = \frac{\Delta y}{\Delta x} = \frac{30,000 \ \$}{1480 \ \text{ft}^2} = 20.27 \ \frac{\$}{\text{ft}^2}$$

We conclude that building costs for a house are about $20.27 per square foot. Whenever we have a straight line, we can use its slope to find the rate of increase or decrease in the curve.

Question What if we want to know a rate of increase, but the curve isn't a straight line. For example, what is the rate of increase with these data?

Year	1950	1955	1960	1965	1970	1974
Money spent on education (billions $)	6.7	11.2	17.6	28.1	50.9	72.7

ANSWER First, we graph the data as shown. We see that this isn't a straight line, so we can't find its slope. Still, suppose that we want to know how fast education expenditures were rising in 1965. We will need a slope.

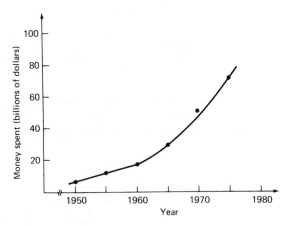

Since the curve itself isn't a straight line, we do the next best thing. Through the point (1965, 28.1 billion) on the curve, we draw a straight line that just barely touches the curve, as illustrated.

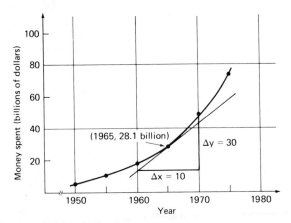

Why do we do this? Because we need a line to find the slope, and this line is the closest to the curve at the point (1965, 28.1 billion). Now we can calculate the slope of the line.

$$m = \frac{\Delta y}{\Delta x} = \frac{30 \text{ billion}}{10 \text{ yr}} = 3.0 \frac{\text{billion}}{\text{yr}}$$

We conclude that in 1965, education expenditures were increasing by about $3.0 billion per year. Notice that, if we did this at different points, we would get different slopes. This is because the curve is not straight.

Whenever we do not have a straight line curve to calculate a slope, we draw a straight line that just barely touches the curve at one point. This line is called a **tangent line**.

Problem   Graph the following data:

Year	1950	1955	1960	1965	1970	1974
Gross National Product (in $ billions)	285	398	504	685	977	1397

Draw a line tangent to the curve at (1970, 977). What is the slope of the tangent line? What does this slope mean?

ANSWER   The figure shows the graph with the tangent line. Notice how the tangent line just barely touches the curve at the point (1970, 977).

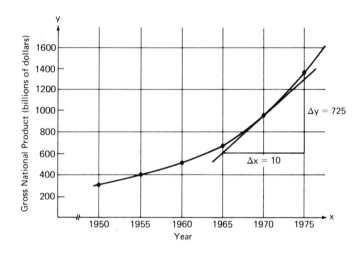

To compute the slope, we conveniently make a triangle with $\Delta x = 10$ years. We find that $\Delta y = 1345 - 620 = 725$. Therefore,

$$m = \frac{\Delta y}{\Delta x} = \frac{725 \text{ billion}}{10 \text{ yr}} = 72.5 \frac{\text{billion}}{\text{yr}}$$

What does this mean? This tells us that in 1970 the Gross National Product was increasing at a rate of about $72.5 billion per year. If we had computed the slope for a tangent at (1955, 398), our slope would have been smaller. Why? Because the GNP was not growing as fast then. The slope of a tangent line always tells us how fast the curve is changing (up or down) at that point.

1. Define or discuss:
   (a) Interpolation.
   (b) Extrapolation.
   (c) Tangent line.

x	0	100	200	300	400	500
y	2000	4500	7000	9500	12,000	14,500

2. Graph the data shown in the table.

3. Find y when $x = 50$; $x = 160$; $x = 325$.

4. Find y when $x = 600$; $x = 1000$.

5. Find the slope.

6. Find the equation of the line.

x	0	1	8	27	64	125	216	343	512	729	1000
$y = \sqrt[3]{x}$	0	1	2	3	4	5	6	7	8	9	10

$\sqrt[3]{x}$ is called the cube root of x.

7. Graph this table.

8. Use the graph to find $\sqrt[3]{15}$, $\sqrt[3]{100}$, $\sqrt[3]{450}$, and $\sqrt[3]{820}$.

9. Use the graph to find $\sqrt[3]{1400}$ and $\sqrt[3]{2000}$.

10. Draw a tangent line to the curve at (125, 5).

11. Find the slope of this tangent line.

Age (in years)	10	20	30	40	50	60	70
Expected remaining years to live	62	53	44	34	26	18	12

12. Graph this table.

13. Use the graph to determine the expected remaining years to live if one is 15, 23, 37, 45, or 65.

14. Use the graph to determine the expected remaining years to live if one is 75, 80, 85, 90 or 100.

15. Draw a tangent line to the curve at (70, 12).

16. Find the slope of the tangent line.

17. What does this slope mean?

Car weight (lb)	2000	2500	3000	3500	4000	4500	5000
Miles per gallon	28	22.5	19	16	14	12	11

18. Graph this table.

19. Use this graph to determine the miles per gallon of cars weighing 2250, 2800, 3700, and 4400 pounds.

20. Use the graph to determine the miles per gallon of cars weighing 1500, 6000, and 7000 pounds.

21. Draw a tangent line to the curve at (4000, 12).

22. Find the slope of the tangent line.

23. What does this slope mean?

Turn back to graphs (a) and (b) on page 219.

24. Use part (a) to project the number of bachelor's degrees in 1980 and 1990.

25. Use part (b) to project the number of bachelor's degrees in 1980 and 1990.

26. Do you get the same or approximate answers?

27. Which graph was easier to extrapolate from?

The following are actual data of American life in the mid-1900's. Graph each case. (a) Extrapolate to the year 2000. (b) Determine the rate of growth (or decline) in 1970. In other words, find the slope of the tangent line.

28.
Year	1900	1920	1930	1950	1960	1970
Illiteracy (%)	11.3	6.5	4.7	3.3	2.4	1.2

29.
Year	1950	1955	1960	1965	1968	1969	1970	1971	1972	1973
Hospital expense (per day)	8	11.20	16.50	25.30	37.80	45.00	54.00	63.80	73.90	83.70

30.
Year	1920	1930	1940	1950	1960	1965	1970	1974
Percent of population divorced	0.6	1.1	1.2	1.7	1.9	2.5	2.5	3.5

31.
Year	1910	1920	1930	1940	1950	1960	1970
Number of males to one female (age 25–44)	1.102	1.051	1.018	0.985	0.964	0.957	0.955

32.
Year	1940	1950	1955	1960	1965	1970	1972
Union membership (millions)	8.9	15.0	17.7	18.1	18.5	20.8	20.9

33.

Year	1950	1955	1960	1965	1969	1970	1971	1972	1973
Gonorrhea cases (thousands)	287	239	259	325	535	600	670	767	843

34.

Congress number	82	83	84	85	86	87	88	89	90	91	92	93	94	95
Democrats in House	234	211	232	233	283	263	258	295	247	243	254	243	290	291
Democrats in Senate	49	47	48	49	64	65	67	68	64	57	54	56	62	62

(In 1970 the ninety-first Congress convened. In 2000, the one hundred sixth will convene.)

Misuses of Graphs

Question Can graphs be misused?

ANSWER Yes. Graphs carry a powerful visual message, so a distorted graph can carry a distorted message.

Question What are the major ways to misuse a graph?

ANSWER One way is by distorting the scaling.

Example We could distort the following data to make them look "bad" or "good."

Month	Jan.	Feb.	Mar.	Apr.	May	June	July	Aug.	Sept.	Oct.	Nov.	Dec.
Sales volume (in millions)	24.10	24.11	24.13	24.16	24.20	24.25	24.31	24.38	24.46	24.55	24.65	24.76

First, let us make this sales growth look as slow as possible. We do this by choosing a scale on which the data are lost. It appears that nothing is happening.

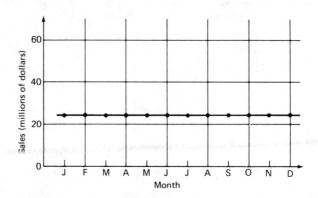

Now let us stretch out the scale so that the data dominate it. It appears that the company is growing like crazy. Remember, these are the *same* data. Notice that the key to distorting the graph's appearances is to stretch or shrink the axes to suit your needs.

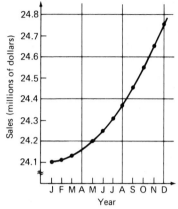

Question   How else can a graph be misused?

ANSWER   A graph can also be misused by improper extrapolation. Recall that extrapolation is taking information from beyond the data points. This requires a lot of caution. It also has the great potential for error or abuse.

Problem   What differences can come about from extrapolation of the following data?

Year	1976	1977	1978	1979	1980
Price of ABC Yo-Yo stock ($)	85	120	80	55	100

ANSWER   Let us first graph only the data that are given.

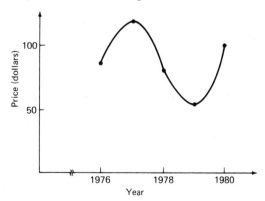

What can we conclude about the future? It depends on who interprets the graph. First, it is possible to see a great upward trend. In this case, we would get the graph in part (a) (with extrapolation in a dotted line).

On the other hand, someone else might conclude that the stock was just going up and down like a Yo-Yo. Now we would get the graph in part (b).

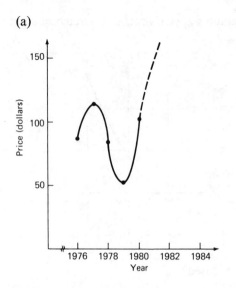

(a)

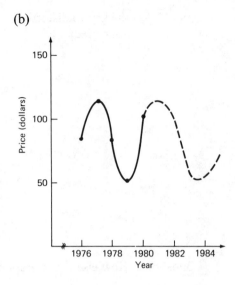

(b)

Question Which graph is right?

ANSWER Both graphs look right in their own way. Nobody will really know which is right until the future comes. These graphs show that extrapolation can sometimes be used to prove anything a person wants.

PROBLEM SET 4–4–2

1. Discuss some of the possible misuses of graphing.

Make two graphs for each of the following sets of data. In one graph, make the data look bad. In the other, make the data look good. (If you can't tell which is bad or good, just make the graphs look very different.)

2.

Month (Jan. is 1)	1	2	3	4	5	6	7	8	9	10	11	12
Price of 1 lb of hamburger	78	80	81	81	83	85	86	87	89	91	92	94

3.

Car weight (lb)	2000	2500	3000	3500	4000	4500	5000
Miles per gallon	25	22.5	19	16	14	12	11

4.

Year	1910	1920	1930	1940	1950	1960	1970	1974
Egg consumption (per person)	306	299	331	319	389	335	311	287

5.

Annual income	2000	4000	6000	8000	10,000	12,000	14,000
Income tax (one person)	0	117	432	763	1135	1547	2029

6.

Year	1973	1974	1975	1976	1977	1978
Average price of a share of Slinki Guitar, Inc.	41	52	43	36	48	59

7–10. Make two graphs of the data given in Problems 28–31 of Problem Set 4–4–1 look very different.

11. Make two extrapolations of the data given in Problem 28 in Problem Set 4–4–1 look very different.

12, 13. Make the two extrapolations of the data given in Problems 4 and 6 above look different.

IMPORTANT WORDS

axis (4–1)
bar graph (4–2)
broken scale (4–1)
Cartesian coordinate system (4–1)
coordinate (4–1)
correlation (4–2)
extrapolation (4–4)
frequency graph (4–2)
general relation graph (4–2)

graph (4–1)
interpolation (4–4)
log scale (4–1)
origin (4–1)
scattergram (4–2)
scientific graph (4–2)
slope (4–3)
tangent line (4–4)
time graph (4–2)

REVIEW EXERCISES

Using the given graph, find the coordinates of the following:

1. Point A

2. Point B

3. Point C

4. Point D

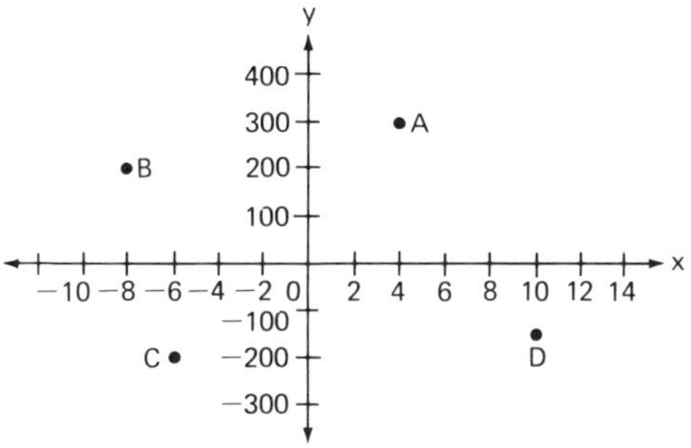

Give a range for the following situations, and draw an appropriate axis.

5. A student's grade-point average.

6. The number of spectators at a ball game.

7. Plot the frequency graph for the following data.

Hours studied per night	$\frac{1}{2}$	1	$1\frac{1}{2}$	2	$2\frac{1}{2}$	3	$3\frac{1}{2}$	4
Percent of students	15	22	25	18	11	6	2	$\frac{1}{2}$

8. Make a bar graph and a line graph for the following data.

Grade point average	0–0.5	0.5–1.0	1.0–1.5	1.5–2.0	2.0–2.5	2.5–3.0	3.0–3.5	3.5–4.0
Number of students	13	35	101	189	423	631	301	89

9. Place the following stock prices into convenient groups or ranges, and then make a bar graph: $33\frac{3}{8}$, $11\frac{1}{2}$, $21\frac{3}{4}$, $3\frac{1}{4}$, 9, 98, $16\frac{1}{2}$, $49\frac{1}{8}$, 34, $50\frac{3}{8}$, $15\frac{3}{8}$, 47, $32\frac{1}{2}$, $20\frac{1}{4}$, $24\frac{3}{4}$, $33\frac{3}{8}$, $1\frac{1}{2}$, 21, $25\frac{1}{4}$, $20\frac{3}{4}$, $11\frac{1}{4}$, $7\frac{3}{4}$, 18, $2\frac{5}{8}$, $19\frac{5}{8}$, $16\frac{1}{4}$, $4\frac{1}{4}$, $34\frac{1}{4}$, $14\frac{1}{4}$, $24\frac{3}{8}$, $30\frac{7}{8}$, 82.

10. Graph the following data.

Year	1960	1965	1968	1969	1970	1972	1973
TV broadcast revenue ($ million)	1504	2328	2995	3325	3337	3770	4107

11. Graph the following data.

Month (Jan. is 1)	1	2	3	4	5	6	7	8	9	10	11	12
Temperature in Chicago (°F)	24	27	37	50	60	71	75	74	66	55	40	29

12. Graph the following data.

Radius of circle	2	4	6	8	10	12	14
Area of circle	12.57	50.27	113.10	201.06	314.16	452.39	615.75

13. Make a scattergram for the following data.

Animal	Ass	Cat	Chicken	Cow	Dog	Duck
Gestation (days)	360	58	21	280	63	28
Average life (yr)	19	11	7	10	11	10

Animal	Elephant	Goat	Hamster	Hippopotamus
Gestation (days)	628	150	17	237
Average life (yr)	35	12	2	30

Animal	Lion	Mare	Monkey	Rabbit	Wolf	Woman
Gestation (days)	108	336	164	31	63	270
Average life (yr)	10	22	13	7	11	72

Complete the following table.

	m	Δx	Δy
14.	?	14	-3
15.	37	21	?
16.	-12	?	-3

Compute the slope of the line segment between the following points.

17. $A = (4, 9)$, $B = (8, 3)$ **18.** $C = (7, 100)$, $D = (12, 250)$

19. Find the equation of the line with slope 4 and through the point (3, 78).

20. Find the equation of the line through the point (2, 4) and (4, 10).

The cost of running a certain car is $1200 plus 6.5 cents per mile.

21. Find the equation for this rule.

22. Graph the equation of the line.

Year	1900	1910	1920	1930	1940	1950	1960	1970
Death rate (per 1000)	17.4	14.7	13.0	11.3	10.8	9.6	9.5	9.5

23. Graph this table.

24. Use the graph to determine the death rate in 1915 and 1935.

25. Draw a tangent line at the point (1920, 13.0).

26. Find the slope of that tangent line.

27. What does the slope mean?

28. Make two graphs that make the data look very different.

29. Make two possible extrapolations of the data.

A family struggling to get their money's worth. (© 1976 by Christa Armstrong from Rapho/Photo Researchers)

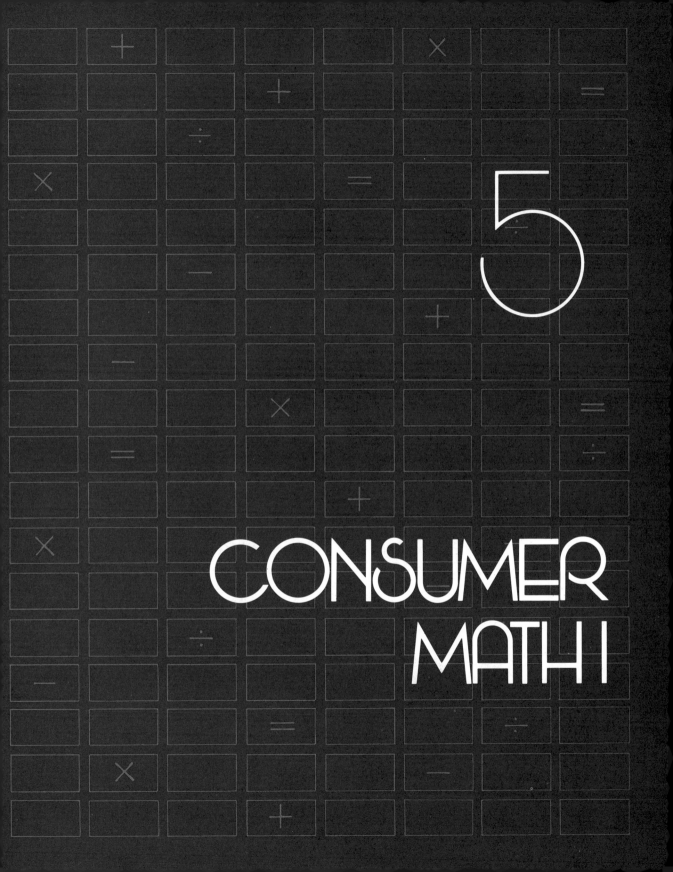

5

CONSUMER MATH I

5-1 FOOD AND HEALTH

Question   Susan and Tim are interested in saving money at the grocery store. What do they need to know?

ANSWER   They should understand pricing, vitamins, minerals, protein, calories, and the like.

Pricing

Problem   Which is the better deal: peanut butter at 83 cents for 15 ounces, or $1.19 for 22 ounces?

ANSWER   The most systematic way to solve this is to reduce both to a common unit, like cents per ounce. The first jar is

$$\frac{83\cancel{c}}{15 \text{ oz}} = 5.53 \frac{\cancel{c}}{\text{oz}}$$

The second jar is

$$\frac{119\cancel{c}}{22 \text{ oz}} = 5.41 \frac{\cancel{c}}{\text{oz}}$$

So the 22-ounce jar gives the peanut butter at a slightly cheaper price.

(David Strickler/Monkmeyer Press Photo Service)

To the shopper it must appear that prices and weights are purposely chosen to be confusing. This is probably partly true.

> To compare prices:
>
> 1. Reduce all the prices to a common unit of price per unit measure (such as cents per ounce or cents per gallon).
> 2. Compare the ratios for the lowest.

Problem Milk costs $1.65 per gallon or 47 cents per quart. Which is cheaper?

ANSWER We reduce both of these to a common ratio, cents per quart. Remember that a gallon is 4 quarts. So the first case is 165 cents per 4 quarts = 41.25 cents per quart. The second case is already 47 cents per quart. So the gallon container is a cheaper buy.

Problem Suppose that Barb and Art are buying milk at the above prices, and they only drink $\frac{3}{4}$ gallon and the rest sours. Now which is cheaper?

ANSWER If they buy a gallon for 165 cents and use only 3 quarts of it, they are paying 165 cents for 3 quarts of milk, or 165 cents per 3 quart = 55 cents per quart. Compare this to 47 cents for a quart container. If spoilage or waste is possible, often the larger size isn't the cheaper. When you calculate the ratio of cents to sizes, you must use only the quantity you actually consume.

Question Why do smaller quantities usually cost more per unit size than larger quantities?

ANSWER Probably packaging. Whenever you buy anything, you are paying for two things: the actual merchandise and the package design that caught your eye.

Problem Smiles Toothpaste comes in three sizes: 6 ounces for 50 cents, 12 ounces for 80 cents, and 20 ounces for $1.20. Can we determine the price of the packaging?

ANSWER Yes, we can get an idea of the cost of the packaging by graphing the price versus the size, as shown. Notice that this line crosses the cost axis at about 20 cents. This means 0 ounces of toothpaste would cost 20 cents! What is this 20 cents for? It

Size (ounces)	Price (cents)
6	50
12	80
20	120

pays for the fancy colorful box, the special plastic tube, and the testimony of actors that Smiles gives them the whitest teeth. In general, we can get a rough idea of a product's packaging and advertisement costs by graphing several prices and sizes. Then we find the curve through the points, and extend it to see how much a size of 0 would cost.

PROBLEM SET 5–1–1 Determine the cheaper quantity by computing and comparing the cost per unit size ratio.

1. Peanut butter at 83 cents for 15 ounces or $1.45 for 28 ounces?

2. Bread at 59 cents for 16 ounces or 73 cents for a 20-ounce loaf?

3. Soup at 25 cents for a can that makes 21 ounces of soup, or 19 cents for a package that makes 3 cups of soup?

4. Tomatoes at 59 cents for a 28-ounce can or $1.00 for three 15-ounce cans?

5. (a) Cottage cheese at 88 cents for 16 ounces or $1.19 for 24 ounces?
   (b) What if only $\frac{3}{4}$ of the 24-ounce container is eaten, and the rest is thrown out?

6. Detergent at $1.49 for 5 pounds or $5.49 for 20 pounds?

7. Apples sold at 25 cents per pound loose, or at $4 for a 30-pound basket when only 80% of the basket will be eaten?

In the following cases, use a graph to approximate the cost of the packaging and advertisement.

8. Soda pop (no return)

Size (oz)	16	32	64
Price (¢)	25	45	75

9. TV dinner

Size (oz)	8	15	32
Price (¢)	38	55	99

10. Milk

Size	1 qt	$\frac{1}{2}$ gal	1 gal
Price (¢)	51	91	165

11. Use the graphing method to analyze the packaging costs of other products, such as beer, candy bars, stationery, and aspirin.

Protein

Problem   Peanut butter is 105 cents for a 20-ounce jar. According to nutrition tables, peanut butter provides 13 grams of protein for each 50 grams of peanut butter. How many grams of protein per dollar is this?

ANSWER   This is a unit conversion problem of the type seen in Chapter 2. We want to end up with units of grams of protein per dollar (gP/$). We put the quantities so that grams of protein is in the numerator and dollars is in the denominator, as follows.

$$\frac{gP}{g} \cdot \frac{oz}{\$} \cdot \boxed{?} = \frac{gP}{\$}$$

To find the appropriate units for $\boxed{?}$, we see that we must cancel the ounces on the top and the grams on the bottom. So the right conversion factor is grams per ounce. Since 1 ounce = 28.3 grams, we have the conversion factor 28.3 grams per ounce. Now we have

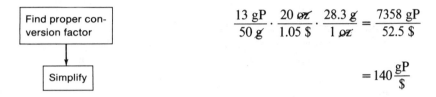

$$\frac{13\ gP}{50\ g} \cdot \frac{20\ oz}{1.05\ \$} \cdot \frac{28.3\ g}{1\ oz} = \frac{7358\ gP}{52.5\ \$}$$

$$= 140 \frac{gP}{\$}$$

So peanut butter provides about 140 grams of protein for $1.

Hand Calculator Instant Replay

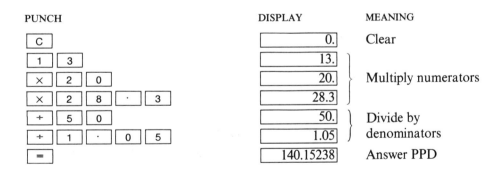

Protein is one of the most important nutrients in our diet. The average adult needs about 70 grams of protein (gP) per day. The chief sources of protein are dairy foods, meats, fishes, nuts, and grains. Different foods provide different quantities and qualities of protein at different prices. To compare foods, we must have a standard unit. We call this unit **PPD (protein per dollar)**. This naturally has the units grams of protein per dollar (gP/$).

To compute the protein per dollar (PPD):

1. Determine the protein content of the food in units of grams of protein per size_1. (Get this information from health tables or the container itself.)

2. Write the price in units of size_2 per dollar.

3. Choose conversion factors so that size_2 and size_1 cancel, leaving only grams of protein per dollar.

4. Multiply across as in any unit conversion.

The student should realize that there are many other important nutrients in foods besides protein. We are using protein as one example. We could do this same analysis with vitamin B, vitamin C, niacin, fiber, and so on.

Problem   Whole milk is $1.65 per gallon. It provides 32 grams of protein per quart. What is the PPD?

ANSWER   Again, we want the answer to be in grams of protein per dollar. So we set up the product as follows:

$$\text{PPD} = \frac{\text{gP}}{\text{qt}} \cdot \frac{\text{gal}}{\$} \cdot \boxed{?} = \frac{\text{gP}}{\$}$$

To cancel gallons on top and quarts on the bottom, we need a conversion factor, quarts per gallon. This is 4 quarts per gallon. Now we have

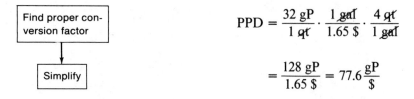

$$\text{PPD} = \frac{32 \text{ gP}}{1 \text{ qt}} \cdot \frac{1 \text{ gal}}{1.65 \$} \cdot \frac{4 \text{ qt}}{1 \text{ gal}}$$

$$= \frac{128 \text{ gP}}{1.65 \$} = 77.6 \frac{\text{gP}}{\$}$$

Question   Which is a more economical source of protein, milk or peanut butter?

ANSWER   To compare foods, we check to see how much protein we can buy for $1 (this is the PPD). We can buy 140 grams of protein with peanut butter, and 77.6 grams of protein with whole milk. So peanut butter is clearly more economical. (It should be noted that some foods like peanut butter are an *incomplete protein*, which should be balanced with a *complete protein*, like milk.)

Problem   On the side of a box of macaroni and cheese it says that each serving provides 8 grams of protein. There are 4 servings per package, and the box costs 29 cents. What is the PPD?

ANSWER   Again, we set up the product so that the units will cancel properly.

$$PPD = \frac{8 \text{ gP}}{1 \text{ serv.}} \cdot \frac{4 \text{ serv.}}{1 \text{ box}} \cdot \frac{1 \text{ box}}{0.29 \text{ \$}}$$

$$= \frac{32 \text{ gP}}{0.29 \text{ \$}} = 110 \frac{\text{gP}}{\text{\$}}$$

PROBLEM SET 5–1–2

1. Define or discuss PPD (protein per dollar).

Find the PPD for the following:

	Protein	Cost	Conversion factor	PPD
2.	$\frac{7 \text{ gP}}{1 \text{ oz}}$	$\frac{1 \text{ lb}}{1.95 \text{ \$}}$	$\frac{16 \text{ oz}}{1 \text{ lb}}$	? *57.4*
3.	$\frac{18 \text{ gP}}{8 \text{ oz}}$	$\frac{8 \text{ oz}}{0.35 \text{ \$}}$	?	?
4.	$\frac{14 \text{ gP}}{100 \text{ g}}$	$\frac{1 \text{ lb}}{1.29 \text{ \$}}$	?	?

5. A certain food claims to provide 8% of the daily recommended allowance of protein, which is 70 grams. How much protein does this food provide?

6. Another food claims to provide 2% of the daily recommended allowance of protein. It only provides 1 gram of protein. Use this to find what they are using for the daily recommended allowance?

7. Go to a local supermarket and get the prices and sizes necessary to complete the following table. For some the protein is given; for others it is on the label.

Food	Cost (size/$)	Protein (gP/size)	Conversion factors	PPD
Milk (whole)		32 gP/1 qt		
Milk (skim)		36 gP/1 qt		
Powdered milk (whole)		27 gP/100 g		
Powdered milk (skim)		30 gP/100 g		
American cheese		7 gP/1 oz		
Cottage cheese		30 gP/1 cup		
Ice cream		6 gP/1 cup		
Eggs (large)		6 gP/egg		
Bacon		2 gP/1 slice		
Ground beef		7 gP/1 oz		
Sirloin steak		20 gP/3 oz		
Corned beef hash		4 gP/1 oz		
Beef pot pie		18 gP/8 oz		
Chicken		23 gP/3 oz		
Lamb chop		6 gP/1 oz		

Food	Cost (size/$)	Protein (gP/size)	Conversion factors	PPD
Pork chop		16 gP/100 g		
Ham		16 gP/3 oz		
Turkey		27 gP/100 g		
Chile con carne (with beans)		19 gP/250 g		
Chile con carne (no beans)		26 gP/255 g		
Bologna		7 gP/50 gm		
Hot dogs		14 gP/100 g		
Crabmeat		14 gP/3 oz		
Fishsticks		19 gP/112 g		
Halibut		26 gP/100 g		
Lobster		18 gP/100 g		
Salmon		17 gP/3 oz		
Shrimp		23 gP/3 oz		
Tuna		25 gP/3 oz		
Lima beans (dry)		16 gP/192 g		
Red kidney beans (canned)		15 gP/260 g		
Corn		5 gP/200 g		
Soybeans		22 gP/200 g		
Bread (white)		39 gP/1-lb loaf		
Bread (wheat)		48 gP/1-lb loaf		
Rice (brown)		15 gP/208 g		
Rice (white)		14 gP/190 g		
Cheese pizza		64 gP/14-in. pie		
Wheat germ		20 gP/65 g		
Peanut butter		13 gP/50 g		
Cashews		12 gP/70 g		
Oatmeal		5 gP/236 g		
Candy bar		[on label]		
Space food		[on label]		
Corn flakes		[on label]		

Weight and Calories

Problem Alice is 5 feet, 6 inches = 66 inches tall, and has a medium frame. What is her desirable weight range?

ANSWER The accompanying figure shows desirable weight ranges. For Alice we find the curve, woman-medium frame. Since Alice is 5 feet 6 inches (66 inches) tall, this corresponds to a weight of 125 pounds. Notice that the graph says the weight is ±6 pounds. So Alice's weight range should be 119 to 131 pounds. She can narrow down this range by determining if she is well-muscled or not. For example, if Alice is unmuscled, her weight should be about 120. This is called the **ideal weight**.

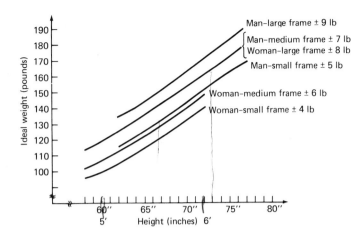

Problem Bob is 6 feet, 2 inches = 74 inches tall. He has a large frame but only average muscle development. What is his desirable weight?

ANSWER According to the figure man–large frame, Bob's weight should be about 180 ± 9 pounds. This range is 171 to 189 pounds. Since Bob has an average muscle development, he should probably be in the middle at about 180 pounds.

To find the ideal weight for a person:

1. Determine the sex, height, frame, and musculature.
2. Find the proper sex–frame curve on the graph.
3. Find the weight that corresponds to the height.
4. Add pounds for a well-muscled body; subtract pounds for a poorly muscled body.

Problem Ron's ideal weight is 175 pounds. He is 22 years old and slightly more active than average. What should his daily calorie intake be?

ANSWER We use the accompanying figure, which gives the **daily calorie needs**. We follow 22 years up to the men curve, and we find that the calorie needs are about 3000. But

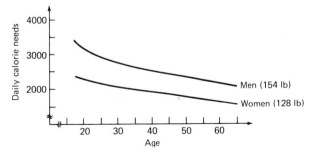

this is for a 154-pound man; so we have to adjust this for Ron's weight. We use proportions.

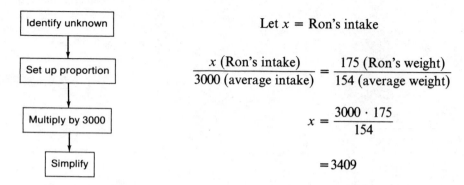

Let x = Ron's intake

| Identify unknown |

| Set up proportion |

$$\frac{x \text{ (Ron's intake)}}{3000 \text{ (average intake)}} = \frac{175 \text{ (Ron's weight)}}{154 \text{ (average weight)}}$$

| Multiply by 3000 |

$$x = \frac{3000 \cdot 175}{154}$$

| Simplify |

$$= 3409$$

This says that the average 22-year-old man whose ideal weight is 175 should eat about 3409 calories per day. But Ron is slightly more active than average; so he might need 5% more calories per day. This is 5% of 3409 or $(0.05) \cdot (3409) = 170$ calories. So we have

$$
\begin{aligned}
\text{Regular calories} &= 3409 \\
\text{Extra calories} &= +170 \\
\hline
\text{Total for Ron} &= 3579
\end{aligned}
$$

Ron needs about 3579 calories per day. As Ron grows older and less active, this number will go down.

To compute a person's daily calorie needs:

1. Find the proper men/women curve on the graph.
2. Determine the calorie needs that correspond to the age.
3. Set up the following proportion
$$\frac{x \text{ (calorie need)}}{\text{intake (from graph)}} = \frac{\text{ideal weight}}{154 \text{ lb (men) or } 128 \text{ lb (women)}}$$
4. Solve for x.
5. Add 5 or 10% for an active person; subtract 5 or 10% for an inactive person.

Problem   Linda is 29 years old. Her ideal weight is 115 pounds, and she is very inactive. What is her proper daily calorie intake?

ANSWER We first use the graph. We read up 29 years to the women curve to 2050 calories. Since this is for a 128-pound woman, we must adjust it proportionately to Linda's weight.

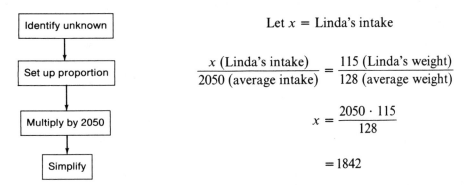

Let x = Linda's intake

$$\frac{x \text{ (Linda's intake)}}{2050 \text{ (average intake)}} = \frac{115 \text{ (Linda's weight)}}{128 \text{ (average weight)}}$$

$$x = \frac{2050 \cdot 115}{128}$$

$$= 1842$$

Finally, since Linda is inactive, we should probably subtract 10% of the 1842, which is $(0.10) \cdot (1842) = 184$. Then we have

$$
\begin{aligned}
\text{Regular calories} &= 1842 \\
\text{Less calories} &= -184 \\
\hline
\text{Total for Linda} &= 1658
\end{aligned}
$$

Linda needs about 1658 calories a day. As she ages, this will go down further. If she becomes more active, it will go up.

We can put these two procedures together to find a person's ideal weight and daily calorie needs. Notice that we need to know the person's sex, height, frame, and musculature to find the ideal weight, and then we need sex, age, and activity level to determine calorie needs.

Problem Fran is 25 years old. She is 5 feet, 6 inches = 66 inches tall, and she has a medium frame and average musculature. She is slightly inactive, since she has a desk job. What is her ideal weight and daily calorie intake?

ANSWER First, we find her ideal weight. We read the height 66 inches up to the women–medium frame curve, and see that her weight should be 126 ± 6 pounds. Since her muscle tone is average, she should probably weigh about 126.

We now consult the second graph and see that a 25-year-old woman should take in about 2100 calories. We have to adjust for Fran's ideal weight of 126, since the chart is for a 128-pound woman.

Let x = Fran's intake

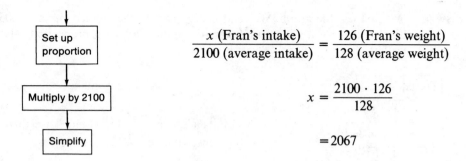

| Set up proportion | $\dfrac{x \text{ (Fran's intake)}}{2100 \text{ (average intake)}} = \dfrac{126 \text{ (Fran's weight)}}{128 \text{ (average weight)}}$ |

| Multiply by 2100 | $x = \dfrac{2100 \cdot 126}{128}$ |

| Simplify | $= 2067$ |

Since Fran is slightly inactive, we should deduct about 5% from this. This is 5% of 2067, which is $(0.05) \cdot (2067) = 103$.

$$
\begin{array}{rl}
\text{Regular calories} = & 2067 \\
\underline{\text{Less calories} =} & \underline{-103} \\
\text{Total for Fran} = & 1964
\end{array}
$$

Fran should weigh about 126 pounds and eat about 1964 calories per day.

PROBLEM SET 5–1–3

1. Define or discuss:
   (a) Ideal weight.
   (b) Daily calorie needs.

Use the given graph to determine the ideal weight for the following people.

	Name	Height	Frame	Musculature	Ideal weight
2.	Howie	6 ft., 1 in.	Medium	Average	
3.	Susan	5 ft., 6 in.	Large	Strong	
4.	Tim	5 ft., 11 in.	Small	Weak	
5.	Helene	5 ft., 2 in.	Small	Average	
6.	Kevin	6 ft., 4 in.	Large	Strong	
7.	Pam	5 ft., 10 in.	Large	Average	

Use the given graph to determine the daily calorie needs of the following people.

	Name	Age	Ideal weight	Activity level	Calorie needs
8.	Marion	48	135	Inactive (-15%)	
9.	Bonnie	34	130	Active ($+10\%$)	
10.	Allan	23	137	Average	
11.	Sid	57	157	Active ($+5\%$)	
12.	Barbara	19	145	Active ($+15\%$)	
13.	Marv	28	165	Inactive (-10%)	

14. Compute your own ideal weight.

15. Compute your own daily calorie needs.

There is a joke, "I'm not overweight; I should just be 8 inches taller for my weight." Seriously, though, use the given graph to determine the heights necessary for the following weights to be ideal. Assume average activity level. (Some extrapolation may be necessary.)

	Name	Frame	Weight	Ideal height
16.	Barb	Medium	167	
17.	Kent	Large	235	
18.	Betty	Small	225	
19.	Larry	Small	140	

Weight Loss

Question   What is the best way to lose weight?

ANSWER   Push yourself away from the table three times a day.

There are estimates that 50% of all Americans are overweight to some degree. Most of this is caused by bad eating habits and lack of exercise. As a consequence, there are 7×10^8 different books on dieting on the market. Basically, weight loss is related to calorie intake by the following formula.

Formula 5–1–1

$$\boxed{\text{1 lb} = \text{3500 calories}}$$

This formula states that to lose 1 pound one must take in 3500 less calories. Also, if one overeats by 3500 calories, he or she will gain a pound. Formula 5–1–1 really represents just an average since weight rarely comes off at a steady rate.

Problem   Lisa should eat 1700 calories a day. She wants to lose 2 pounds a week. What should her calorie intake be?

ANSWER   To lose 1 pound, she must reduce her intake by 3500 calories; so to reduce by 2 pounds, she must reduce her intake by 7000 calories a week. This is 1000 calories a day. Since her normal intake should be 1700, this means that a diet of 1700 − 1000 = 700 calories per day will help her lose 2 pounds a week. (This diet may not be healthy. Also, the weight loss may not come off at a steady pace.)

Problem   Becky eats a chocolate chip cookie (100 calories) every day. How much weight could she lose in a year by giving up this one cookie each day?

ANSWER   In a year, Becky eats 365 cookies, which is 365 · 100 = 36,500 extra calories. To see how many pounds this is, we set up the proportion.

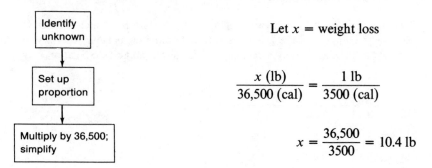

Let x = weight loss

$$\frac{x \ (\text{lb})}{36,500 \ (\text{cal})} = \frac{1 \ \text{lb}}{3500 \ (\text{cal})}$$

$$x = \frac{36,500}{3500} = 10.4 \ \text{lb}$$

Becky might lose 10.4 pounds per year by giving up a cookie a day.

PROBLEM SET 5–1–4   Use Formula 5–1–1 to complete the following table.

	Calories	Pounds
1.	10,000	
2.	18,000	
3.		20
4.		5
5.	800	
6.		35

Fran should eat 1950 calories per day.

7. If she goes on a 1600 calorie per day diet, how long will it take her to lose 1 pound? 5 pounds? 20 pounds?

8. If she goes on a 1200 calorie per day diet, how long will it take her to lose 1 pound? 5 pounds? 20 pounds?

9. Bob and Betty are both 15 pounds overweight. Bob's daily calorie needs should be 2900, and Betty's 1800. They both decide to cut their calorie intake by 25%. How long will it take each of them to lose 15 pounds?

10. Mary, whose daily calorie needs are 1900 calories, goes on Dr. Quack's Idiot Diet. This is 100 calories per day. If she lives, how much weight can she lose in 30 days?

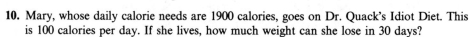

Walking up stairs burns about 1000 calories per hour.

11. If Renee walks up the stairs 5 minutes a day instead of taking an elevator, how many extra calories will she burn off in a day?

12. How many extra calories will she burn off in a year?

13. How many pounds will she lose in a year by taking the stairs instead of the elevator?

14. Walking slowly burns about 130 calories per hour. Rich takes an extra piece of apple pie (300 calories), and he feels a short walk will burn off the pie. How long must Rich walk?

15. Jane should eat 1950 calories a day. She decides to cut this by 20% and do 250 calories of exercise a day. How long will it take her to lose 20 pounds?

16. Use Formula 5–1–1 to figure how many calories per day equals 1 pound per year.

5-2 CREDIT AND INSTALLMENT BUYING

Question Bob and Marsha need $1000. How much interest will they have to pay to borrow $1000?

ANSWER That depends on the interest rate, and on how long they need the money. It depends on whether the interest is simple, add-on, or discounted. It also depends on whether or not there is a carrying charge.

Simple Interest

Problem Sam borrows $600 at 8% interest for $\frac{1}{2}$ year. How much interest does he pay?

ANSWER To use the money for a whole year it would cost 8% of 600, or $(0.08)(600) = 48$. Since he only uses the money for $\frac{1}{2}$ year, the cost is $\frac{1}{2}(48) = \$24$.

When we use someone else's money, we must pay them a premium. This extra payment is called **interest**, or **finance charge**. This depends on the amount we borrow, the interest rate, and the length of time that we keep the money. The formula for interest is

Formula 5-2-1

$$\boxed{I = P \cdot r \cdot t}$$

where I = interest or finance charge, P = **principal** (the amount borrowed), r = **rate** (written as a decimal), and t = **time** of the loan. (*Note:* Time may be any whole or fractional amount of years.)

Problem Bob and Marsha borrow $1000 at 10% interest. After 1 year, they pay back $600 of the loan. At the end of the second year, they pay back the remaining $400. How much interest do they pay?

ANSWER During the first year they borrowed $1000 at 10% for 1 year. So we can use Formula 5–2–1 with $P = 1000$, $r = 0.10$, and $t = 1$. Then we get

$$I = P \cdot r \cdot t$$

Formula 5–2–1

↓

Substitute

$$I = (1000)(0.10)(1)$$

↓

Simplify

$$= \$100$$

When they pay back \$600, they only owe \$400 during the second year. So now $P = 400$, and

Substitute

$$I = (400)(0.10)(1)$$

↓

Simplify

$$= \$40$$

Therefore, in addition to the \$1000 borrowed, they pay \$140 in interest. In this example, notice that Bob and Marsha only paid interest on the money that they still owed. We call this **simple interest** when the interest is charged only on the unpaid balance, or the amount of money that is still owed. Revolving charge accounts, credit unions, and mortgages are the most common examples of simple interest. We will also study the other types of loans later.

Problem A charge card has a $1\frac{1}{2}\%$ per month finance charge (interest). Jerry owes \$400. At the end of 1 month he pays \$100. How much does he owe at the end of the second month?

ANSWER After 1 month Jerry is charged $1\frac{1}{2}\%$ of 400 in interest. This is $(0.015)(400) = \$6.00$. So now Jerry owes \$406. When he pays back \$100, he owes \$306.

After another month he is charged $1\frac{1}{2}\%$ of 306, which is $(0.015)(306) = \$4.59$. So now he owes $306 + 4.59 = \$310.59$.

Problem   Sandy borrows $240 from her credit union, which charges 12% per year. If she pays back $20 a month plus the interest charges, what are her monthly payments and how much total interest does she pay?

ANSWER   Twelve percent per year is the same as 1% per month. The first month she owes 1% of 240, which is $(0.01)(240) = \$2.40$. So her first payment is $20.00 + 2.40 = \$22.40$. In the second month, she only owes $220, so her interest is $(0.01)(220) = \$2.20$. Then her second-month payment is $20.00 + 2.20 = \$22.20$. For the entire year, we get the following table.

Months	Unpaid balance	Interest (1% of unpaid balance)	Monthly payment
1	$240	$2.40	$22.40
2	220	2.20	22.20
3	200	2.00	22.00
4	180	1.80	21.80
5	160	1.60	21.60
6	140	1.40	21.40
7	120	1.20	21.20
8	100	1.00	21.00
9	80	0.80	20.80
10	60	0.60	20.60
11	40	0.40	20.40
12	20	0.20	20.20
		Total = $15.60	$255.60

For the year, it cost Sandy $15.60 to use the $240. Let us compute this percent.

$$r = \frac{15.60}{240} = 0.065 = 6.5\%$$

Question   But the credit union charges 12%. Did Sandy pay only 6.5%?

ANSWER   This is where the confusion about credit begins. We say that the true annual rate of interest is 12%. If Sandy had kept the *entire* $240 for the *entire* year, she would have had to pay 12% of 240 or $(0.12)(240) = \$28.80$.

But Sandy didn't keep all the money for the whole year. By 6 months, she had returned almost half of it. In the first month, she owed $240, and in the last month she owed $20. We can then think of her average unpaid balance as $\frac{1}{2}(240 + 20) = \frac{1}{2}(260) = \$130$. This means that over the year the average amount that she owed was $130.

We see that the $15.60 interest she paid is

$$r = \frac{15.60}{130} = 0.12 = 12\%$$

So her true interest rate is 12%, as it should be.

We will consider simple interest as a standard of comparison. All other forms of interest will be converted to a rate equivalent to simple interest. This is called **annual percentage rate (APR)**.

Add-on Interest

Question Roger is going to buy a car. The dealer says he can give Roger 7% add-on interest. This is a good deal, isn't it?

ANSWER The best credit risks in the nation, like General Motors and Sears, must pay 8 or 9% to borrow money. Do you really think a nobody like Roger is going to get out of the showroom only paying 7%?

Problem Roger buys a $4000 car. He puts down $1000, and finances the remaining $3000 at 7% add-on interest for 3 years. How much interest does he pay? What are his monthly payments?

ANSWER The finance company figures the interest as follows: $3000 at 7% for 3 years is (3000)(0.07)(3) = $630. So his interest is $630.

This $630 is added on to the $3000 to give a total debt of $3630. (That is why this is called **add-on**.) Over 3 years, he will make 36 monthly payments. So each payment is $\dfrac{3630}{36}$ = $100.83.

Question This seems legitimate. What's the catch?

ANSWER Remember that Roger borrows $3000, but he doesn't keep all of it for the whole 3 years. Every month he pays a little of it back. We can figure that approximately $\dfrac{3000}{36}$ = $83.33 each month goes toward paying back the original loan of 3000. As in Sandy's case, we can average his first month's debt and his last month's debt to get an average debt over the 3-year period. We average 3000 with 83.33 to get

$$\frac{1}{2}(3000 + 83.33) = \frac{1}{2}(3083.33) = \$1541.67.$$

This is a more realistic number to use for his principal or debt over the 3-year period. Now we can use Formula 5–2–1 to compute the true interest rate.

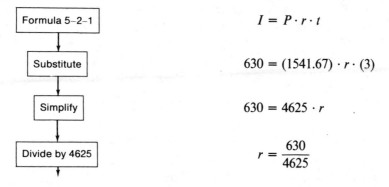

| Formula 5–2–1 | $I = P \cdot r \cdot t$ |

| Substitute | $630 = (1541.67) \cdot r \cdot (3)$ |

| Simplify | $630 = 4625 \cdot r$ |

| Divide by 4625 | $r = \dfrac{630}{4625}$ |

$$=0.136 = 13.6\%$$

Thus 7% add-on interest is approximately equivalent to 13.6% simple interest, where you only pay interest on the unpaid balance.

Let us now use some simple algebra to find a formula to compute the approximate APR for situations like this. The information that we will need is I = total interest paid, B = amount borrowed, and n = number of monthly payments.

Since we borrow B dollars for n months, we can figure that each month we are paying back about $\dfrac{B}{n}$ dollars of the debt. As before, the *average unpaid balance* (AUB) is the average of the first and last month's debt. So

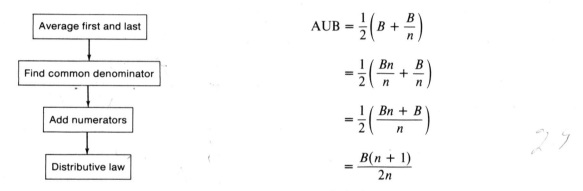

$$AUB = \frac{1}{2}\left(B + \frac{B}{n}\right)$$

$$= \frac{1}{2}\left(\frac{Bn}{n} + \frac{B}{n}\right)$$

$$= \frac{1}{2}\left(\frac{Bn + B}{n}\right)$$

$$= \frac{B(n + 1)}{2n}$$

In Formula 5–2–1, we will use AUB as the principal, P, since this is really the average debt over the period of the loan. We must now compute t, the time in years. Since there are 12 months in a year, and the money is borrowed for n months, we get $t = \dfrac{n}{12}$. For example, if $n = 24$ months, then $t = \dfrac{24}{12} = 2$ years.

If we substitute all this into Formula 5–2–1, we get

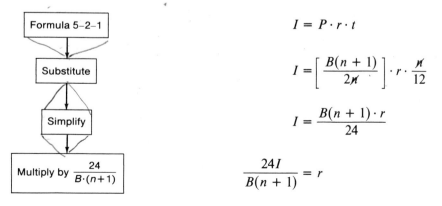

$$I = P \cdot r \cdot t$$

$$I = \left[\frac{B(n + 1)}{2\cancel{n}}\right] \cdot r \cdot \frac{\cancel{n}}{12}$$

$$I = \frac{B(n + 1) \cdot r}{24}$$

$$\frac{24I}{B(n + 1)} = r$$

This formula *approximates* the true rate of interest equivalent to simple interest. Summarizing,

Formula 5–2–2

$$APR = \frac{24I}{B(n + 1)}$$

where I = total finance charges, B = amount borrowed, and n = number of month payments.

Problem Allan and Bonnie buy a $400 TV. They are given a payment plan of $20 per month for 2 years. This seems reasonable to them. What is the APR?

ANSWER To use Formula 5–2–2, we need I, B, and n. The last two are easy: $B = 400$ borrowed and $n = 24$ monthly payments. To find I, we compute $20 per month for 24 months to get $480 in total payments. Since 400 was for the original debt, $I = 480 - 400 = \$80$. So

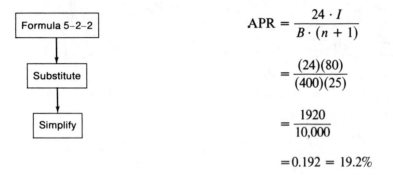

$$APR = \frac{24 \cdot I}{B \cdot (n + 1)}$$

$$= \frac{(24)(80)}{(400)(25)}$$

$$= \frac{1920}{10,000}$$

$$= 0.192 = 19.2\%$$

Those "reasonable" payments turn out to be about 19.2% in terms of simple interest.

Hand Calculator Instant Replay

PUNCH	DISPLAY	MEANING
C	0.	Clear
2 4	24.	Multiply numerators
× 8 0	80.	
+ 4 0 0	400.	Divide by denominators
+ 2 5	25.	
=	0.192	APR (Answer)

Discounted Interest

Question   Larry must borrow some money. HFG Loan Co. will loan him money at a 10% discounted rate. This sounds like a good deal, doesn't it?

ANSWER   The term, **discounted interest**, makes the interest sound like a department store bargain. Unfortunately for Larry and many others, it is not only not a bargain, but is worse than add-on.

Problem   Larry borrows $1000 at 10% discount interest for a year. How does this work? What is the APR?

ANSWER   First, HFG figures the interest, which is 10% of $1000, or $100. HFG then hands Larry a check for $900. That's the discount: he gets $900, but he has to pay back $1000 in 12 monthly payments of $\frac{1000}{12}$ = $83.33. To compute the APR, we use Formula 5–2–2 with $I = 100$, $B = 900$ (that's all he really got), and $n = 12$. Thus

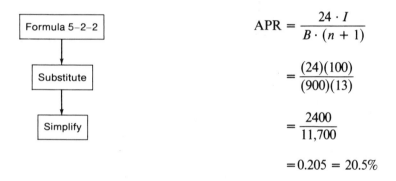

$$APR = \frac{24 \cdot I}{B \cdot (n + 1)}$$

$$= \frac{(24)(100)}{(900)(13)}$$

$$= \frac{2400}{11,700}$$

$$= 0.205 = 20.5\%$$

The 10% discounted interest turns out to be about 20.5%.

PROBLEM   1. Define or discuss:
SET 5-2-1   (a) Principal. *amount borrowed*
                 (b) Interest. *extra amount paid*
                 (c) Simple interest. *only pay I on what you still owe*
                 (d) APR. *Average percentage rate*
                 (e) Add-on interest. *interest paid no matter what you owe*
                 for early ? pay back
                 (f) Discounted interest.

Use Formula 5–2–1 to complete the following table.

	t (months)	r (per year)	P (amount borrowed)	I (interest)
2.	36	10%	1500	?
3.	42	?	2500	1050
4.	?	15%	2000	200

5. Eric borrows $600 at 9% simple interest. If he repays it all at the end of 9 months, how much interest does he pay?

6. Marge borrowed $500 for college expenses. She paid it back 4 years later with $80 interest. What was the interest rate?

7. Tony borrowed $150 and 1 month later paid back $152. What is this interest rate?

8. Terry and Nancy have a charge account, which charges $1\frac{1}{2}$% per month interest. No matter how hard they try, they can't seem to reduce their balance below $300. If their balance always stays at $300, what will their yearly interest be?

Use Formula 5–2–2 to complete the following table.

	n (months)	APR (per year)	B (amount borrowed)	I (interest)
9.	36	18%	2000	?
10.	24	?	1500	300
11.	36	?	3000	800
12.	30	?	4000	1000

Susan buys a $5000 car. She puts down $800 and finances the rest at 6.5% add-on interest for $2\frac{1}{2}$ years.

13. How much is she borrowing to buy the car? 4200

14. How much interest is added on?

15. What is her total debt? 4882.50

16. How many months is the term of the loan?

17. What are her total monthly payments?

18. Use Formula 5–2–2 to compute the APR.

Cliff and Sandy buy a new bedroom set that costs $1000. The store arranges the following deal with them: They pay $200 down and $45 a month for the next 24 months.

19. How much are they financing?

20. How much will they pay in over the next 24 months?

21. How much interest are they paying?

22. What is the APR?

23. Ken buys a stereo outfit for $700. He agrees to pay $42 a month for the next 20 months. What is the APR?

24. Judy buys a formal dress for $200. She agrees to pay $20 down and $20 a month for the next year. What is the APR?

17.3

19200

111000

The XYZ Loan Co. has an interest rate policy of 24% on the first $500, 18% on the next $500, and 12% on the amount over $1000.

25. Find 1 year's interest on $240.

26. Find 1 year's interest on $750.

27. Find 1 year's interest on $1400.

Gene goes to a new car dealer who advertises "No Money Down!" The car he likes is $6000. The finance company will finance $4000 of this at 7% add-on rate for 3 years. To finance the rest, Gene is taken to the PDQ Loan Co., which finances the $2000 down payment at a 10% add-on rate for 2 years. (Gene now senses something is fishy, but he is too much in love with the car to back out.)

28. How much interest will he pay the first finance company?

29. What will his monthly payment be to the first finance company?

30. How much interest will he pay the PDQ Loan Co?

31. What will his monthly payments be to the PDQ Loan Co.?

32. What will his total monthly payments be for the first 2 years?

Dick borrowed $800 at a 9% discounted rate for one year.

33. How much of a loan will he actually receive?

34. What is the APR?

Mary needs $1200 in a hurry, so she goes to the ABC Loan Co. They charge her a $90 "bookkeeping fee," in addition to an 11% add-on interest rate for 2 years.

35. How much interest does she pay, including the bookkeeping fee?

36. How much is her total debt?

37. What are her monthly payments?

38. What is the APR?

A slightly more accurate formula for APR is given by

$$\boxed{\text{APR} = \frac{72I}{3B(n+1) + I(n-1)}}$$

39–46. Use this improved formula to recompute Problems 10, 12, 18, 22, 23, 24, 34, and 38.

5–3 INFLATION

Question   The cost of a loaf of bread today is 65 cents. Is this a high price?

ANSWER   By 1939 standards, this is a high price. By 1999 standards, this will no doubt be considered cheap.

All prices are relative. We need a scale to measure the change in prices. This scale is called the **consumer price index (CPI)**.

Consumer Price Index

Question   How does the CPI work?

ANSWER   First, a standard of comparison has to be set. For instance, the average price level of 1967 is set equal to 100.

Question   Why 100?

ANSWER   There is no special reason other than the fact that most people are comfortable with the number 100.

Question   What does the CPI mean?

ANSWER   The best way to understand this number is to imagine a big basket full of everyday goods such as bread, shoes, gasoline, doctor bills, rent receipts, and so on. See the accompanying figure. In 1967, this imaginary basket cost $100.

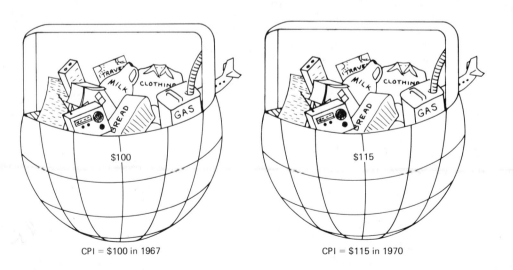

CPI = $100 in 1967   CPI = $115 in 1970

Question Suppose that the CPI for 1970 was 115. What does this mean?

ANSWER This means that exactly the same imaginary basket of goods now costs $115. The numbers 100 and 115 are simply guides that we use to see how prices have risen or fallen. We say that $115 in 1970 has the same purchasing power as $100 did in 1967. We will use 1967 as a standard. *All dollars in any year can be converted into 1967 dollars by using the CPI.*

Problem In 1976, the CPI was 159. Diane made $600 a month. How much was this in 1967 dollars?

ANSWER This is a proportion problem. The problem is this: If $159 in 1976 buys what $100 did in 1967, how much will $600 buy?

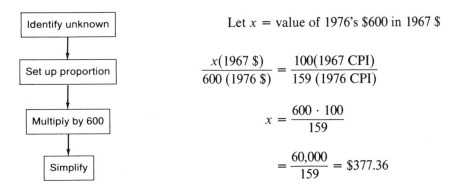

Let x = value of 1976's $600 in 1967 $

$$\frac{x\,(1967\ \$)}{600\,(1976\ \$)} = \frac{100\,(1967\ \text{CPI})}{159\,(1976\ \text{CPI})}$$

$$x = \frac{600 \cdot 100}{159}$$

$$= \frac{60{,}000}{159} = \$377.36$$

So her $600 in 1976 could only buy what $377.36 would buy in 1967. Since all purchasing power is proportional to the CPI for the year, we have the formula.

These prices were high when this photograph was taken, but now they seem low.

Formula 5–3–1

$$\frac{\text{Year } A \ \$}{\text{Year } B \ \$} = \frac{\text{Year } A \ \text{CPI}}{\text{Year } B \ \text{CPI}}$$

Problem The 23rd National Bank of Kishnev advertises that $1000 left on deposit for 10 years will grow to $2000. The CPI is estimated to grow from 162 today to 275 in 10 years. What will the $2000 really be worth in terms of today's purchasing power?

ANSWER We will use the proportion in Formula 5–3–1.

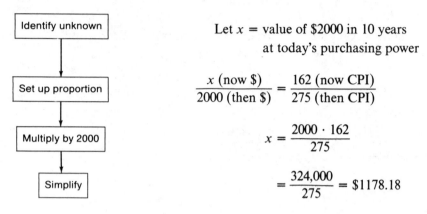

Let x = value of $2000 in 10 years
at today's purchasing power

$$\frac{x \ (\text{now } \$)}{2000 \ (\text{then } \$)} = \frac{162 \ (\text{now CPI})}{275 \ (\text{then CPI})}$$

$$x = \frac{2000 \cdot 162}{275}$$

$$= \frac{324{,}000}{275} = \$1178.18$$

Sad to say, the $1000 has really only grown to $1178.18 by today's standards. However, inflation has the opposite effect on money that you borrow.

Problem Barry borrows $500 at 12% interest for 1 year. How much does he owe at the end of the year?

ANSWER By Formula 5–2–1, the interest is 12% of $500 or $(0.12) \cdot (500) = \$60$. Adding this to the original principal of $500, he owes a total of $560.

Problem Suppose in that year that the CPI went from 162 to 172. How much is that $560 worth in terms of last years dollars?

ANSWER We again set up the proportion.

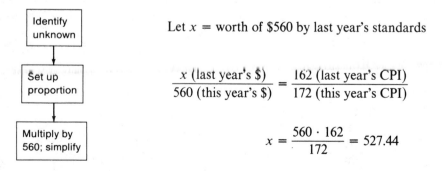

Let x = worth of $560 by last year's standards

$$\frac{x \ (\text{last year's } \$)}{560 \ (\text{this year's } \$)} = \frac{162 \ (\text{last year's CPI})}{172 \ (\text{this year's CPI})}$$

$$x = \frac{560 \cdot 162}{172} = 527.44$$

Barry is paying back $560, but it is really only worth about $527 by the standards of the year he borrowed it. This is often called paying back a loan with cheap dollars.

Problem What interest rate does Barry really pay at?

ANSWER Since he only pays back $527 in last year's dollars, the interest is really only $27. So, after inflation, the interest rate is $\frac{27}{500} = 0.054 = 5.4\%$. (Do not confuse this with APR from the last section. The APR refers to actual dollars paid in interest. This after-inflation interest refers only to purchasing power, not actual dollars.)

Problem Suppose that this year's CPI is 167, and next year it will be 177. What is the rate of inflation (as a percent)?

ANSWER The CPI went up 10 points. This is *not* the same as 10%, since the base is 167.

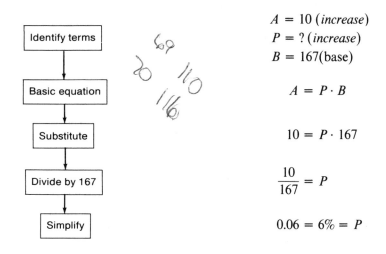

$A = 10$ (*increase*)

$P = ?$ (*increase*)

$B = 167$ (base)

$$A = P \cdot B$$

$$10 = P \cdot 167$$

$$\frac{10}{167} = P$$

$$0.06 = 6\% = P$$

Problem Gary earns $250 per week. In 1 year there was 9% inflation. What is Gary's salary worth if he gets no raise? Is this a 9% decrease in purchasing power?

ANSWER Since the inflation rate is 9%, we can again imagine a basket of everyday goods that last year cost $100 and this year cost $109. Unfortunately, Gary is still making $250 per week this year. We want to know what this is worth in terms of last year's dollars.

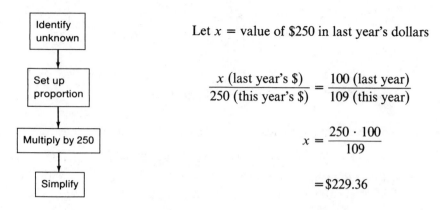

Let x = value of $250 in last year's dollars

$$\frac{x \text{ (last year's \$)}}{250 \text{ (this year's \$)}} = \frac{100 \text{ (last year)}}{109 \text{ (this year)}}$$

$$x = \frac{250 \cdot 100}{109}$$

$$= \$229.36$$

The fact that Gary did not get a raise means that his $250 can only buy $229.36 worth of goods. He has lost $20.64 of purchasing power. From the $250 base, this is $\frac{20.64}{250} = 0.083 = 8.3\%$. So Gary's purchasing power has dropped by 8.3%. Notice that even though prices went up 9%, his purchasing power went down only 8.3%. Why do you think this happens?

PROBLEM SET 5–3–1

1. Define or discuss CPI.

Complete the following table. Use Formula 5–3–1.

	Year A		Year B	
	$	CPI	$	CPI
2.	300	136	?	156
3.	1500	?	2500	187
4.	?	125	500	169
5.	25,000	131	?	168
6.	450	144	595	?
7.	?	108	1200	158

For the following problems, use the following table of CPI's.

Year	1967	1968	1969	1970	1971	1972	1973	1974	1975	1976
CPI	100	103	110	116	120	125	134	147	164	175

Find the percent of inflation in the following cases. (Be sure to use the right base.)

8. From 1969 to 1970.

9. From 1971 to 1972.

10. From 1974 to 1975.

11. From 1968 to 1976.

12. Mrs. Plotski got $13,000 in a divorce settlement in 1975. How much is this in 1967 dollars?

Bob and Marilyn bought a house in 1968 for $25,000.

13. How much is the house worth in 1975? *39,805*

14. How much profit did they make because of inflation? *83*

A certain 1974 model car cost $3350. A comparably equipped 1975 model cost $3660.

15. How much does the 1975 model cost in 1974 dollars? *3280.61*

16. Which car is cheaper (in terms of 1974 dollars)?

Mr. Sern earned $7500 in 1970. In 1975, he earned $9700.

17. What percent of increase is this?

18. How much is $9700 in 1970 dollars?

19. What is the percent of increase (or decrease) in Mr. Sern's purchasing power (of 1970 dollars)?

In 1975, Rich earned $11,000. He got a 9% raise in 1976.

20. How much does he earn in 1976?

21. How much is this in 1975 dollars? *11,236.34*

22. What is the percent of increase (or decrease) in purchasing power (of 1975 dollars)?

Carol borrowed $2000 for school in 1973 at 15% interest.

300

23. If she keeps the money until 1974 (1 year), how much interest does she owe?

206.60

24. What is her total debt (principal + interest)?

25. How much is this debt in 1973 dollars?

26. In terms of 1973 dollars, how much interest does she pay?

27. What is this rate of interest?

28. Suppose that inflation is 10% in one year. How much purchasing power does $1000 have in terms of last year's dollars?

5-4 TAXES

Question Is there any way to avoid taxes?

ANSWER Yes, if you are very rich, very poor, or very dishonest. Otherwise, no.

Question What do those of us in the middle have to know?

ANSWER The tax laws are so complicated that it is almost impossible for any book under 1000 pages to discuss every aspect of taxes. All we will try to do is to discuss the most important and common facts of the U.S. federal income tax. (For more details, the reader is urged to buy and read one of the many tax guides that come out every January.)

The main terms (along with our abbreviations) are as follows: (These definitions are very simplified.)

1. **Exemptions (EXEM)**: This is basically the number of people that we support.
2. **Adjusted gross income (AGI)**: This is the total of all wages, tips, bank interest, and prizes, less certain expenses.
3. **Tax tables**: These tables give the shortest calculation of tax, where AGI is less than $20,000, or $40,000 for a married joint return.
4. **Tax (TAX)**: This is the amount you actually owe the government.
5. **Tax withheld (TW)**: This is the total amount withheld from all your paychecks throughout the year.
6. **Balance due (BALANCE)**: If your tax bill is more than the amount withheld throughout the year, you pay the balance.
7. **Refund (REFUND)**: If the amount that was withheld was more than your tax, you get a refund.
8. **Deductions (DEDUCT)**: This includes money spent on such things as medical expenses, interest on loans, taxes, donations, and union dues. You can deduct much of this from your income and not pay tax on it.

The front and back of the standard 1040 form is shown on pages 288 and 289. Although there are many other terms on the 1040 form, we will concentrate our attention on the ones listed above.

Refer to the flowchart, which indicates how taxes are computed. First, the EXEM and AGI are computed. Then we must decide whether to itemize our deductions or not. We will see shortly how both of these methods compute the final TAX. Once we have the TAX, we compare it with the tax withheld, TW. If TAX is more than TW, we have to pay the balance. If TW is more than the TAX, we get a refund. Then we're through.

Every year the tax forms, laws, and tables change slightly. Thus it would be impossible to write a text in one year that would be 100% correct the next. But taxes are a very important topic. Thus we have included them even though the forms and tables will change slightly. The general rules are the same. Also, in this section, we will only do the simple cases.

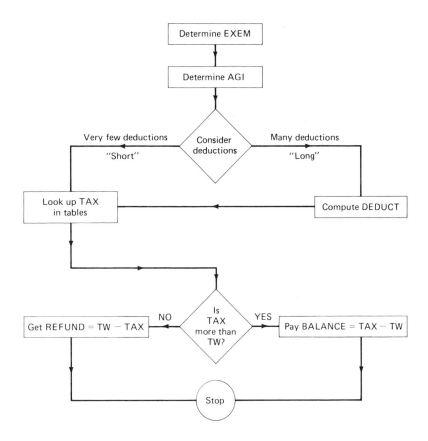

Short Method

Question   Is there a quick way to compute income tax?

ANSWER   Yes, if you earn less than $20,000 (or $40,000 married joint return) a year, you can use Tax Tables A–D. These tables start on page 290. To save space, we have included only the portions of the tables needed for the examples and exercises.

If you are single, use Table A; if you are married filing jointly, use Table B; if you are married filing separately, use Table C; if you are the head of a household, use Table D.

Problem   Paul and Judy earn $13,412 together. Can they use the tax tables? If so, what is their tax?

ANSWER   Since they made less than $40,000, they can use the tax tables. Assuming that Paul and Judy are married, they would use Table B: married filing joint return. Since there are two of them, they would use column 2. Their income is between 13,400 and 13,450. So we see in the figure that their tax is $1365.

Form **1040**　Department of the Treasury—Internal Revenue Service
U.S. Individual Income Tax Return 19**78**

For Privacy Act Notice, see page 3 of Instructions | For the year January 1–December 31, 1978, or other tax year beginning 1978, ending , 19

Use IRS label. Otherwise, please print or type.	Your first name and initial (if joint return, also give spouse's name and initial)	Last name	Your social security number
	Present home address (Number and street, including apartment number, or rural route)		Spouse's social security no.
	City, town or post office, State and ZIP code		Your occupation

▶ Do you want $1 to go to the Presidential Election Campaign Fund? ... Yes [ ] No [ ]
If joint return, does your spouse want $1 to go to this fund? .. Yes [ ] No [ ]
Note: Checking Yes will not increase your tax or reduce your refund.

Spouse's occupation

Filing Status
Check only one box.

1 [ ] Single
2 [ ] Married filing joint return (even if only one had income)
3 [ ] Married filing separate return. If spouse is also filing, give spouse's social security number in the space above and enter full name here ▶
4 [ ] Unmarried head of household. Enter qualifying name ▶ .... See page 6 of Instructions.
5 [ ] Qualifying widow(er) with dependent child (Year spouse died ▶ 19). See page 6 of Instructions.

Exemptions
Always check the box labeled Yourself.
Check other boxes if they apply.

6a [ ] Yourself　[ ] 65 or over　[ ] Blind
b [ ] Spouse　[ ] 65 or over　[ ] Blind

} Enter number of boxes checked on 6a and b ▶ [ ]

c First names of your dependent children who lived with you ▶ ....
Enter number of children listed ▶ [ ]

d Other dependents:

(1) Name	(2) Relationship	(3) Number of months lived in your home	(4) Did dependent have income of $750 or more?	(5) Did you provide more than one-half of dependent's support?

Enter number of other dependents ▶ [ ]

Add numbers entered in boxes above ▶ [ ]

7 Total number of exemptions claimed ....

Income
Please attach Copy B of your Forms W–2 here.

If you do not have a W–2, see page 5 of Instructions.

8 Wages, salaries, tips, and other employee compensation .... | 8 |
9 Interest income (If over $400, attach Schedule B) .... | 9 |
10a Dividends (If over $400, attach Schedule B) ...., 10b Exclusion. |
10c Subtract line 10b from line 10a .... | 10c |
11 State and local income tax refunds (does not apply unless refund is for year you itemized deductions) .... | 11 |
12 Alimony received .... | 12 |
13 Business income or (loss) (attach Schedule C) .... | 13 |
14 Capital gain or (loss) (attach Schedule D) .... | 14 |
15 Taxable part of capital gain distributions not reported on Schedule D (see page 9 of Instructions) . | 15 |
16 Net gain or (loss) from Supplemental Schedule of Gains and Losses (attach Form 4797) .... | 16 |
17 Fully taxable pensions and annuities not reported on Schedule E .... | 17 |
18 Pensions, annuities, rents, royalties, partnerships, estates or trusts, etc. (attach Schedule E) .... | 18 |
19 Farm income or (loss) (attach Schedule F) .... | 19 |
20 Other income (state nature and source—see page 10 of Instructions) ▶ .... | 20 |

Please attach check or money order here.

21 Total income. Add lines 8, 9, and 10c through 20 .... ▶ | 21 |

Adjustments to Income

22 Moving expense (attach Form 3903) .... | 22 |
23 Employee business expenses (attach Form 2106) .. | 23 |
24 Payments to an IRA (see page 10 of Instructions) .... | 24 |
25 Payments to a Keogh (H.R. 10) retirement plan ... | 25 |
26 Interest penalty due to early withdrawal of savings | 26 |
27 Alimony paid (see page 10 of Instructions) .... | 27 |
28 Total adjustments. Add lines 22 through 27 .... ▶ | 28 |

Adjusted Gross Income

29 Subtract line 28 from line 21 .... | 29 |
30 Disability income exclusion (attach Form 2440) ...... | 30 |
31 Adjusted gross income. Subtract line 30 from line 29. If this line is less than $8,000, see page 2 of Instructions. If you want IRS to figure your tax, see page 4 of Instructions .... ▶ | 31 |

263-053-2

☆ U.S. GOVERNMENT PRINTING OFFICE : 1978 O-263-053

Form 1040 (1978)

Tax Compu-tation

32 Amount from line 31	32	
33 If you do not itemize deductions, enter zero . . . . }		
If you itemize, complete Schedule A (Form 1040) and enter the amount from Schedule A, line 41 . . . . }	33	

Caution: If you have unearned income and can be claimed as a dependent on your parent's return, check here ▶ ☐ and see page 11 of the Instructions. Also see page 11 of the Instructions if:
- You are married filing a separate return and your spouse itemizes deductions, OR
- You file Form 4563, OR
- You are a dual-status alien.

34 Subtract line 33 from line 32. Use the amount on line 34 to find your tax from the Tax Tables, or to figure your tax on Schedule TC, Part I	34	

Use Schedule TC, Part I, and the Tax Rate Schedules ONLY if:
- The amount on line 34 is more than $20,000 ($40,000 if you checked Filing Status Box 2 or 5), OR
- You have more exemptions than those covered in the Tax Table for your filing status, OR
- You use any of these forms to figure your tax: Schedule D, Schedule G, or Form 4726.

Otherwise, you MUST use the Tax Tables to find your tax.

35 Tax. Enter tax here and check if from ☐ Tax Tables or ☐ Schedule TC	35	
36 Additional taxes. (See page 11 of Instructions.) Enter total and check if from ☐ Form 4970, } ☐ Form 4972, ☐ Form 5544, ☐ Form 5405, or ☐ Section 72(m)(5) penalty tax . . . }	36	
37 **Total.** Add lines 35 and 36 . . . . ▶	37	

Credits

38 Credit for contributions to candidates for public office . .	38	
39 Credit for the elderly (attach Schedules R&RP)	39	
40 Credit for child and dependent care expenses (attach Form 2441) .	40	
41 Investment credit (attach Form 3468)	41	
42 Foreign tax credit (attach Form 1116)	42	
43 Work Incentive (WIN) Credit (attach Form 4874)	43	
44 New jobs credit (attach Form 5884)	44	
45 Residential energy credits (see page 12 of Instructions, attach Form 5695) . . .	45	

46 **Total credits.** Add lines 38 through 45	46	
47 **Balance.** Subtract line 46 from line 37 and enter difference (but not less than zero) . ▶	47	

Other Taxes

48 Self-employment tax (attach Schedule SE)	48	
49 Minimum tax. Check here ▶ ☐ and attach Form 4625	49	
50 Tax from recomputing prior-year investment credit (attach Form 4255)	50	
51 Social security (FICA) tax on tip income not reported to employer (attach Form 4137) . .	51	
52 Uncollected employee FICA and RRTA tax on tips (from Form W–2)	52	
53 Tax on an IRA (attach Form 5329)	53	
54 **Total tax.** Add lines 47 through 53 . . . . ▶	54	

Payments

Attach Forms W–2, W–2G, and W–2P to front.

55 Total Federal income tax withheld	55	
56 1978 estimated tax payments and credit from 1977 return .	56	
57 Earned income credit. If line 31 is under $8,000, see page 2 of Instructions. If eligible, enter child's name ▶........................	57	
58 Amount paid with Form 4868	58	
59 Excess FICA and RRTA tax withheld (two or more employers)	59	
60 Credit for Federal tax on special fuels and oils (attach Form 4136) .	60	
61 Regulated Investment Company credit (attach Form 2439)	61	

62 **Total.** Add lines 55 through 61 . . . . ▶	62	

Refund or Due

63 If line 62 is larger than line 54, enter amount **OVERPAID** . . . . ▶	63	
64 Amount of line 63 to be **REFUNDED TO YOU** . . . . ▶	64	
65 Amount of line 63 to be credited on 1979 estimated tax . ▶	65	
66 If line 54 is larger than line 62, enter **BALANCE DUE.** Attach check or money order for full amount payable to "Internal Revenue Service." Write your social security number on check or money order . . ▶ (Check ▶ ☐ if Form 2210 (2210F) is attached. See page 14 of instructions.) ▶ $	66	

Please Sign Here

Under penalties of perjury, I declare that I have examined this return, including accompanying schedules and statements, and to the best of my knowledge and belief, it is true, correct, and complete. Declaration of preparer (other than taxpayer) is based on all information of which preparer has any knowledge.

▶ Your signature	Date	▶ Spouse's signature (if filing jointly, BOTH must sign even if only one had income)

Paid Preparer's Information

Preparer's signature ▶		Preparer's social security no.	Check if self-employed ▶ ☐
Firm's name (or yours, if self-employed), address and ZIP code ▶		E.I. No. ▶	
		Date ▶	

263-053-2

Form 1040 (1978)

Table A – Single Taxpayers (Continued)

If Tax Table Income is—Over—	But not over—	1	2	3	If Tax Table Income is—Over—	But not over—	1	2	3	If Tax Table Income is—Over—	But not over—	1	2	3
		Your tax is—					Your tax is—					Your tax is—		
11,000	11,050	1,447	1,282	1,117	14,000	14,050	2,200	1,998	1,804	17,000	17,050	3,053	2,834	2,617
11,050	11,100	1,459	1,293	1,128	14,050	14,100	2,214	2,011	1,816	17,050	17,100	3,069	2,849	2,631
11,100	11,150	1,470	1,304	1,139	14,100	14,150	2,227	2,025	1,829	17,100	17,150	3,084	2,863	2,646
11,150	11,200	1,482	1,315	1,150	14,150	14,200	2,241	2,038	1,841	17,150	17,200	3,100	2,878	2,660
11,200	11,250	1,493	1,326	1,161	14,200	14,250	2,254	2,052	1,854	17,200	17,250	3,115	2,892	2,675
11,250	11,300	1,505	1,337	1,172	14,250	14,300	2,268	2,065	1,866	17,250	17,300	3,131	2,907	2,689
11,300	11,350	1,516	1,348	1,183	14,300	14,350	2,281	2,079	1,879	17,300	17,350	3,146	2,921	2,704
11,350	11,400	1,528	1,359	1,194	14,350	14,400	2,295	2,092	1,891	17,350	17,400	3,162	2,936	2,718
11,400	11,450	1,539	1,370	1,205	14,400	14,450	2,308	2,106	1,904	17,400	17,450	3,177	2,950	2,733
11,450	11,500	1,551	1,381	1,216	14,450	14,500	2,322	2,119	1,917	17,450	17,500	3,193	2,965	2,747
11,500	11,550	1,562	1,392	1,227	14,500	14,550	2,335	2,133	1,930	17,500	17,550	3,208	2,979	2,762
11,550	11,600	1,574	1,403	1,238	14,550	14,600	2,349	2,146	1,944	17,550	17,600	3,224	2,994	2,776
11,600	11,650	1,585	1,414	1,249	14,600	14,650	2,362	2,160	1,957	17,600	17,650	3,239	3,008	2,791
11,650	11,700	1,597	1,425	1,260	14,650	14,700	2,376	2,173	1,971	17,650	17,700	3,255	3,023	2,805
11,700	11,750	1,608	1,436	1,271	14,700	14,750	2,389	2,187	1,984	17,700	17,750	3,270	3,038	2,820
11,750	11,800	1,620	1,447	1,282	14,750	14,800	2,403	2,200	1,998	17,750	17,800	3,286	3,053	2,834
11,800	11,850	1,631	1,459	1,293	14,800	14,850	2,416	2,214	2,011	17,800	17,850	3,301	3,069	2,849
11,850	11,900	1,643	1,470	1,304	14,850	14,900	2,430	2,227	2,025	17,850	17,900	3,317	3,084	2,863
11,900	11,950	1,654	1,482	1,315	14,900	14,950	2,443	2,241	2,038	17,900	17,950	3,332	3,100	2,878
11,950	12,000	1,666	1,493	1,326	14,950	15,000	2,457	2,254	2,052	17,950	18,000	3,348	3,115	2,892
12,000	12,050	1,679	1,505	1,337	15,000	15,050	2,472	2,268	2,065	18,000	18,050	3,363	3,131	2,907
12,050	12,100	1,691	1,516	1,348	15,050	15,100	2,486	2,281	2,079	18,050	18,100	3,379	3,146	2,921
12,100	12,150	1,704	1,528	1,359	15,100	15,150	2,501	2,295	2,092	18,100	18,150	3,394	3,162	2,936
12,150	12,200	1,716	1,539	1,370	15,150	15,200	2,515	2,308	2,106	18,150	18,200	3,410	3,177	2,950

Table B – Married Taxpayers Filing Jointly

If Tax Table Income is—Over—	But not over—	2	3	4	5	6	7	8	9	If Tax Table Income is—Over—	But not over—	2	3	4	5	6	7	8	9
		Your tax is—										Your tax is—							
If $5,200 or less your tax is 0										11,600	11,650	1,037	910	751	573	397	235	78	0
5,200	5,250	4	0	0	0	0	0	0	0	11,650	11,700	1,046	918	760	583	406	243	86	0
5,250	5,300	11	0	0	0	0	0	0	0	11,700	11,750	1,054	927	770	592	415	252	94	0
5,300	5,350	18	0	0	0	0	0	0	0	11,750	11,800	1,063	935	779	602	424	260	102	0
5,350	5,400	25	0	0	0	0	0	0	0	11,800	11,850	1,071	944	789	611	434	269	110	0
										11,850	11,900	1,080	952	798	621	443	277	118	0
										11,900	11,950	1,088	961	808	630	453	286	126	0
7,800	7,850	401	245	94	0	0	0	0	0	11,950	12,000	1,097	969	817	640	462	294	134	0
7,850	7,900	410	253	101	0	0	0	0	0	12,000	12,050	1,105	978	827	649	472	303	142	0
7,900	7,950	418	261	109	0	0	0	0	0	12,050	12,100	1,114	986	836	659	481	311	150	0
7,950	8,000	427	269	116	0	0	0	0	0	12,100	12,150	1,122	995	846	668	491	320	158	3
8,000	8,050	435	277	124	0	0	0	0	0	12,150	12,200	1,131	1,003	855	678	500	328	166	11
8,050	8,100	444	285	131	0	0	0	0	0	12,200	12,250	1,139	1,012	865	687	510	337	174	19
8,100	8,150	452	293	139	0	0	0	0	0	12,250	12,300	1,148	1,020	874	697	519	345	183	27
8,150	8,200	461	301	146	0	0	0	0	0	12,300	12,350	1,156	1,029	884	706	529	354	191	35
										12,350	12,400	1,165	1,037	893	716	538	362	200	43
										12,400	12,450	1,173	1,046	903	725	548	371	208	51
9,000	9,050	595	443	282	127	0	0	0	0	12,450	12,500	1,182	1,054	912	735	557	380	217	59
9,050	9,100	604	451	290	135	0	0	0	0										
9,100	9,150	612	460	298	143	0	0	0	0										
9,150	9,200	621	468	306	151	1	0	0	0	13,000	13,050	1,285	1,148	1,017	839	662	484	310	148
9,200	9,250	629	477	314	159	9	0	0	0	13,050	13,100	1,295	1,156	1,026	849	671	494	319	156
9,250	9,300	638	485	323	167	16	0	0	0	13,100	13,150	1,305	1,165	1,036	858	681	503	327	165
9,300	9,350	646	494	331	175	24	0	0	0	13,150	13,200	1,315	1,173	1,045	868	690	513	336	173
9,350	9,400	655	502	340	183	31	0	0	0	13,200	13,250	1,325	1,182	1,054	877	700	522	345	182
9,400	9,450	663	511	348	191	39	0	0	0	13,250	13,300	1,335	1,190	1,063	887	709	532	354	190
9,450	9,500	672	520	357	199	46	0	0	0	13,300	13,350	1,345	1,199	1,071	896	719	541	364	199
9,500	9,550	680	529	365	207	54	0	0	0	13,350	13,400	1,355	1,207	1,080	906	728	551	373	207
9,550	9,600	689	539	374	215	61	0	0	0	13,400	13,450	1,365	1,216	1,088	915	738	560	383	216
9,600	9,650	697	548	382	223	69	0	0	0	13,450	13,500	1,375	1,225	1,097	925	747	570	392	224
9,650	9,700	706	558	391	231	76	0	0	0	13,500	13,550	1,385	1,235	1,105	934	757	579	402	233
9,700	9,750	714	567	399	239	84	0	0	0	13,550	13,600	1,395	1,245	1,114	944	766	589	411	241
9,750	9,800	723	577	408	247	92	0	0	0	13,600	13,650	1,405	1,255	1,122	953	776	598	421	250
9,800	9,850	731	586	416	255	100	0	0	0	13,650	13,700	1,415	1,265	1,131	963	785	608	430	258
9,850	9,900	740	596	425	263	108	0	0	0	13,700	13,750	1,426	1,275	1,139	972	795	617	440	267
9,900	9,950	748	605	433	271	116	0	0	0	13,750	13,800	1,437	1,285	1,148	982	804	627	449	275
9,950	10,000	757	615	442	279	124	0	0	0	13,800	13,850	1,448	1,296	1,156	991	814	636	459	284
10,000	10,050	765	624	450	288	132	0	0	0	13,850	13,900	1,459	1,305	1,165	1,001	823	646	468	292

Table B—Married Taxpayers Filing Jointly (Continued)

If Tax Table Income is— Over—	But not over—	And the number of exemptions is— 2	3	4	5	6	7	8	9
		Your tax is—							
14,200	14,250	1,536	1,375	1,225	1,067	890	712	535	357
14,250	14,300	1,547	1,385	1,235	1,077	899	722	544	367
14,300	14,350	1,558	1,395	1,245	1,086	909	731	554	376
14,350	14,400	1,569	1,405	1,255	1,096	918	741	563	386
14,400	14,450	1,580	1,415	1,265	1,105	928	750	573	395
14,450	14,500	1,591	1,426	1,275	1,115	937	760	582	405
14,500	14,550	1,602	1,437	1,285	1,124	947	769	592	414
14,550	14,600	1,613	1,448	1,295	1,134	956	779	601	424
14,600	14,650	1,624	1,459	1,305	1,143	966	788	611	433
14,650	14,700	1,635	1,470	1,315	1,153	975	798	620	443
14,700	14,750	1,646	1,481	1,325	1,162	985	807	630	452
14,750	14,800	1,657	1,492	1,335	1,172	994	817	639	462
15,600	15,650	1,844	1,679	1,514	1,354	1,156	978	801	623
15,650	15,700	1,855	1,690	1,525	1,365	1,165	988	810	633
15,700	15,750	1,866	1,701	1,536	1,375	1,176	997	820	642
15,750	15,800	1,877	1,712	1,547	1,385	1,187	1,007	829	652
15,800	15,850	1,888	1,723	1,558	1,395	1,198	1,016	839	661
15,850	15,900	1,899	1,734	1,569	1,405	1,209	1,026	848	671
15,900	15,950	1,910	1,745	1,580	1,415	1,220	1,035	858	680
15,950	16,000	1,921	1,756	1,591	1,426	1,231	1,045	867	690
16,000	16,050	1,932	1,767	1,602	1,437	1,242	1,054	877	699
16,050	16,100	1,943	1,778	1,613	1,448	1,253	1,064	886	709
16,100	16,150	1,954	1,789	1,624	1,459	1,264	1,073	896	718
16,150	16,200	1,965	1,800	1,635	1,470	1,275	1,083	905	728
16,200	16,250	1,976	1,811	1,646	1,481	1,286	1,092	915	737
16,250	16,300	1,987	1,822	1,657	1,492	1,297	1,102	924	747
16,300	16,350	1,998	1,833	1,668	1,503	1,308	1,111	934	756
16,350	16,400	2,009	1,844	1,679	1,514	1,319	1,121	943	766
18,800	18,850	2,611	2,424	2,236	2,053	1,858	1,658	1,458	1,258
18,850	18,900	2,624	2,436	2,249	2,064	1,869	1,669	1,469	1,269
18,900	18,950	2,636	2,449	2,261	2,075	1,880	1,680	1,480	1,280
18,950	19,000	2,649	2,461	2,274	2,086	1,891	1,691	1,491	1,291
19,000	19,050	2,661	2,474	2,286	2,099	1,902	1,702	1,502	1,302
19,050	19,100	2,674	2,486	2,299	2,111	1,913	1,713	1,513	1,313
19,100	19,150	2,686	2,499	2,311	2,124	1,924	1,724	1,524	1,324
19,150	19,200	2,699	2,511	2,324	2,136	1,935	1,735	1,535	1,335
19,200	19,250	2,711	2,524	2,336	2,149	1,946	1,746	1,546	1,346
19,250	19,300	2,724	2,536	2,349	2,161	1,957	1,757	1,557	1,357
19,300	19,350	2,736	2,549	2,361	2,174	1,968	1,768	1,568	1,368
19,350	19,400	2,749	2,561	2,374	2,186	1,979	1,779	1,579	1,379
19,400	19,450	2,761	2,574	2,386	2,199	1,990	1,790	1,590	1,390
19,450	19,500	2,774	2,586	2,399	2,211	2,001	1,801	1,601	1,401
19,500	19,550	2,786	2,599	2,411	2,224	2,012	1,812	1,612	1,412
19,550	19,600	2,799	2,611	2,424	2,236	2,023	1,823	1,623	1,423
19,600	19,650	2,811	2,624	2,436	2,249	2,034	1,834	1,634	1,434
19,650	19,700	2,824	2,636	2,449	2,261	2,045	1,845	1,645	1,445
19,700	19,750	2,836	2,649	2,461	2,274	2,056	1,856	1,656	1,456
19,750	19,800	2,849	2,661	2,474	2,286	2,069	1,867	1,667	1,467
19,800	19,850	2,861	2,674	2,486	2,299	2,081	1,878	1,678	1,478
19,850	19,900	2,874	2,686	2,499	2,311	2,094	1,889	1,689	1,489
19,900	19,950	2,886	2,699	2,511	2,324	2,106	1,900	1,700	1,500
19,950	20,000	2,899	2,711	2,524	2,336	2,119	1,911	1,711	1,511
20,000	20,050	2,911	2,724	2,536	2,349	2,131	1,922	1,722	1,522
20,050	20,100	2,924	2,736	2,549	2,361	2,144	1,933	1,733	1,533
20,100	20,150	2,936	2,749	2,561	2,374	2,156	1,944	1,744	1,544
20,150	20,200	2,949	2,761	2,574	2,386	2,169	1,955	1,755	1,555
20,200	20,250	2,961	2,774	2,586	2,399	2,181	1,966	1,766	1,566
20,250	20,300	2,974	2,786	2,599	2,411	2,194	1,977	1,777	1,577
20,300	20,350	2,986	2,799	2,611	2,424	2,206	1,988	1,788	1,588
20,350	20,400	2,999	2,811	2,624	2,436	2,219	1,999	1,799	1,599
20,400	20,450	3,011	2,824	2,636	2,449	2,231	2,010	1,810	1,610
20,450	20,500	3,024	2,836	2,649	2,461	2,244	2,021	1,821	1,621
20,500	20,550	3,036	2,849	2,661	2,474	2,256	2,034	1,832	1,632
20,550	20,600	3,049	2,861	2,674	2,486	2,269	2,046	1,843	1,643
20,600	20,650	3,061	2,874	2,686	2,499	2,281	2,059	1,854	1,654
20,650	20,700	3,074	2,886	2,699	2,511	2,294	2,071	1,865	1,665
20,700	20,750	3,087	2,899	2,711	2,524	2,306	2,084	1,876	1,676
20,750	20,800	3,101	2,911	2,724	2,536	2,319	2,096	1,887	1,687
20,800	20,850	3,115	2,924	2,736	2,549	2,331	2,109	1,898	1,698
20,850	20,900	3,129	2,936	2,749	2,561	2,344	2,121	1,909	1,709
20,900	20,950	3,143	2,949	2,761	2,574	2,356	2,134	1,920	1,720
20,950	21,000	3,157	2,961	2,774	2,586	2,369	2,146	1,931	1,731
21,000	21,050	3,171	2,974	2,786	2,599	2,381	2,159	1,942	1,742
21,050	21,100	3,185	2,986	2,799	2,611	2,394	2,171	1,953	1,753
21,100	21,150	3,199	2,999	2,811	2,624	2,406	2,184	1,964	1,764
21,150	21,200	3,213	3,011	2,824	2,636	2,419	2,196	1,975	1,775

If Tax Table Income is— Over—	But not over—	And the number of exemptions is— 2	3	4	5	6	7	8	9
		Your tax is—							
23,800	23,850	3,955	3,745	3,535	3,325	3,085	2,859	2,636	2,414
23,850	23,900	3,969	3,759	3,549	3,339	3,099	2,871	2,649	2,426
23,900	23,950	3,983	3,773	3,563	3,353	3,113	2,884	2,661	2,439
23,950	24,000	3,997	3,787	3,577	3,367	3,127	2,896	2,674	2,451
24,000	24,050	4,011	3,801	3,591	3,381	3,141	2,909	2,686	2,464
24,050	24,100	4,025	3,815	3,605	3,395	3,155	2,921	2,699	2,476
24,100	24,150	4,039	3,829	3,619	3,409	3,169	2,934	2,711	2,489
24,150	24,200	4,053	3,843	3,633	3,423	3,183	2,946	2,724	2,501
24,200	24,250	4,067	3,857	3,647	3,437	3,197	2,959	2,736	2,514
24,250	24,300	4,081	3,871	3,661	3,451	3,211	2,971	2,749	2,526
24,300	24,350	4,095	3,885	3,675	3,465	3,225	2,984	2,761	2,539
24,350	24,400	4,109	3,899	3,689	3,479	3,239	2,996	2,774	2,551
24,400	24,450	4,123	3,913	3,703	3,493	3,253	3,009	2,786	2,564
24,450	24,500	4,137	3,927	3,717	3,507	3,267	3,022	2,799	2,576
24,500	24,550	4,151	3,941	3,731	3,521	3,281	3,036	2,811	2,589
24,550	24,600	4,165	3,955	3,745	3,535	3,295	3,050	2,824	2,601
24,600	24,650	4,179	3,969	3,759	3,549	3,309	3,064	2,836	2,614
24,650	24,700	4,193	3,983	3,773	3,563	3,323	3,078	2,849	2,626
24,700	24,750	4,208	3,997	3,787	3,577	3,337	3,092	2,861	2,639
24,750	24,800	4,224	4,011	3,801	3,591	3,351	3,106	2,874	2,651
24,800	24,850	4,240	4,025	3,815	3,605	3,365	3,120	2,886	2,664
24,850	24,900	4,256	4,039	3,829	3,619	3,379	3,134	2,899	2,676
24,900	24,950	4,272	4,053	3,843	3,633	3,393	3,148	2,911	2,689
24,950	25,000	4,288	4,067	3,857	3,647	3,407	3,162	2,924	2,701
25,000	25,050	4,304	4,081	3,871	3,661	3,421	3,176	2,936	2,714
25,050	25,100	4,320	4,095	3,885	3,675	3,435	3,190	2,949	2,726
25,100	25,150	4,336	4,109	3,899	3,689	3,449	3,204	2,961	2,739
25,150	25,200	4,352	4,123	3,913	3,703	3,463	3,218	2,974	2,751
27,800	27,850	5,200	4,960	4,720	4,480	4,210	3,960	3,715	3,470
27,850	27,900	5,216	4,976	4,736	4,496	4,226	3,974	3,729	3,484
27,900	27,950	5,232	4,992	4,752	4,512	4,242	3,988	3,743	3,498
27,950	28,000	5,248	5,008	4,768	4,528	4,258	4,002	3,757	3,512
28,000	28,050	5,264	5,024	4,784	4,544	4,274	4,016	3,771	3,526
28,050	28,100	5,280	5,040	4,800	4,560	4,290	4,030	3,785	3,540
28,100	28,150	5,296	5,056	4,816	4,576	4,306	4,044	3,799	3,554
28,150	28,200	5,312	5,072	4,832	4,592	4,322	4,058	3,813	3,568
28,200	28,250	5,328	5,088	4,848	4,608	4,338	4,072	3,827	3,582
28,250	28,300	5,344	5,104	4,864	4,624	4,354	4,086	3,841	3,596
28,300	28,350	5,360	5,120	4,880	4,640	4,370	4,100	3,855	3,610
31,800	31,850	6,605	6,335	6,065	5,795	5,495	5,215	4,940	4,665
31,850	31,900	6,623	6,353	6,083	5,813	5,513	5,231	4,956	4,681
31,900	31,950	6,641	6,371	6,101	5,831	5,531	5,247	4,972	4,697
31,950	32,000	6,659	6,389	6,119	5,849	5,549	5,263	4,988	4,713
32,000	32,050	6,677	6,407	6,137	5,867	5,567	5,279	5,004	4,729
32,050	32,100	6,695	6,425	6,155	5,885	5,585	5,295	5,020	4,745
32,100	32,150	6,713	6,443	6,173	5,903	5,603	5,311	5,036	4,761
32,150	32,200	6,731	6,461	6,191	5,921	5,621	5,327	5,052	4,777
32,200	32,250	6,749	6,479	6,209	5,939	5,639	5,343	5,068	4,793
32,250	32,300	6,767	6,497	6,227	5,957	5,657	5,359	5,084	4,809
32,300	32,350	6,785	6,515	6,245	5,975	5,675	5,375	5,100	4,825
35,800	35,850	8,139	7,846	7,554	7,261	6,939	6,630	6,325	6,020
35,850	35,900	8,158	7,866	7,573	7,281	6,958	6,648	6,343	6,038
35,900	35,950	8,178	7,885	7,593	7,300	6,978	6,666	6,361	6,056
35,950	36,000	8,197	7,905	7,612	7,320	6,997	6,684	6,379	6,074
36,000	36,050	8,217	7,924	7,632	7,339	7,017	6,702	6,397	6,092
36,050	36,100	8,236	7,944	7,651	7,359	7,036	6,720	6,415	6,110
36,100	36,150	8,256	7,963	7,671	7,378	7,056	6,738	6,433	6,128
36,150	36,200	8,275	7,983	7,690	7,398	7,075	6,756	6,451	6,146
36,200	36,250	8,295	8,002	7,710	7,417	7,095	6,774	6,469	6,164
36,250	36,300	8,314	8,022	7,729	7,437	7,114	6,792	6,487	6,182
36,300	36,350	8,334	8,041	7,749	7,456	7,134	6,810	6,505	6,200
36,350	36,400	8,353	8,061	7,768	7,476	7,153	6,828	6,523	6,218
36,400	36,450	8,373	8,080	7,788	7,495	7,173	6,846	6,541	6,236
39,600	39,650	9,709	9,394	9,079	8,764	8,421	8,093	7,766	7,438
39,650	39,700	9,730	9,415	9,100	8,785	8,440	8,113	7,785	7,458
39,700	39,750	9,751	9,436	9,121	8,806	8,461	8,132	7,805	7,477
39,750	39,800	9,772	9,457	9,142	8,827	8,482	8,152	7,824	7,497
39,800	39,850	9,793	9,478	9,163	8,848	8,503	8,171	7,844	7,516
39,850	39,900	9,814	9,499	9,184	8,869	8,524	8,191	7,863	7,536
39,900	39,950	9,835	9,520	9,205	8,890	8,545	8,210	7,883	7,555
39,950	40,000	9,856	9,541	9,226	8,911	8,566	8,230	7,902	7,575

Table C—Married Taxpayers Filing Separately (Continued)

If Tax Table Income is— Over—	But not over—	1	2	3
		Your tax is—		
6,400	6,450	672	492	314
6,450	6,500	683	501	324
6,500	6,550	694	511	333
6,550	6,600	705	520	343
6,600	6,650	716	530	352
6,650	6,700	727	539	362
6,700	6,750	738	549	371
6,750	6,800	749	558	381
6,800	6,850	760	568	390
6,850	6,900	771	577	400
6,900	6,950	782	587	409
6,950	7,000	793	596	419
7,000	7,050	804	606	428
7,050	7,100	815	615	438
7,100	7,150	826	626	447
7,150	7,200	837	637	457
7,200	7,250	848	648	466
7,250	7,300	859	659	476
7,300	7,350	870	670	485
7,350	7,400	881	681	495
7,400	7,450	892	692	504
7,450	7,500	903	703	514
7,500	7,550	914	714	523
7,550	7,600	925	725	533
7,600	7,650	936	736	542
7,650	7,700	947	747	552
7,700	7,750	958	758	561
7,750	7,800	969	769	571
7,800	7,850	980	780	580
7,850	7,900	991	791	591
7,900	7,950	1,002	802	602
7,950	8,000	1,013	813	613
8,000	8,050	1,024	824	624
8,050	8,100	1,035	835	635
8,100	8,150	1,046	846	646
8,150	8,200	1,057	857	657
8,200	8,250	1,068	868	668
8,250	8,300	1,079	879	679
8,300	8,350	1,090	890	690
8,350	8,400	1,101	901	701
8,400	8,450	1,114	912	712
8,450	8,500	1,126	923	723
8,500	8,550	1,139	934	734
8,550	8,600	1,151	945	745
8,600	8,650	1,164	956	756
8,650	8,700	1,176	967	767
8,700	8,750	1,189	978	778
8,750	8,800	1,201	989	789
8,800	8,850	1,214	1,000	800
8,850	8,900	1,226	1,011	811
8,900	8,950	1,239	1,022	822
8,950	9,000	1,251	1,033	833
9,000	9,050	1,264	1,044	844
9,050	9,100	1,276	1,055	855
9,100	9,150	1,289	1,066	866
9,150	9,200	1,301	1,079	877
9,200	9,250	1,314	1,091	888
9,250	9,300	1,326	1,104	899
9,300	9,350	1,339	1,116	910
9,350	9,400	1,351	1,129	921
9,400	9,450	1,364	1,141	932
9,450	9,500	1,376	1,154	943
9,500	9,550	1,389	1,166	954
9,550	9,600	1,401	1,179	965
9,600	9,650	1,414	1,191	976
9,650	9,700	1,426	1,204	987
9,700	9,750	1,439	1,216	998
9,750	9,800	1,451	1,229	1,009
9,800	9,850	1,464	1,241	1,020
9,850	9,900	1,476	1,254	1,031
9,900	9,950	1,489	1,266	1,044
9,950	10,000	1,501	1,279	1,056
10,000	10,050	1,514	1,291	1,069
10,050	10,100	1,526	1,304	1,081
10,100	10,150	1,539	1,316	1,094
10,150	10,200	1,551	1,329	1,106
10,200	10,250	1,564	1,341	1,119
10,250	10,300	1,576	1,354	1,131
10,300	10,350	1,589	1,366	1,144
10,350	10,400	1,602	1,379	1,156
10,400	10,450	1,616	1,391	1,169
10,450	10,500	1,630	1,404	1,181
10,500	10,550	1,644	1,416	1,194
10,550	10,600	1,658	1,429	1,206
10,600	10,650	1,672	1,441	1,219
10,650	10,700	1,686	1,454	1,231
10,700	10,750	1,700	1,466	1,244
10,750	10,800	1,714	1,479	1,256
10,800	10,850	1,728	1,491	1,269
10,850	10,900	1,742	1,504	1,281
10,900	10,950	1,756	1,516	1,294
10,950	11,000	1,770	1,529	1,306
11,000	11,050	1,784	1,541	1,319
11,050	11,100	1,798	1,554	1,331
11,100	11,150	1,812	1,567	1,344
11,150	11,200	1,826	1,581	1,356
11,200	11,250	1,840	1,595	1,369
11,250	11,300	1,854	1,609	1,381
11,300	11,350	1,868	1,623	1,394
11,350	11,400	1,882	1,637	1,406
11,400	11,450	1,896	1,651	1,419
11,450	11,500	1,910	1,665	1,431
11,500	11,550	1,924	1,679	1,444
11,550	11,600	1,938	1,693	1,456
11,600	11,650	1,952	1,707	1,469
11,650	11,700	1,966	1,721	1,481
11,700	11,750	1,980	1,735	1,494
11,750	11,800	1,994	1,749	1,506
11,800	11,850	2,008	1,763	1,519
11,850	11,900	2,022	1,777	1,532
11,900	11,950	2,036	1,791	1,546
11,950	12,000	2,050	1,805	1,560
12,000	12,050	2,064	1,819	1,574
12,050	12,100	2,078	1,833	1,588
12,100	12,150	2,092	1,847	1,602
12,150	12,200	2,106	1,861	1,616
12,200	12,250	2,120	1,875	1,630
12,250	12,300	2,134	1,889	1,644
12,300	12,350	2,148	1,903	1,658
12,350	12,400	2,163	1,917	1,672
12,400	12,450	2,179	1,931	1,686
12,450	12,500	2,195	1,945	1,700
12,500	12,550	2,211	1,959	1,714
12,550	12,600	2,227	1,973	1,728
12,600	12,650	2,243	1,987	1,742
12,650	12,700	2,259	2,001	1,756
12,700	12,750	2,275	2,015	1,770
12,750	12,800	2,291	2,029	1,784
12,800	12,850	2,307	2,043	1,798
12,850	12,900	2,323	2,057	1,812
12,900	12,950	2,339	2,071	1,826
12,950	13,000	2,355	2,085	1,840
13,000	13,050	2,371	2,099	1,854
13,050	13,100	2,387	2,113	1,868
13,100	13,150	2,403	2,128	1,882
13,150	13,200	2,419	2,144	1,896
13,200	13,250	2,435	2,160	1,910
13,250	13,300	2,451	2,176	1,924
13,300	13,350	2,467	2,192	1,938
13,350	13,400	2,483	2,208	1,952
13,400	13,450	2,499	2,224	1,966
13,450	13,500	2,515	2,240	1,980
13,500	13,550	2,531	2,256	1,994
13,550	13,600	2,547	2,272	2,008
13,600	13,650	2,563	2,288	2,022
13,650	13,700	2,579	2,304	2,036
13,700	13,750	2,595	2,320	2,050
13,750	13,800	2,611	2,336	2,064
13,800	13,850	2,627	2,352	2,078
13,850	13,900	2,643	2,368	2,093
13,900	13,950	2,659	2,384	2,109
13,950	14,000	2,675	2,400	2,125
14,000	14,050	2,691	2,416	2,141
14,050	14,100	2,707	2,432	2,157
14,100	14,150	2,723	2,448	2,173
14,150	14,200	2,739	2,464	2,189
14,200	14,250	2,755	2,480	2,205
14,250	14,300	2,771	2,496	2,221
14,300	14,350	2,787	2,512	2,237
14,350	14,400	2,804	2,528	2,253
14,400	14,450	2,822	2,544	2,269
14,450	14,500	2,840	2,560	2,285
14,500	14,550	2,858	2,576	2,301
14,550	14,600	2,876	2,592	2,317
14,600	14,650	2,894	2,608	2,333
14,650	14,700	2,912	2,624	2,349
14,700	14,750	2,930	2,640	2,365
14,750	14,800	2,948	2,656	2,381
14,800	14,850	2,966	2,672	2,397
14,850	14,900	2,984	2,688	2,413
14,900	14,950	3,002	2,704	2,429
14,950	15,000	3,020	2,720	2,445
15,000	15,050	3,038	2,736	2,461
15,050	15,100	3,056	2,752	2,477
15,100	15,150	3,074	2,769	2,493
15,150	15,200	3,092	2,787	2,509
15,200	15,250	3,110	2,805	2,525
15,250	15,300	3,128	2,823	2,541
15,300	15,350	3,146	2,841	2,557
15,350	15,400	3,164	2,859	2,573
15,400	15,450	3,182	2,877	2,589
15,450	15,500	3,200	2,895	2,605
15,500	15,550	3,218	2,913	2,621
15,550	15,600	3,236	2,931	2,637
15,600	15,650	3,254	2,949	2,653
15,650	15,700	3,272	2,967	2,669
15,700	15,750	3,290	2,985	2,685
15,750	15,800	3,308	3,003	2,701
15,800	15,850	3,326	3,021	2,717
15,850	15,900	3,344	3,039	2,734
15,900	15,950	3,362	3,057	2,752
15,950	16,000	3,380	3,075	2,770
16,000	16,050	3,398	3,093	2,788
16,050	16,100	3,416	3,111	2,806
16,100	16,150	3,434	3,129	2,824
16,150	16,200	3,452	3,147	2,842
16,200	16,250	3,470	3,165	2,860
16,250	16,300	3,488	3,183	2,878
16,300	16,350	3,506	3,201	2,896
16,350	16,400	3,525	3,219	2,914
16,400	16,450	3,544	3,237	2,932
16,450	16,500	3,564	3,255	2,950
16,500	16,550	3,583	3,273	2,968
16,550	16,600	3,603	3,291	2,986
16,600	16,650	3,622	3,309	3,004
16,650	16,700	3,642	3,327	3,022
16,700	16,750	3,661	3,345	3,040
16,750	16,800	3,681	3,363	3,058
16,800	16,850	3,700	3,381	3,076
16,850	16,900	3,720	3,399	3,094
16,900	16,950	3,739	3,417	3,112
16,950	17,000	3,759	3,435	3,130

Table D—Heads of Household (Continued)

If Tax Table Income is— Over—	But not over—	1	2	3	4	5	6	7	8
		Your tax is—							
7,300	7,350	644	520	353	183	25	0	0	0
7,350	7,400	652	528	362	192	33	0	0	0
7,400	7,450	661	536	371	201	41	0	0	0
7,450	7,500	669	544	380	210	49	0	0	0
7,500	7,550	678	552	389	219	57	0	0	0
7,550	7,600	686	560	398	228	65	0	0	0
7,600	7,650	695	568	407	237	73	0	0	0
7,650	7,700	703	576	416	246	81	0	0	0
7,700	7,750	712	584	425	255	89	0	0	0
7,750	7,800	720	593	434	264	97	0	0	0
7,800	7,850	729	601	443	273	105	0	0	0
7,850	7,900	737	610	452	282	113	0	0	0
7,900	7,950	746	618	461	291	121	0	0	0
7,950	8,000	754	627	470	300	130	0	0	0
8,000	8,050	763	635	479	309	139	0	0	0
8,050	8,100	771	644	488	318	148	0	0	0
8,100	8,150	780	652	497	327	157	0	0	0
8,150	8,200	788	661	506	336	166	6	0	0
8,200	8,250	797	669	515	345	175	14	0	0
8,250	8,300	805	678	524	354	184	22	0	0
8,300	8,350	814	686	533	363	193	30	0	0
8,350	8,400	822	695	542	372	202	38	0	0
8,400	8,450	831	703	551	381	211	46	0	0
8,450	8,500	839	712	560	390	220	54	0	0
8,500	8,550	848	720	569	399	229	62	0	0
8,550	8,600	856	729	579	408	238	70	0	0
9,200	9,250	975	839	702	525	355	185	19	0
9,250	9,300	985	848	712	534	364	194	27	0
9,300	9,350	995	856	721	544	373	203	35	0
9,350	9,400	1,005	865	731	553	382	212	43	0
9,400	9,450	1,015	873	740	563	391	221	51	0
9,450	9,500	1,025	882	750	572	400	230	60	0
9,500	9,550	1,035	890	759	582	409	239	69	0
9,550	9,600	1,045	899	769	591	418	248	78	0
9,600	9,650	1,055	907	778	601	427	257	87	0
9,650	9,700	1,065	916	788	610	436	266	96	0
9,700	9,750	1,075	925	797	620	445	275	105	0
9,750	9,800	1,085	935	805	629	454	284	114	0
9,800	9,850	1,095	945	814	639	463	293	123	0
9,850	9,900	1,105	955	822	648	472	302	132	0
9,900	9,950	1,115	965	831	658	481	311	141	0
9,950	10,000	1,125	975	839	667	490	320	150	0
10,000	10,050	1,135	985	848	677	499	329	159	0
10,050	10,100	1,145	995	856	686	509	338	168	0
10,100	10,150	1,155	1,005	865	696	518	347	177	8
10,150	10,200	1,165	1,015	873	705	528	356	186	16
10,200	10,250	1,175	1,025	882	715	537	365	195	25
10,250	10,300	1,185	1,035	890	724	547	374	204	34
10,300	10,350	1,195	1,045	899	734	556	383	213	43
10,350	10,400	1,205	1,055	907	743	566	392	222	52
10,400	10,450	1,215	1,065	916	753	575	401	231	61
10,450	10,500	1,225	1,075	925	762	585	410	240	70
10,500	10,550	1,235	1,085	935	772	594	419	249	79
10,550	10,600	1,245	1,095	945	781	604	428	258	88
11,200	11,250	1,378	1,225	1,075	906	727	550	375	205
11,250	11,300	1,388	1,235	1,085	917	737	559	384	214
11,300	11,350	1,399	1,245	1,095	928	746	569	393	223
11,350	11,400	1,409	1,255	1,105	939	756	578	402	232
11,400	11,450	1,420	1,265	1,115	950	765	588	411	241
11,450	11,500	1,430	1,275	1,125	961	775	597	420	250
11,500	11,550	1,441	1,285	1,135	972	784	607	429	259
11,550	11,600	1,451	1,295	1,145	983	794	616	439	268

If Tax Table Income is— Over—	But not over—	1	2	3	4	5	6	7	8
		Your tax is—							
11,800	11,850	1,504	1,346	1,195	1,038	841	664	486	313
11,850	11,900	1,514	1,357	1,205	1,049	851	673	496	322
11,900	11,950	1,525	1,367	1,215	1,060	860	683	505	331
11,950	12,000	1,536	1,378	1,225	1,071	871	692	515	340
12,000	12,050	1,547	1,388	1,235	1,082	882	702	524	349
12,050	12,100	1,559	1,399	1,245	1,093	893	711	534	358
12,100	12,150	1,570	1,409	1,255	1,104	904	721	543	367
12,150	12,200	1,582	1,420	1,265	1,115	915	730	553	376
12,200	12,250	1,593	1,430	1,275	1,125	926	740	562	385
12,250	12,300	1,605	1,441	1,285	1,135	937	749	572	394
12,300	12,350	1,616	1,451	1,295	1,145	948	759	581	404
12,350	12,400	1,628	1,462	1,305	1,155	959	768	591	413
12,400	12,450	1,639	1,472	1,315	1,165	970	778	600	423
12,450	12,500	1,651	1,483	1,325	1,175	981	787	610	432
12,500	12,550	1,662	1,493	1,336	1,185	992	797	619	442
12,550	12,600	1,674	1,504	1,346	1,195	1,003	806	629	451
13,200	13,250	1,829	1,651	1,483	1,325	1,146	946	752	575
13,250	13,300	1,841	1,662	1,493	1,336	1,157	957	762	584
13,300	13,350	1,854	1,674	1,504	1,346	1,168	968	771	594
13,350	13,400	1,866	1,685	1,514	1,357	1,179	979	781	603
13,400	13,450	1,879	1,697	1,525	1,367	1,190	990	790	613
13,450	13,500	1,891	1,708	1,536	1,378	1,201	1,001	801	622
13,500	13,550	1,904	1,720	1,547	1,388	1,212	1,012	812	632
13,550	13,600	1,916	1,731	1,559	1,399	1,223	1,023	823	641
13,600	13,650	1,929	1,743	1,570	1,409	1,234	1,034	834	651
13,650	13,700	1,941	1,754	1,582	1,420	1,245	1,045	845	660
13,700	13,750	1,954	1,766	1,593	1,430	1,256	1,056	856	670
13,750	13,800	1,966	1,779	1,605	1,441	1,267	1,067	867	679
14,400	14,450	2,129	1,941	1,754	1,582	1,414	1,210	1,010	810
14,450	14,500	2,141	1,954	1,766	1,593	1,426	1,221	1,021	821
14,500	14,550	2,154	1,966	1,779	1,605	1,437	1,232	1,032	832
14,550	14,600	2,166	1,979	1,791	1,616	1,449	1,243	1,043	843
14,600	14,650	2,179	1,991	1,804	1,628	1,460	1,254	1,054	854
14,650	14,700	2,191	2,004	1,816	1,639	1,472	1,265	1,065	865
14,700	14,750	2,204	2,016	1,829	1,651	1,483	1,276	1,076	876
14,750	14,800	2,216	2,029	1,841	1,662	1,493	1,287	1,087	887
15,400	15,450	2,388	2,191	2,004	1,816	1,639	1,437	1,230	1,030
15,450	15,500	2,402	2,204	2,016	1,829	1,651	1,448	1,241	1,041
15,500	15,550	2,415	2,216	2,029	1,841	1,662	1,460	1,252	1,052
15,550	15,600	2,429	2,229	2,041	1,854	1,674	1,471	1,264	1,063
15,600	15,650	2,442	2,241	2,054	1,866	1,685	1,483	1,275	1,074
15,650	15,700	2,456	2,254	2,066	1,879	1,697	1,494	1,287	1,085
15,700	15,750	2,469	2,267	2,079	1,891	1,708	1,506	1,298	1,096
15,750	15,800	2,483	2,280	2,091	1,904	1,720	1,517	1,310	1,107
15,800	15,850	2,496	2,294	2,104	1,916	1,731	1,529	1,321	1,118
15,850	15,900	2,510	2,307	2,116	1,929	1,743	1,540	1,333	1,129
15,900	15,950	2,523	2,321	2,129	1,941	1,754	1,552	1,344	1,140
15,950	16,000	2,537	2,334	2,141	1,954	1,766	1,563	1,356	1,151
16,000	16,050	2,550	2,348	2,154	1,966	1,779	1,575	1,367	1,162
16,050	16,100	2,564	2,361	2,166	1,979	1,791	1,586	1,379	1,173
16,100	16,150	2,577	2,375	2,179	1,991	1,804	1,598	1,390	1,184
16,150	16,200	2,591	2,388	2,191	2,004	1,816	1,609	1,402	1,195
16,200	16,250	2,604	2,402	2,204	2,016	1,829	1,621	1,413	1,206
16,250	16,300	2,618	2,415	2,216	2,029	1,841	1,632	1,425	1,217
16,300	16,350	2,631	2,429	2,229	2,041	1,854	1,644	1,436	1,229
16,350	16,400	2,645	2,442	2,241	2,054	1,866	1,655	1,448	1,240
16,400	16,450	2,658	2,456	2,254	2,066	1,879	1,667	1,459	1,252
16,450	16,500	2,672	2,469	2,267	2,079	1,891	1,678	1,471	1,263
16,500	16,550	2,685	2,483	2,280	2,091	1,904	1,690	1,482	1,275
16,550	16,600	2,699	2,496	2,294	2,104	1,916	1,701	1,494	1,286

Question Who pays more tax, single people or married people?

ANSWER That depends on how the question is phrased.

Problem Al and Bill both earn $11,835. Al is single, and Bill is married. Who pays more tax?

ANSWER Since Al is single, we find his tax in Table A. If he has no other exemptions, he looks under column 1. Since his income is between 11,800 and 11,850, his tax is $1631.

    Since Bill is married, we use Table B, under column 2: married filing joint return. Their tax is $1071. So it would appear that single people do pay more tax. But read on.

Problem Amy and Bob are married, and each of them earns $12,013. Cindy and Dan are single, but live together; each of them also earns $12,013. Which couple pays more tax?

ANSWER For Amy and Bob, we add up their income 12,013 + 12,013 = 24,026. We look this up in Table B, column 2. This is between 24,000 and 24,050, so their tax is $4011.

    Cindy and Dan each look up their tax in Table A, column 1, since they are each single. Since each has an income between 12,000 and 12,050, each must pay $1679 in taxes. Together their tax is 1679 + 1679 = $3358. So Amy and Bob pay $653 more in taxes just because they are married.

Problem Rodney has a wife and five children. He earns $9840. What are their taxes?

ANSWER We use column 7 (since there are seven of them) of Table B. For $9840, we see that the tax is $0. So Rodney pays no tax.

PROBLEM SET 5–4–1

1. Define or discuss:
   (a) Tax.
   (b) Exemptions.
   (c) Deductions.
   (d) Refund.
   (e) Adjusted gross income.
   (f) Tax tables.
   (g) Tax withheld.
   (h) Balance due.

Use the tax tables (Table A–Table D) to complete the following.

	Name	Exemptions	Situation	Income ($)	Tax
2.	Mary	1	Head of household	12,347	
3.	Smiths	4	Joint return	16,169	
4.	Jack	1	Single	11,235	
5.	Rubins	3	Joint return	24,678	
6.	Rileys	8	Joint return	19,482	
7.	Sandy	3	Head of household	8,327	
8.	Tom	1	Married-separate	13,246	

Jim earns $11,960 and Kathy earns $12,360. Compute their taxes if:

9. They are both single filing as head of household.

10. They are married filing a joint return.

11. They are married filing separate returns.

12. The Roses have six exemptions and an income of $32,067. What is their tax?

13. Frank has a wife and two children, and his mother lives with him. He earns $13,211 per year. What are his taxes?

Dick and Jane have four children. Both of them earn $9432 per year. Compute their total tax bill if:

14. They file a joint return with all the children as exemptions.

15. They each claim two children on separate returns.

16. Reggie earns $14,443. Make a graph showing his tax when he is single, married, has one child, two children, three children, . . . , seven children. Put exemptions on the x-axis and tax on the y-axis.

17. Bob has a wife and two children. Make a graph showing his tax if he earns $4000, $8000, $12000, . . . , $36,000, $40,000. Put AGI on the x-axis and tax on the y-axis.

Complete the following table for Steve and Sherri's *total* tax.

	Steve ($)	Sherri ($)	Married, joint	Married, separate	Both head of household
18.	12,435	12,435			
19.	13,435	11,435			
20.	14,435	10,435			
21.	15,435	9,435			
22.	16,435	8,435			

Danny earns $14,321. He has a wife and a child.

23. What is their AGI?

24. How much tax does he pay?

25. What percent of his income is this?

26. Suppose that he earns an extra $100. What is his tax now?

27. How much extra tax does he pay?

28. What percent is the extra tax of the extra $100 pay?

Diane is single head of household and earns $15,921 a year.

29. How much tax does she pay?

30. What percent of her income is this?

31. Suppose that she earns an extra $100. What is her tax now?

32. How much extra tax is this?

33. What percent is the extra tax of the extra $100 pay?

Deductions

Question   What do we do if we have very large deductions?

ANSWER   We can compute the **itemized deduction (ID)**. This includes deductions for medical expenses, certain taxes, interest on loans, donations, certain losses, and other miscellaneous expenses. Table 5-1 gives a reasonable idea of what can and cannot be deducted. (This table does not include everything. See a tax guide for more details.) There is also a *tax credit* for child care. This is different from a deduction, so we will not discuss it here.

Question   Can we deduct every cent of the expenses listed in the "Can deduct" column of Table 5-1?

ANSWER   Yes, for taxes, interest, donations, and miscellaneous. For losses, we cannot deduct the first $100. For medicines, we cannot deduct the first 1% of our AGI. For doctor and hospital bills we cannot deduct the first 3% of our AGI.

Problem   Mr. Teeters suffers a $650 loss. How much of this can he deduct?

ANSWER   He cannot deduct the first $100. So he can deduct $650 - 100 = 550.

Problem   The Bakers have an AGI of $15,000. They spent $125 on medicines and $640 in doctor and hospital bills. How much can they deduct?

ANSWER   One percent of their AGI of $15,000 is $150. Since they only spent $125 on medicine, they cannot deduct anything for medicine. Three percent of their AGI is $(0.03) \cdot (15,000) = \$450$. They spent $640. They cannot deduct the first 450, but they can deduct $640 - 450 = \$190$.

Table 5-1   Itemized Deductions

Can deduct	Cannot deduct
Medical Expenses	
Medicine	Illegal drugs
Hospital bills	Vitamins
Doctor and dentist bills	Cosmetics
Medical supplies (eyeglasses, braces, etc.)	Health clubs
Laboratory tests	
Transportation to and from	
Taxes	
State tax	Social security tax
Local tax	Licenses (auto, dog, hunting, etc.)
Gasoline tax	Cigarette and alcohol tax
Sales tax	Federal excise tax
Real estate tax	
Property tax	
Interest on	
Personal loans	Life insurance loan
Mortgage	Another person's debt
Late taxes	Loans for tax-free bonds
Revolving charge	
Bank credit card	
Installment loans	
Contributions to	
Religious organizations	Friends and relatives
Charitable organizations	Foreign or propaganda organizations
(Salvation Army, Boy Scouts,	Civic leagues (Chamber of Commerce)
CARE, YMCA, Red Cross, and so on)	Social clubs
Nonprofit schools and hospitals	Tuition to private school
Medical research organizations	Labor unions
Losses	
Property damage by natural causes	Property damage by animals
Airplane crashes	Gradual natural deterioration
Car crashes	Accidental property loss
Theft and vandalism	
Miscellaneous	
Certain job expenses	Transportation to and from work
Schooling to improve job skills	Schooling to change jobs
Alimony	Day-care payments to relatives
Contribution to political candidates	
Union dues	

Question   How do we compute DEDUCT?

ANSWER   We are not allowed to use the entire ID.

To compute DEDUCT:

1. *Married filing joint return.* DEDUCT = ID − 3200 (unless this is negative; then DEDUCT = 0).

2. *Single or unmarried head of household.* DEDUCT = ID − 2200 (unless this is negative; then DEDUCT = 0).

3. *Married filing separate returns.* DEDUCT = ID − 1600 (unless this is negative; then DEDUCT = 0).

Example 1 Tina and John are married filing a joint return, and their ID is $4150. Therefore, their DEDUCT = ID − 3200 = 4150 − 3200 = $950.

Example 2 Susan is a single woman with an ID of $1400. Since ID − 2200 = 1400 − 2200 = − 800, DEDUCT = 0. In this case, she is probably better off not itemizing at all.

Problem Tony and Donna have an AGI of $22,000. These are their expenses: $95 for medicine, $800 for doctors and hospitals, $1400 in various taxes, $2100 in interest (mostly mortgage), $250 for charities and churches, $55 in losses, and $300 in union dues. What is their itemized deduction, ID, and their DEDUCT?

ANSWER To compute their itemized deduction (ID), we compute the medical expenses first. One percent of 22,000 is 220. Since they only spent $95 for medicine they cannot deduct the medicine expenses. Three percent of their AGI is (0.03) · (22,000) = $660. They spent $800 for doctor and hospital bills, so they can deduct 800 − 660 = $140. Since their loss was under $100 they cannot deduct any of it. Summarizing,

Medical	$ 140
Taxes	1400
Interest	2100
Contributions	250
Losses	0
Miscellaneous	300
Total ID	= $4190

Their ID = $4190. Since they are married filing a joint return, we have DEDUCT = ID − 3200 = 4190 − 3200 = $990.

Now we use the formula

Formula 5–4–1

$$\boxed{\text{Tax Table Income} = \text{AGI} - \text{DEDUCT}}$$

Problem If Tony and Donna have one child, what is their tax?

ANSWER We take AGI − DEDUCT = 22,000 − 990 = \$21,010. This is their Tax Table Income. Since they are married filing a joint return, we used Table B, Column 3. Their TAX = \$2974.

```
MATH FLASH
1 2 3 4 5 6 7 8
/  %  +/-  ÷  MC
7  8  9  X  MR
4  5  6  −  M-
1  2  3  +  M+
CE 0  .  =  C
```

Hand Calculator Instant Replay

PUNCH	DISPLAY	MEANING
C	0.	Clear
1 4 0	140.	⎫
+ 1 4 0 0	1400.	⎪
+ 2 1 0 0	2100.	⎬ Add deductions
+ 2 5 0	250.	⎪
+ 3 0 0	300.	⎭
=	4190.	ID
− 3 2 0 0	3200.	
=	990.	DEDUCT
2 2 0 0 0	22000.	AGI
− 9 9 0	990.	Minus DEDUCT
=	21010.	Tax Table Income

This is a fairly simple procedure. However, if your income is over \$20,000 (or \$40,000, if married filing joint return), you must use a different procedure which we will not discuss here.

PROBLEM SET 5-4-2

1. Define or discuss itemized deduction.

In the following cases, compute the itemized deduction (ID) and DEDUCT. (mj = married, joint return; shh = single, head of household; snhh = single, not head of household.)

	Situation	AGI	Medicine	Doctor	Taxes	Interest	Gifts	Losses	Misc.	ID	DEDUCT
2.	mj	\$23,500	\$110	\$790	\$2000	\$2900	\$350	\$130	\$275	?	?
3.	shh	14,500	55	250	600	100	150	25	150	?	?
4.	mj	9,250	250	1250	700	750	200	45	150	?	?
5.	mj	19,609	65	300	550	50	100	25	55	?	?

Situation	AGI	Medicine	Doctor	Taxes	Interest	Gifts	Losses	Misc.	ID	DEDUCT
6. snhh	17,400	45	150	400	30	30	40	100	?	?
7. shh	10,500	150	670	690	650	100	40	1200	?	?
8. mj	42,223	160	850	1800	3500	500	300	3500	?	?
9. mj	15,500	140	600	1500	1800	250	105	300	?	?
10. snhh	18,000	35	320	600	100	100	20	140	?	?

Use the above data to compute the following taxes: (Problem 11 refers back to Problem 2, Problem 12 to Problem 3, and so on).

	EXEM	Tax Table Income	TAX	TW	BALANCE or REFUND?
11.	4	?	?	$3,500	?
12.	1	?	?	1,900	?
13.	2	?	?	1,035	?
14.	3	?	?	2,400	?
15.	1	?	?	1,020	?
16.	2	?	?	1,200	?
17.	5	?	?	10,000	?
18.	4	?	?	1,400	?
19.	1	?	?	2,100	?

Marginal Tax Rate

Problem John and Mary Pappas have an AGI of $15,910. What is their tax?

ANSWER We look this up in Tax Table B, column 2. Their TAX is $1910.

Problem Suppose that their AGI goes *up* by $100. Now what is the TAX? How much more tax is this?

ANSWER Since AGI is now $16,010, their tax is $1932. Notice that this is $22 more.

Problem Suppose that their AGI goes *down* by $100. Now what is their TAX? How much less tax is this?

ANSWER Now their AGI is $15,810, and their TAX is $1888. Notice that this is $22 less tax than it was originally.

In these cases we were looking at what happened to *extra* or *less* income. We saw that if the AGI went *up* $100, TAX went *up* $22; and when the AGI went *down* $100, TAX went *down* $22. In both cases this is a 22% change.

This percentage change is called the **marginal tax rate**.

> To compute the marginal tax rate:
>
> 1. Compute TAX for the given AGI.
>
> 2. Add $100 to AGI.
>
> 3. Compute TAX for this new AGI. The *extra* TAX is the marginal tax rate as a percent.

Problem   Tim and Marybeth Deaner have an AGI of $25,000. What is their marginal tax rate?

ANSWER   The TAX on $25,000 is $4288. If they earned an extra $100, their AGI would be $25,100, and their TAX, $4320. This is $32 more, or 32% more. So, their marginal tax rate is 32%.

Question   What does this 32% mean?

ANSWER   This means that for every *extra* $1 that they earn they will pay an *extra* 32 cents in tax. Also, if they can find another $1 to deduct ($1 *less* AGI), then they will pay 32 cents *less* tax.

Sometimes, people will say this is the 32% **tax bracket**. This marginal rate tells what the tax rate is doing to the top of your income. (We will use both terms: *tax bracket* and *marginal rate*.)

Problem   Bonnie is in the 22% tax bracket. Suppose that she donates $150 to charity. How much tax does she save? How much of the donation actually comes out of her pocket?

ANSWER   The donation has reduced her AGI by $150. Since she is in the 22% bracket, this will reduce her tax by 22% of 150 = $33. To see how much she actually donated out of her own pocket, we have

$$
\begin{array}{ll}
\text{Original donation} = & \$150 \\
\text{Reduction in tax} = & -33 \\
\hline
\text{Her donation} = & \$117
\end{array}
$$

Question   But the charity got $150. If only $117 came out of Bonnie's pocket, where did the other $33 come from?

ANSWER   The U.S. government donated the other $33.

Problem Bert is in the 25% tax bracket. He is considering a part-time second job to make some extra money. He figures he can earn an extra $225 a month on this second job. How much of this extra money will he actually get?

ANSWER The extra job will increase his AGI by 225 a month. Since he is in the 25% tax bracket, the extra tax will be 25% of 225 every month. This is an extra $56.25 in taxes every month. So

$$\text{Extra income per month} = \$225.00$$
$$\text{Extra taxes per month} = -56.25$$
$$\overline{\text{Net extra pay per month} = \$168.75}$$

Another way to figure this is to say that, if his tax rate is 25%, then he gets to *keep* 75% of every extra dollar. Sure enough, 75% of 225 = $168.25.

Problem Ted and Linda are in the 22% tax bracket. They buy an $800 TV/stereo on time at 10% add-on interest. How much interest do they pay in a year? How much will this lower their taxes? What is the *after-tax* rate of interest?

ANSWER Their interest is 10% of 800, which is $80. Since interest is tax deductible, this lowers their AGI by 80. Since they are in a 22% tax bracket, this lowers their tax by 22% of 80, which is $17.60.

So, of the $80 interest, Ted and Linda pay $62.40, and the U.S. government pays the other $17.60. Then it really only cost them $62.40 to borrow $800. This is $\frac{62.40}{800} = 0.078 = 7.8\%$. Thanks to tax deductions, they really paid 7.8% add-on interest.

Suppose that a person is in the *T*% tax bracket. If he or she earns more money, they pay *T*% of it in taxes and keep the other (100 − *T*)%. If they deduct some money from their AGI, they lower their taxes by *T*%, and then they must pay (100 − *T*)% of the deduction. The number (100 − *T*)% is called the **after-tax rate** because this is what is left of the extra money or the deduction after taxes.

Problem The Hendersons are in the 32% tax bracket. What is their after-tax bracket?

ANSWER (100 − 32)% = 68%.

Question What does this 68% mean?

ANSWER Suppose that the Hendersons earn an extra $200. Then they can keep 68% of it, or $136, after taxes. If the Hendersons pay an extra $300 in interest, then they really pay 68% of it, or $204, after taxes.

PROBLEM SET 5–4–3
1. Define or discuss:
   (a) Marginal rate.
   (b) After-tax rate.

Find the tax bracket and after-tax rate for the following cases. (Note: hh = head of household.)

	AGI	Situation	Tax bracket	After-tax rate
2.	$ 9,456	Married, joint		
3.	24,111	Married, joint		
4.	7,604	Single, hh		
5.	39,772	Married, joint		
6.	13,567	Single, hh		
7.	19,087	Married, joint		
8.	27,915	Married, joint		
9.	16,900	Married, separate		
10.	13,450	Married, joint		

11. Marty and Sue are in the 25% tax bracket. Marty gets a second job and earns an extra $1500. How much extra tax does he pay?

12. Greg buys a $5000 car. He puts $1000 down and finances the remaining $4000 at 7% interest. How much interest will he pay? How much will his interest deduction save him in taxes if he is in the 28% tax bracket?

13. Frank and Karen are in the 32% bracket. They donate $500 to their church. After taxes, how much of the donation came out of their pockets?

14. Elton Throat is a famous rock music star. He is in the 70% tax bracket. He is offered $60,000 to do a concert. After taxes how much of this will he get?

15. Fred and Judy Smith are in the 42% tax bracket. They have just bought a $100,000 home. Every month they pay about $950 just in interest and taxes on the house. After taxes, how much of this do they pay, and how much does the U.S. government pay?

16. Peter put $1500 into a bank account that earns 7% interest. How much interest does he earn in a year? If Peter's AGI is $17,890 and he is single, what is his tax bracket? How much interest does he get after taxes? How much has the $1500 really grown to?

17. In Problem 16, suppose that inflation is 8% a year. How much is the money he has this year after taxes worth in terms of last year's dollars? How much has he gained or lost from the original $1500?

18. Jerry and Tammy had a $350 loss that was not covered by insurance. How much of this will they be allowed to deduct? If they are in the 22% tax bracket, how much will they get back after taxes?

19. Judy has a garage and attic full of junk. She donates it to the Salvation Army and assesses the worth at $200. If she and her husband are in the 28% tax bracket, how much money has she "earned" just by cleaning out her garage and attic?

20. Stan is in the 36% tax bracket. He has $3000 to invest. He can get 8% interest with a certain stock, but he will lose some of it to taxes. Or he can get $5\frac{1}{2}\%$ tax-free interest with a certain municipal bond. Which is a better deal?

21. Ron borrows $4500 at 15% a year to buy a car. How much interest does he pay in the first year? If Ron is in the 25% tax bracket, how much interest does he pay after taxes? What is his after-taxes rate of interest?

5-5 AUTOMOBILE OWNERSHIP

Question What is the best way to beat the high cost of owning an automobile?

ANSWER Sell your car; then walk, mooch rides, or use public transportation.

Unfortunately, this solution isn't too practical for most people. The next best thing is to understand the costs of owning a car.

Operating Costs

Question What are the types of costs of operating a car?

ANSWER Fixed and variable expenses. **Fixed expenses** are the costs for the car that you pay whether or not you ever drive the car at all. These include licenses, interest, insurance, depreciation, and the like. **Variable expenses** are the costs that depend on how much you drive the car. These include gas, oil, tires, repairs, and the like.

Problem Fred figures his fixed auto expenses are $1200 per year and his variable expenses are 5 cents per mile. How much does it cost Fred to operate his car?

ANSWER Fred's total cost depends on how much he drives. Suppose that he drives the car m miles every year. Then his variable expenses are $(0.05)m$. For example, if $m = 10,000$, then the variable cost is $(0.05) \cdot (10,000) = \$500$. His fixed cost is $1200 no matter what m is. So we get

$$\text{Fixed costs} = \$1200$$
$$\underline{\text{Variable costs} = (0.05)m}$$
$$C = \text{total cost} = \$1200 + (0.05)m$$

Recall from Chapter 4 that the equation $C = 1200 + 0.05m$ is the equation of

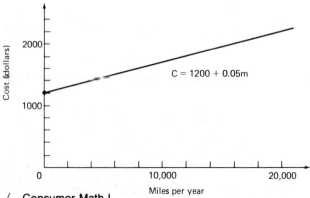

$C = 1200 + 0.05m$

a straight line, where 0.05 is the slope and 1200 is the *y*-intercept. If we graph this, we get the graph shown.

Depreciation

Question   What is the most invisible cost of owning a car?

ANSWER   Depreciation.

Question   What is the most expensive cost of owning a car?

ANSWER   Again the answer is depreciation. It is the most invisible and costly. You never write a check or pay cash for it. But it is there. If you buy a $5000 car and sell it 3 years later for $2100, you have lost $2900 somewhere. You never saw it go. This loss of value is called **depreciation**.

Depreciation can be measured in terms of the percent of the car's value that is lost every year. For instance, the average car depreciates about 25% of its value each year. This does not mean 25% of its original value, but 25% of its current value.

Problem   Linda buys a $4000 car that depreciates 25% per year. How much does it depreciate each year?

I like you, and I'm going to make you a deal. (Burk Uzzle/© Magnum Photos)

ANSWER   The first year it loses 25% of its original 4000, or $1000. The car is now worth
4000 − 1000 = $3000. The second year the car loses 25% of the 3000, which is
$750. Now the car is worth 3000 − 750 = $2250. The third year the car loses 25%
of the 2250, which is $562.50. The car is now worth $1687.50. So, in these 3 years
Linda has lost $2312.50 in depreciation.

Depreciation is a lot like discount. Remember that a 25% discount can be
computed as 75% paid. We can do the same thing with depreciation. If a car's
value is V, then after a year it has depreciated 25% of V. So its new value is 75% of
V. We can write this new value as $(0.75)V$. After a second year, it is down to 75%
of that, or $(0.75)(0.75)V$. After a third year it is down to 75% of that, or
$(0.75)(0.75)(0.75)V = (0.75)^3V$.

In general, after n years the car's value is given by

Formula 5–5–1

$$\boxed{\textbf{Value after } n \textbf{ years} = (\mathbf{0.75})^n V}$$

where V is the original value.

We saw how to compute $(0.75)^n$ in Chapter 1. Table 5-2 is a table of values for
$(0.75)^n$ for 10 years.

Table 5–2

n (year)	$(0.75)^n = \%$	Percent of loss off original
0	1.00 = 100	—
1	0.75 = 75	25
2	0.56 = 56	19
3	0.42 = 42	14
4	0.32 = 32	10
5	0.24 = 24	8
6	0.18 = 18	6
7	0.13 = 13	5
8	0.10 = 10	3
9	0.08 = 8	2
10	0.06 = 6	2

The percent of loss column tells what percent of the value was lost that year.
For instance, in year 5, the car lost 8% of its original value. The middle column
tells the percent of the original value that is left. For example, after year 5, the car
is only worth 24% of its original value. (If you use a hand calculator with Formula
5–5–1, the answer will usually be slightly different because the values in Table 5–2
are only approximate.)

Problem   Barry and Kathy buy a $4500 car. Assuming 25% depreciation per year, what will
the car be worth in 5 years?

ANSWER   According to Formula 5–5–1, the value after 5 years is $(0.75)^5(4500)$. Here $V = 4500$, the original worth of the car. Now we use Table 5–2 to find $(0.75)^5 = 0.24$. So the car is now worth $(0.24) \cdot (4500) = \$1080$.

Hand Calculator Instant Replay

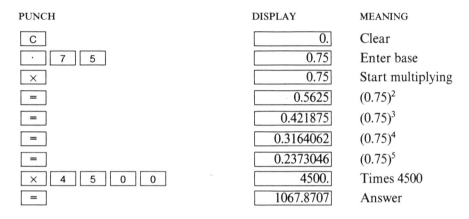

PUNCH	DISPLAY	MEANING
C	0.	Clear
. 7 5	0.75	Enter base
×	0.75	Start multiplying
=	0.5625	$(0.75)^2$
=	0.421875	$(0.75)^3$
=	0.3164062	$(0.75)^4$
=	0.2373046	$(0.75)^5$
× 4 5 0 0	4500.	Times 4500
=	1067.8707	Answer

Problem   Ed buys a 6-year-old car for $750. Assuming a 25% depreciation, what was the car worth when it was new?

ANSWER   From Table 5–2, a 6-year-old car is worth only 18% of its original value. To find the original value, we set this up as a percentage problem.

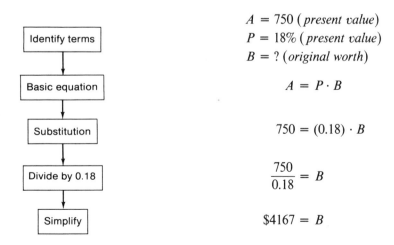

$A = 750$ (*present value*)
$P = 18\%$ (*present value*)
$B = ?$ (*original worth*)

$$A = P \cdot B$$

$$750 = (0.18) \cdot B$$

$$\frac{750}{0.18} = B$$

$$\$4167 = B$$

The original value of the car was about $4167.

Problem   Al buys a $7000 car. Assuming 25% depreciation, how much money does Al lose the first year? the second year? in each of the next eight years?

ANSWER   To compute loss, we use the percent of loss column of Table 5–2. In year 1, he loses 25% of 7000, which is $1750. In year 2, he loses 19% of the original 7000, which is $1330. To find the other years' losses, we multiply the number in the percent of loss column by the original 7000. This is given below.

Year	1	2	3	4	5	6	7	8	9	10
Loss	1750	1330	980	700	560	420	350	210	140	140

**PROBLEM
SET 5–5–1**

1. Define or discuss:
   (a) Fixed costs.
   (b) Variable costs.
   (c) Depreciation.

Mike figures his fixed auto expenses as $1500 per year; his variable expenses are 6 cents per mile. (This is a new car.)

2. Write an equation for his total car costs, C, in terms of m, the number of miles he drives per year.

3. Graph this equation.

Ron figures his fixed expenses are $1000 per year; his variable expenses are 8 cents per mile. (This is a used car.)

4. Why do you think Ron's fixed costs for a used car are lower than Mike's for a new car?

5. Why do you think Ron's variable costs for a used car are higher than Mike's for a new car?

6. Write an equation for Ron's total cost, C, in terms of m, the total number of miles he drives per year.

7. Graph this equation on the graph of Problem 3.

8. Where do these lines intersect?

9. What does the intersection point mean?

10. Ms. Folie buys a $12,000 car. Assuming 25% depreciation per year, what is her car worth in 1 year? in 2 years? in 5 years?

11. Sid buys a $5000 car. With 25% depreciation per year, how much does he lose the first year? the second year? the fifth year? the tenth year?

The Petersons buy a $6000 car. Assuming 25% per year depreciation, complete the following table.

	Year	Worth of car	Loss that year
	0	$6000	$ —
	1	4500	1500
12.	2		
13.	3		
14.	4		
15.	5		
16.	6		
17.	7		
18.	8		
19.	9		
20.	10		

21. John buys a 4-year-old car that has been depreciating at 25% per year. He pays $1600. What was the original price?

Harvey buys a $4000 subcompact. This car only depreciates at 20% per year.

22. What is the car worth after 1 year?

23. What is the car worth after 2 years?

24. What is the car worth after 10 years? (Use your calculator.)

Suppose that a car depreciates at $D\%$ a year, and it is originally worth V dollars.

25. Give an expression for what the car is worth after 1 year; after 2 years; after n years.

26. Give an expression for how much money is lost in year n.

Other Fixed Auto Expenses

Question What are the other fixed expenses of a car?

ANSWER Basically, they are taxes, licenses, insurance, and interest on the car loan.

Problem Roger buys a $4000 car. He puts $1000 down and finances the rest. What might his fixed expenses be?

ANSWER If the car depreciates at 25% per year, we know how much depreciation he will lose. If the interest is 7% add-on, we can compute the interest. His insurance will depend on his age, driving record, town, sex, and other factors. We can approximate the insurance and taxes. Table 5–3 (page 310) lists Roger's fixed expenses.

The first year, Roger is paying $1560 before he drives 1 mile! Notice the big costs are depreciation and interest, which are either gone or small by the fourth year. But this is the time that most people trade their car in and start over at year 1. Clearly, one good way to lower your fixed costs is to buy a used car (low depreciation) and pay cash (no interest).

Table 5-3

Expense	Year 1	2	3	4	5
Depreciation (25% loss)	$1000	$ 760	$ 560	$400	$320
Interest (7% of 3000)	210	210	210	0	0
Taxes, license	50	50	50	50	50
Insurance	300	300	300	300	300
Total fixed costs	$1560	$1320	$1120	$750	$670

Variable Costs

Question What are the variable costs in operating a car?

ANSWER Essentially, these are gas, oil, tires, and repairs. These expenses are roughly proportional to how much we drive. The more we drive, the more gas we use, and the sooner we have to change oil and tires and make repairs.

Problem Suppose that gas is 70 cents per gallon, and Carol gets 18 miles per gallon. What are her gasoline costs?

ANSWER Let GC be **gasoline costs**. This is in units of cents per mile. Using the units method of Chapter 2, we get

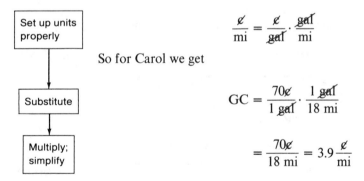

$$\frac{\cancel{c}}{\text{mi}} = \frac{\cancel{c}}{\cancel{\text{gal}}} \cdot \frac{\cancel{\text{gal}}}{\text{mi}}$$

So for Carol we get

$$GC = \frac{70\cancel{c}}{1 \cancel{\text{gal}}} \cdot \frac{1 \cancel{\text{gal}}}{18 \text{ mi}}$$

$$= \frac{70\cancel{c}}{18 \text{ mi}} = 3.9 \frac{\cancel{c}}{\text{mi}}$$

Problem Ken gets an $8 oil change and lubrication every 3000 miles. He also adds $1 worth of oil between changes. What are his oil costs?

ANSWER Let OC be the **oil costs**. Every 3000 miles he spends $8 + $1 = $9 = 900 cents. So we have

$$OC = \frac{900\cancel{c}}{3000 \text{ mi}} = 0.3 \frac{\cancel{c}}{\text{mi}}$$

Problem Every 30,000 miles, Eric has to buy 4 tires at $200 for the set. What are his tire costs?

Let TC be the **tire costs**. Then it costs him $200 or 20,000 cents every 30,000 miles:

$$TC = \frac{20,000\cancel{c}}{30,000 \text{ mi}} = 0.67\,\frac{\cancel{c}}{\text{mi}}$$

This expense should not occur until about the third year or 20,000 miles, since the car has its original tires until then.

Notice that *all the variable costs are computed in units of cents per mile.* The last variable cost is repairs. This is much trickier than gas, oil, and tires, which have to be bought at fairly predictable times. But no one knows when the transmission or the shocks will go out. One car may be a lemon, and another might be a gem. Let RC be the **repair costs**. Based on data that have been accumulated, an *approximate* formula might be

Formula 5–5–2

$$\boxed{RC = (0.6) \cdot n\!\left(\frac{\cancel{c}}{\text{mi}}\right)}$$

where *n* is the car's age in years.

Problem Alice buys a 7-year-old car. What sort of repair costs might she expect that year?

(Courtesy of General Motors)

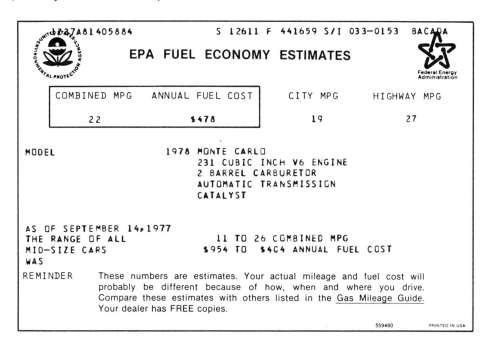

ANSWER We can use Formula 5–5–2 with $n = 7$, since this is the age of the car:

$$RC = (0.6) \cdot (7) = 4.2 \frac{\cancel{c}}{\text{mi}}$$

Problem How much is this if Alice drives 8000 miles a year?

ANSWER At 4.2 cents per mile, the cost is

$$8000 \cancel{\text{mi}} \cdot \frac{4.2\cancel{c}}{\cancel{\text{mi}}} = 33,600\cancel{c} = \$336$$

Problem If Roger drives his car 10,000 miles a year for 10 years, what might his expenses be in each of the 10 years?

ANSWER We already have the fixed costs in Table 5–3. The variable costs are

Gas: $(\frac{3.9\cancel{c}}{\cancel{\text{mi}}}) \cdot (10,000 \cancel{\text{mi}}) = 39,000\cancel{c} = \$390$

Oil: $(\frac{0.3\cancel{c}}{\cancel{\text{mi}}}) \cdot (10,000 \cancel{\text{mi}}) = 3000\cancel{c} = \$30$

Tires: $(\frac{0.67\cancel{c}}{\cancel{\text{mi}}}) \cdot (10,000 \cancel{\text{mi}}) = 6700\cancel{c} = \$67$

Repairs: $(\frac{0.6n\cancel{c}}{\cancel{\text{mi}}}) \cdot (10,000 \cancel{\text{mi}}) = 6000n\cancel{c} = \$60n$

Notice that the repair costs will increase about $60 a year. Total costs are shown in Table 5–4.

Table 5–4

	Fixed Costs				Variable Costs				
Year	Interest	Insur.	Taxes	Deprec.	Tires	Gas	Oil	Repairs	Total
1	$210	$300	$50	$1000	$ 0	$390	$30	$ 60	$2040
2	210	300	50	760	0	390	30	120	1860
3	210	300	50	560	67	390	30	180	1787
4	0	300	50	400	67	390	30	240	1477
5	0	300	50	320	67	390	30	300	1457
6	0	300	50	240	67	390	30	360	1437
7	0	300	50	200	67	390	30	420	1457
8	0	300	50	120	67	390	30	480	1437
9	0	300	50	80	67	390	30	540	1457
10	0	300	50	80	67	390	30	600	1517

Problem What is the cost per mile if he keeps the car for 3 years? for 5 years? for 10 years?

ANSWER Using Table 5–4, the cost for the first 3 years is $2040 + 1860 + 1787 = 5687$. Since he drives the car 30,000 miles in this time, his total cost per mile is 568,700 cents per 30,000 miles = 19.0 cents per mile.

For 5 years, his cost is 2040 + 1860 + 1787 + 1477 + 1457 = 8621. Since he drove 50,000 miles, this is 862,100 cents per 50,000 miles = 17.2 cents per mile.

For 10 years, his cost per mile is 1,592,600 cents per 100,000 = 15.9 cents per mile. We can see that the cheapest years to own a car are between the fifth and ninth when the interest is gone and the depreciation is small.

PROBLEM SET 5-5-2

1. Define or discuss:
   (a) Gasoline costs.
   (b) Oil costs.
   (c) Tire costs.
   (d) Repair costs.

Compute the following gas costs.

	Gas price ($\frac{\cancel{c}}{gal}$)	Gas mileage ($\frac{mi}{gal}$)	GC ($\frac{\cancel{c}}{mi}$)
2.	65	24	
3.	75	15	
4.	70	21	
5.	73	32	
6.	82	26	

Lee fills his tank at 30,257. He refills it at 30,505, and it takes 16.2 gallons to refill the tank.

7. How many miles per gallon does Lee get?

8. At 75 cents per gallon, what is Lee's GC?

Let W be the weight of a car in pounds. *Motor Trend* estimates that the gas mileage of a car with automatic transmission can be approximated by $\frac{56,000}{W}$ (miles per gallon). If gas costs 70 cents per gallon, find the following GC's.

	Weight	Gas mileage ($\frac{mi}{gal}$)	GC ($\frac{\cancel{c}}{mi}$)
9.	1800		
10.	2300		
11.	3000		
12.	4000		
13.	4700		

Use Formula 5–5–2 to complete the following table.

	Age of car	Miles per year	RC (repair costs)
14.	1	10,000	
15.	5	20,000	
16.	8	5,000	
17.	6	14,000	
18.	2	25,000	
19.	4	7,500	
20.	3	12,500	

21. Howie's car burns a lot of oil. He uses 1 quart of oil every 200 miles. Every 3000 miles, he drains the oil, and adds 4 quarts. The oil he uses costs 70 cents per quart. Compute his OC.

Danny is in the market for tires. He is confused by the different grades, prices, and lifetimes of the different tires. Complete the following table.

	Grade of tire	Wear-out period (mi)	Cost ($)	TC ($\frac{\cancel{c}}{mi}$)
22.	Radials	40,000	250	
23.	Four-ply belted	30,000	160	
24.	Two-ply belted	25,000	120	
25.	Bias	20,000	95	
26.	Retreads	12,000	55	

Dan decides to buy a set of 4 radials for $250 with a 40,000-mile guarantee (prorated with wear.)

27. How much does each tire cost?

28. Suppose that Dan has a blowout after 16,000 miles. What percent of the tire's life was used?

29. What percent of the tire's life was left?

30. How much will be credited toward Dan's replacement tire? (Use Problems 27 and 29.)

31. Complete tables similar to Table 5–4 if Roger drives 5000 miles per year and 20,000 miles per year.

32. In Problem 31, compute the costs per mile after 1 year, 3 years, 7 years, and 10 years.

Car Buying

Question Can mathematics help in buying a car?

ANSWER As long as Americans buy cars with the attitude, "If my friends could see me now ...," it might be difficult. Mathematics is very factual, and buying a car seems to be very emotional for a lot of Americans. But we will try anyway.

Question Can we figure out the dealer's cost for a new car?

ANSWER An *approximate* rule of thumb can be used. On the window is the list price, L, and the shipping and dealer preparation charge, S. To get the dealer's cost, D, we subtract the shipping charge from the list price. This is $L - S$. We now take 80% of this, or $(0.80) \cdot (L - S)$. Now we add the shipping charge back on. Finally, we get

Formula 5–5–3

$$D = (0.80)(L - S) + S$$

where D = dealer's cost, L = list price (the bottom line), and S = shipping charge.

Problem Donna sees a car that she likes. The list price is $4975, and the shipping and dealer preparation charge is $135. What is the approximate dealer's cost?

ANSWER Here we have $L = 4975$ and $S = 135$. Using Formula 5–5–3,

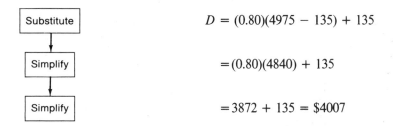

$$D = (0.80)(4975 - 135) + 135$$

$$= (0.80)(4840) + 135$$

$$= 3872 + 135 = \$4007$$

The dealer's cost is about $4000. The 80% figure is just an average figure. For smaller cars, it might be 85%, and for bigger cars it might be 75%.

Question What should Donna offer the salesman?

ANSWER This would depend on how hungry the salesman was. Ten percent over cost might be reasonable. This is 110% of the dealer's cost, which is $(1.10) \cdot (4000) = \$4400$. Donna might offer about $4400.

Question Donna has a trade-in worth $900. The salesman offers her $1100 for it. Is this for real?

ANSWER   This is called a "highball," an exaggerated quote for the trade-in to fool Donna. She should pay $4400 less a $900 trade-in, which is 4400 − 900 = $3500.

No doubt the salesman will come down no lower than $4600. And with the phony $1100 trade-in, she pays 4600 − 1100 = $3500, the same price.

Question   Suppose that the salesman quotes Donna a price of $4150. Is this a good offer?

ANSWER   Yes, but Donna will never get it. This is a "low-ball," a low price used to trick Donna into committing herself to buying the car. She'll end up paying $4400 or more.

Problem   Candy wants to buy a car. She is trying to choose between a $4000 new car and a $2500, 3-year-old used car. Her father tells her that the used car will "nickel and dime" her to death. Does this make sense?

ANSWER   To answer this we look at the different costs that actually depend on the age of the car: depreciation, repairs, and tires. We will also consider interest on a loan. (Assume that taxes, insurance, gas, and oil are always about the same. We won't consider them in order to make the computation easier.)

First, let's look at the $4000 car. If Candy puts $1000 down, she will probably pay 6% add-on interest on the $3000 balance. This is $180 a year for 3 years. If Candy drives about 10,000 miles per year, her repairs will be about $40 the first year, $80 the second, and so on. Also, her tire costs won't begin until the third year. Her big expense will be depreciation.

Now let's look at the used car. If she puts the same $1000 down, Candy will pay 6% add-on interest on $1500, which is $90 a year. Since the car is entering its fourth year, the repairs will be $160 the first year, $200 the next year, and so on.

Both cars depreciate at 25% a year. Table 5–5 summarizes the costs. (Remember that these figures do not include taxes, insurance, gas, and oil.)

Table 5–5

Year	Depreciation	Tires	Repairs	Interest	Total
New Car ($4000) (without gas, tires, and so forth)					
1	$1000	$0	$60	$180	$1240
2	760	0	120	180	1060
3	560	67	180	180	987
4	440	67	240	0	747
5	320	67	300	0	687

Year	Depreciation	Tires	Repairs	Interest	Total	Saving with used car
Used Car ($2500) (without gas, tires, and so forth)						
1	$625	$67	$240	$90	$1022	$ + 218
2	475	67	300	90	932	+ 128
3	350	67	360	90	867	+ 120
4	275	67	420	0	762	− 15
5	200	67	480	0	747	− 60

Question   What is the conclusion?

ANSWER   For the first 3 years, the used car is cheaper because of the high depreciation on the new car. After that, the used car does start to "nickel and dime" her, and the newer car is cheaper.

PROBLEM SET 5–5–3   Use Formula 5–5–3 to compute the following dealer's costs and a price which is 10% over the cost.

	L = list price	S = shipping charge	D = dealer's cost	10% over cost
1.	$5200	$300		
2.	4345	235		
3.	6125	275		
4.	3995	185		
5.	6755	305		

Paul and Rhonna want to buy a big luxury car, which lists for $8200. The shipping and dealer's preparation charge is $295.

6. What is the dealer's cost? (Use 75% instead of 80% in Formula 5–5–3.)

7. If they pay 10% over cost, what should they expect to pay?

8. What is the total bill with a 5% sales tax and $50 tax and license fee?

9. If they put $2000 down and finance the rest at 6% add-on interest for 3 years, what are their monthly payments?

Dave and Nancy are looking at a $3860 subcompact. The shipping and dealer preparation charge is $155.

10. What is the dealer's cost? (Use 85% instead of 80% in Formula 5–5–3.)

11. What is 10% over cost?

12. If they have a trade-in worth $750, what can they expect to pay for the car?

13. Suppose that it is company policy not to give a discount off the list price. How much of a trade-in should Dave and Nancy get so that the actual price is still the same as in Problem 12?

Fred and Linda want to buy a camper van very badly. The list price is $6150, of which $275 is shipping and preparation.

14. What is the dealer's cost? (Use Formula 5–5–3.)

15. The salesman quotes them $5000 for the van. Does this seem like a legitimate quote? Why or why not?

Study the following itemized pricing.

Base price	$3430
Interior package	145
Auto. transmission	279
Air conditioning	381
AM/FM–tape player	126
Power brakes	88
Exterior package	155
Radial tires	106
Subtotal	= $4710
Shipping and preparation	145
Total price	= $4855

16. The final price is what percent of the base price?

17. How much do all the options add up to?

18. By what percent have the options inflated the base price?

19. What is the dealer's cost by Formula 5–5–3?

20. What is a price 10% over the dealer's cost?

IMPORTANT WORDS

add-on interest (5–2)
adjusted gross income (5–4)
after-tax rate (5–4)
annual percentage rate (APR) (5–2)
balance due (5–4)
consumer price index (CPI) (5–3)
daily calorie needs (5–1)
deductions (5–4)
depreciation (5–5)
discounted interest (5–2)
exemptions (5–4)
fixed expenses (5–5)
gasoline costs (5–5)
ideal weight (5–1)

interest (5–2)
itemized deduction (5–4)
marginal tax rate (5–4)
oil costs (5–5)
principal (5–2)
protein per dollar (5–1)
refund (5–4)
repair costs (5–5)
simple interest (5–2)
tax (5–4)
tax tables (5–4)
tax withheld (5–4)
tire costs (5–5)
variable expenses (5–5)

REVIEW EXERCISES

1. Which is a better bargain:
  (a) A cereal at 99 cents for 20 ounces, or 79 cents for 16 ounces?
  (b) A frozen pizza at $1.49 for 14 ounces, or $2.19 for 23 ounces?

2. Approximate the cost of packaging and advertising corn chips.

Size (oz.)	1	4	8	16
Cost(¢)	20	50	75	115

3. Hot dogs provide 14 grams of protein per 100 grams of hot dogs. One pound of hot dogs costs $1.29. Find the protein per dollar (PPD) of this package.

4. Mark is 5 feet, 10 inches tall, has a medium frame, and strong musculature. What is his ideal weight?

5. Carol is 26 years old. Her ideal weight is 122 pounds, and she is slightly inactive (-5%). Determine her daily calorie needs.

6. Playing tennis works off about 400 calories per hour. Kay plays 3 hours of tennis a week. How many calories is this per year? How much weight can she lose (or not gain) in one year by playing tennis?

7. The Kilgores borrowed $2500 for 2 years at 10% interest. How much interest do they pay?

8. Al borrowed $200 and 2 months later he paid back $203. What was the interest rate?

Ben and Amy buy a car for $5000. They put $1000 down and agree to pay $140 a month for the next 3 years. *4000*

9. How much are they financing? *4000*

10. How much will they pay over the next 3 years? *5040*

11. How much interest are they paying? *1040*

12. What is the APR? *16.7*

13. Jack is making $20,000 in a year when the CPI is 175. How much is this income in terms of a year when the CPI is 120?

In 1975 the CPI was 164; in 1976 it was 175. In 1975, Betty earned $12,500. In 1976 she got a $6\frac{1}{2}\%$ raise.

14. How much did she earn in 1976?

15. How much was this in 1975 dollars?

16. By what percent did her purchasing power go up (or down) in 1976?

Ken earns $10,260 a year and Pat earns $10,330. Use the tax tables.

17. Compute their total taxes if they are both filing as single heads of household. (Table D.)

18. Compute their total taxes if they are married, filing a joint return. (Table B.)

Use the tax tables to compute the tax in the following situations.

19. The Hadleys have five exemptions, $3950 for an ID, and an AGI of $13,120. They file a joint return.

20. Mary Lou has an AGI of $16,400. She is single, head of household, with only herself as an exemption. Her itemized deductions are $2100.

The Carlsons file a joint return. They have four exemptions and an AGI of $26,920. Their deductible expenses were medicine, $150; doctor, $838; taxes, $2659; interest, $2859; donations, $320; losses, $250; miscellaneous, $130.

21. Compute the itemized deduction.

22. Compute DEDUCT.

23. What is their tax table income?

24. Compute their tax.

25. If they had $4300 withheld from their paychecks, what is their refund or balance due?

Fred and Marsha have an AGI of $20,355. They file a joint return.

26. What is their tax bracket?

27. What is their after-tax rate?

28. If Marsha gets a $200 bonus, how much will they have left after taxes?

29. If they donate $150 worth of old clothes to the Salvation Army, how much will they get back on their taxes?

Ralph's fixed automobile expenses are $1100 per year. His variable costs are 5.5 cents per mile.

30. Write an equation for his total costs.

31. What are his annual expenses if he drives 5000 miles per year? 10,000 miles? 25,000 miles?

The Sternbergs buy a $7500 car that depreciates at 25% a year. Complete the following table.

Year	Value of car	Loss that year
0	$7500	—
32. 1		
33. 2		
34. 3		

Maria fills her car's gas tank at 39,815 miles. She refills it at 40,103 miles, and it takes 23.6 gallons of gas.

35. How many miles per gallon did she get?

36. If gas is 72.9 cents per gallon, what are her gas costs in cents per mile?

37. George's car is 7 years old. He drives it about 8000 miles per year. What are his approximate repair costs?

Toni buys a set of 4 tires with a 30,000-mile guarantee for $220.

38. What are Toni's tire costs in cents per mile?

39. Suppose that one of the tires has a blow-out after 17,000 miles. How much will she be credited toward a new tire?

Marty is interested in a car with a list price of $4500. The shipping costs are $221.

40. Find the dealer's cost using Formula 5–5–3.

41. What is 10% over the dealer's cost?

The probability of a pregnancy resulting in twins is about one in ninety.

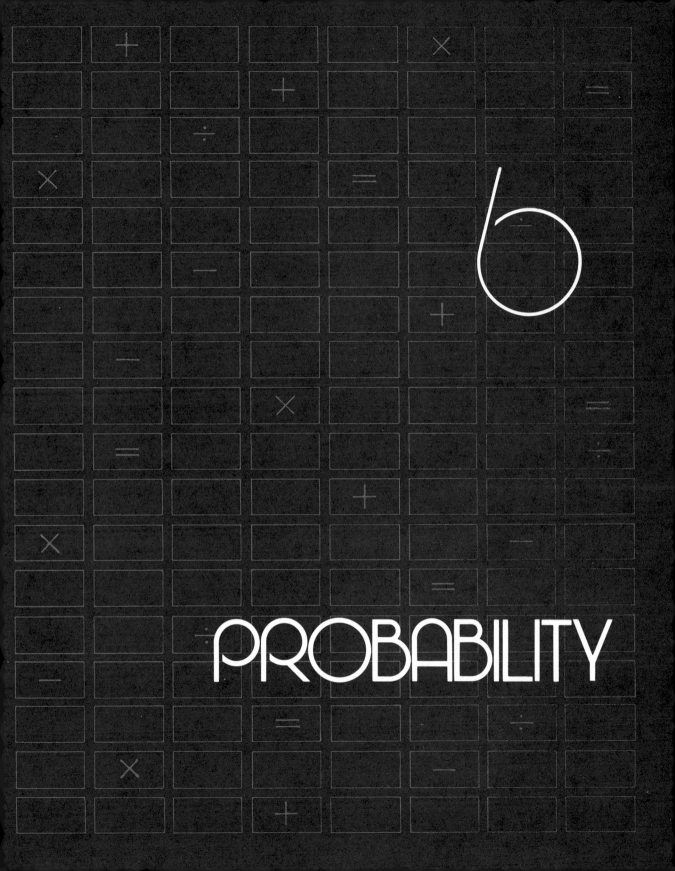

6

PROBABILITY

Question Will it rain tomorrow? Will you be a success in life? Will the world ever find true and lasting peace? Will you be involved in a car accident in the next year? Will mankind ever escape the depths of existential alienation and find full human dignity? Will you get a date for Saturday night?

ANSWER The answers to these questions might be:
(a) Maybe.
(b) Who knows?
(c) It will be one way or the other.

If these answers seem vague to you it is because they are vague. We cannot give a simple yes or no answer to these questions. Life is uncertain. In mathematics, the study of uncertainty is called **probability**.

6-1 BASIC CONCEPTS

Question Will it snow on March 30?

ANSWER We cannot safely say yes or no. We sense the answer is closer to no than to yes. But how close?

To make this precise, we can think of no as 0, and yes as 1. Now that we are using numbers, we can find an answer between yes and no. Suppose that we put no and yes at the opposite ends of a number line.

In English, we have very few expressions between no and yes:

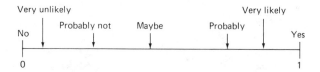

Notice that there are not very many English words to express the likelihood of an event. Also, those that we have are not too precise. Just what does "probably" mean? So, instead of using words between no and yes, we use numbers between 0 and 1. These numbers may be fractions, decimals, or percentages, whichever is most convenient. For example, the likelihood of flipping a head on a coin is $\frac{1}{2}$ or

0.5 or 50%, whatever you prefer. So, in answering the question, "Will it snow on March 30?", we might say: The likelihood of snow on March 30 is 5% or $\frac{1}{20}$. We call **probability** the measure of the likelihood of an event occurring.

Basic Terms

The study of probability has its own special words. The first word that we meet is experiment. Here we mean experiment to be more than something done in a scientific laboratory. We will call an **experiment** anything (scientific or everyday) that has an uncertain outcome.

Question What are a few typical experiments?

ANSWER Remember that an experiment is anything with an uncertain outcome, for instance, flipping a coin, walking alone on a city street at night, conceiving a child, tomorrow's weather, or next year's Super Bowl.

Once we have an experiment, we must then determine the outcomes. The **outcomes** are simply all possible results of an experiment.

Question If we have the experiment of flipping a coin, what are the outcomes?

ANSWER The possible results are {heads, tails}.

Question If an experiment is tomorrow's weather, what are the outcomes?

ANSWER The possible outcomes are {sunny, cloudy, rainy, snowy, and so on}.

Notice that the outcomes are written as the elements of a set. The set of all outcomes is called the **sample space**. Any subset of this sample space is called an **event**.

Problem Bob is taking a mathematics test. Is this an experiment? What are the outcomes and some of the events?

ANSWER For Bob, taking a mathematics test definitely has an uncertain outcome, so it is an experiment. If it is graded the way most tests are graded, the sample space

$$U = \{0, 1, 2, 3, \ldots, 98, 99, 100\}$$

These are the possible outcomes. Some events, or subsets, might be

$$A = \{90, 91, 92, \ldots, 99, 100\} \qquad \text{(grade of A)}$$

$$B = \{80, 81, 82, \ldots, 89\} \qquad \text{(grade of B)}$$

$$C = \{70, 71, 72, \ldots, 79\} \qquad \text{(grade of C)}$$

$$D = \{60, 61, 62, \ldots, 69\} \qquad \text{(grade of D)}$$

$$F = \{0, 1, 2, \ldots, 58, 59\} \qquad \text{(grade of F)}$$

$$P = \{60, 61, \ldots, 99, 100\} \qquad \text{(grade of pass)}$$

We can also get new events that are combinations of other events. If A and B are two events, we have:

1. *A and B* is the event that *both A and B* happen.
2. *A or B* is the event that *either A or B or both* happen.
3. *Not A* is the event that A does *not* happen.

These combinations can be pictured as shown.

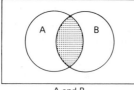
A and B

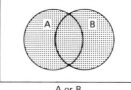
A or B

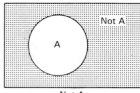
Not A

Problem Let drawing a card from a standard 52-card deck be an experiment. Let $A = \{$red cards$\}$ and $B = \{$face cards$\}$.
  (a) What is the sample space?
  (b) What is *A and B*?
  (c) What is *A or B*?
  (d) What is *not A*?
  (e) What is *not B*?

ANSWER   (a) The sample space is the set of all outcomes. In this case,

$$U = \{2\clubsuit, \ldots, A\spadesuit\} \quad \text{or} \quad \{\text{all 52 cards}\}$$

  (b) *A and B* is the event that *both A and B* happen; in other words, the drawn card is *both* red and a face card. So

$$A \text{ and } B = \{J\blacklozenge, J\heartsuit, Q\blacklozenge, Q\heartsuit, K\blacklozenge, K\heartsuit\}$$

  (c) *A or B* is the event that *either* a red card *or* a face card, *or both* are drawn. So

$$A \text{ or } B = \{\text{all red cards, all face cards}\}$$

(d) *Not A* is the event that a red card is *not* drawn. So

$$not\ A = \{\text{black cards}\}$$

(e) *Not B* is the event that a face card does *not* come up. So

$$not\ B = \{\text{all cards, ace through 10}\}$$

PROBLEM SET 6–1–1

1. Define or discuss:
   (a) Probability.
   (b) Experiment.
   (c) Outcome.
   (d) Sample space.
   (e) Event.
   (f) *And.*
   (g) *Or.*
   (h) *Not.*

For the following events or statements, choose an appropriate English word or phrase to describe its likelihood (for example, probably, most likely, maybe, or never). Then position the phrase on the number line between 0 (no) and 1 (yes).

2. It will snow on Christmas.

3. You can flip a head on a coin.

4. You can flip 10 straight heads on a coin.

5. You can get *exactly* 5 heads and 5 tails on 10 flips of a coin.

6. The Chicago Cubs will win next year's World Series.

7. You will be involved in a car accident this year.

8. You will be on the dean's list this term.

9. You can find the book you need at the library.

10. You will have a great time Saturday night.

11. You will have more than four children.

For the following experiments, write out the sample space of all possible outcomes.

12. Rolling a die.

13. Rolling two dice.

14. Going out on a blind date.

15. Walking across the street.

16. Registering for a difficult course.

17. Buying a new record.

18. A player batting in baseball.

19. Taking your car in for a repair.

20. Your basketball team's final standing in the league.

Let drawing a card from a 52-card deck be an experiment. Let $R = \{$red cards$\}$, $F = \{$face cards$\}$, $A = \{$aces$\}$, $H = \{$hearts$\}$, $B = \{$black cards$\}$, and $E = \{$even-numbered card$\}$.

21. Find *B and F*.

22. Find *A and H*.

23. Find *B and H*.

24. Find *F and H*.

25. Find *not H*.

26. Find *not A*.

27. Find *not B*.

28. Find *F or E*.

29. Find *R or B*.

30. Find *F or A*.

Let $U = \{a, b, c, d, f\}$ be the possible outcomes or grades in a course that Diane is taking. Let $P = \{a, b, c, d\}$, $F = \{f\}$, $S = \{a, b\}$, and $M = \{c, d\}$.

31. Find *P and F*.

32. Find *not S*.

33. Find *S or M*.

34. Find *not F*.

35. Find *not M*.

Definition of Probability

Suppose that we have two events. Sometimes these events have outcomes in common. For instance, drawing a red card and drawing an ace are events that have common outcomes:

$$\{A\heartsuit, A\diamondsuit\}.$$

Sometimes, two events have no outcomes in common. Here we say that they are **mutually exclusive events**. Symbolically, we could say that *A* and *B* are mutually exclusive if *A and B* = { }.

LA TH
ÉORIE
DES P
ROBAB
ILITIÉS
N'EST
QUE LE
BON S
ENS C
ONFIR
MÉ PA
R LE
CALCU
L.

: LA
PLACE
1796 ·

PROBABILITY

Probability distribution model. The balls fall randomly through the maze of pegs. Probability theory predicts that the balls will arrange themselves at the bottom in a normal distribution. (Courtesy of Museum of Science and Industry, Chicago)

Question   Are drawing a black card and a heart mutually exclusive?

ANSWER   Yes, since it is impossible for a card to be a black and a heart at the same time.

Definition   If A is any event in a sample space U, then the ***probability of A***, written $P(A)$, is a number that measures the likelihood of the event A occurring.

The following rules are true for probabilities:

The probability of A is between 0 and 1; symbolically,

Formula 6–1–1

$$0 \leqslant P(A) \leqslant 1$$

The probability of the whole sample space is 1; symbolically,

Formula 6–1–2

$$P(U) = 1$$

If A and B are mutually exclusive events, then the probability of *A or B* occurring is the sum of their probabilities. In symbols, if *A and B* = { }, then

Formula 6–1–3

$$P(A \text{ or } B) = P(A) + P(B)$$

To give probabilities to a sample space:

1. Assign each outcome a likelihood between 0 and 1.
2. Make sure that all probabilities add up to 1.

Question The experiment is rolling a fair die. What are the probabilities?

ANSWER The sample space $U = \{1, 2, 3, 4, 5, 6\}$. We make the following assignments:

Outcome	Probability
1	1/6
2	1/6
3	1/6
4	1/6
5	1/6
6	1/6
U	1

Problem What might the sample space of probabilities be for an adult at a national election?

ANSWER The sample space is $U = \{$vote Democratic, vote Republican, vote another party, vote a write-in, not vote at all$\}$. According to recent statistics, we can say

Outcome	Probability
Democratic	0.288
Republican	0.273
Another party	0.048
Write-in	0.001
No vote	0.390
U	1.000

Problem   What are the probabilities for a woman's marital status?

ANSWER   The sample space is U = {single (never married), married, divorced, widowed}. According to recent statistics,

Outcome	Probability
Single $\longrightarrow$	11.9%
Married $\longrightarrow$	70.4%
Widowed $\longrightarrow$	13.0%
Divorced $\longrightarrow$	4.8%
Total U	100.1%

Can you explain why the total came out to 100.1% instead of 100.0%?

Complementary Events

Problem   There is a 30% chance of rain. What is the probability that it will not rain?

ANSWER   Seventy percent.

Problem   There is a $\frac{2}{3}$ chance that Mary will pass chemistry. What is the probability that she will fail?

ANSWER   One third.

These examples suggest a general rule. We can state this rule as a formula.

Formula 6–1–4

$$\boxed{P(not\ A) = 1 - P(A)}$$

This is exactly what our common sense would have guessed. If $P(A)$ is a percent, then $P(not\ A) = 100\% - P(A)$. The events A and *not* A are called **complementary events**.

Problem   The probability of a 30-year-old man dying before he is 31 is 0.0016. What is the probability that he will live to be 31?

ANSWER   Let D = the event that the man dies before he is 31. Then *not* D is the event that the man lives to be 31. Thus $P(not\ D) = 1 - P(D) = 1 - 0.0016 = 0.9984$, by Formula 6–1–4.

Problem   There is a $\frac{1}{5}$ chance that an arrested person will be convicted. What is the probability that an arrested person will not be convicted?

ANSWER   Let C = the event that an arrested person is convicted. Then

$$P(not\ C) = 1 - \frac{1}{5} = \frac{4}{5}.$$

1. Define or discuss:
   (a) Mutually exclusive.
   (b) Probability.
   (c) Complementary events.

U is an experiment with five outcomes, $\{a, b, c, d, e\}$. Which of the following cases are possible probabilities for U according to the definition? Tell why or why not.

2. $P(a) = 0.1$ 3. $P(a) = 0.2$ 4. $P(a) = \dfrac{1}{4}$ 5. $P(a) = \dfrac{1}{2}$

   $P(b) = 0.2$ $P(b) = 0.3$ $P(b) = \dfrac{5}{16}$ $P(b) = \dfrac{1}{4}$

   $P(c) = 0.3$ $P(c) = 0.3$ $P(c) = \dfrac{1}{8}$ $P(c) = \dfrac{1}{8}$

   $P(d) = 0.2$ $P(d) = 0.2$ $P(d) = \dfrac{1}{4}$ $P(d) = \dfrac{3}{40}$

   $P(e) = 0.1$ $P(e) = 0.1$ $P(e) = \dfrac{3}{32}$ $P(e) = \dfrac{1}{20}$

6. $P(a) = 15\%$ 7. $P(a) = 0.5$ 8. $P(a) = \dfrac{1}{2}$ 9. $P(a) = 1$

   $P(b) = 25\%$ $P(b) = 0.2$ $P(b) = \dfrac{1}{4}$ $P(b) = 0$

   $P(c) = 11\%$ $P(c) = 0.7$ $P(c) = \dfrac{1}{8}$ $P(c) = 0$

   $P(d) = 34\%$ $P(d) = 0.4$ $P(d) = \dfrac{1}{16}$ $P(d) = 0$

   $P(e) = 15\%$ $P(e) = -0.8$ $P(e) = \dfrac{1}{32}$ $P(e) = 0$

For the following experiments or situations, write out the sample space and assign probabilities based on your intuition.

10. The next presidential election.

11. Your marital status in 10 years.

12. The weather in 10 days.

13. Your having children.

14. Your income in 10 years (for example, $0–$5000, $5000–$10,000, and so on).

15. Your rank in the graduating class.

16. Your job after graduation.

Find the following probabilities

17. $P(A) = 0.27$; then $P(not\ A) =$

18. $P(B) = \dfrac{7}{13}$; then $P(not\ B) =$

19. $P(C) = 16.3\%$; then $P(not\ C) =$

20. $P(D) = 0.0076$; then $P(not\ D) =$

21. If $U = \{a, b, c, d\}$, $P(a) = 0.12$, $P(b) = 0.27$, and $P(c) = 0.43$, find $P(d)$.

Complete the following table.

	$P(A)$	$P(not\ A)$
22.	$\frac{1}{78}$	?
23.	0.003	?
24.	0.89	?
25.	?	0.01%
26.	?	79%
27.	1.00	?
28.	?	$\frac{3}{14}$
29.	?	$\frac{78}{79}$
30.	31.8%	?
31.	$\frac{1}{1,000,000}$	?

6–2 COMPUTING AND INTERPRETING PROBABILITIES

Question   How do we compute these probabilities?

ANSWER   That's a good question. Up to now, all we know is that a probability is a number between 0 and 1 that somehow reflects the likelihood of an event occurring. In this section we will see how probabilities are computed and exactly what they mean. We will consider three types of probability:

1. Countable probability
2. Scientific (or historical) probability
3. Guesstimate probability

Countable Probability

Question   The probability of flipping a head on a coin is $\frac{1}{2}$. What does this number mean, and how was it computed?

ANSWER   When we flip a coin, we either get a head or we don't; there is no such thing as $\frac{1}{2}$

head. The number $\frac{1}{2}$ means that if the coin were tossed a million times, for instance, we would expect about half a million heads. Here the probability is based on being able to repeat exactly the same experiment many times in the future.

The probability was computed by counting the number of possible outcomes of a coin flip, which is 2. We then count the number of outcomes for the event heads, which is 1. We assume either of the outcomes, heads or tails, is equally likely. We have one favorable outcome out of two total outcomes, so we say the probability is $\frac{1}{2}$.

Problem There are 27 men and 34 women in a room. One of them will be chosen at random to win a radio for a door prize. What is the probability of a man winning the radio?

ANSWER There are 27 + 34 = 61 people in the room. So the sample space has 61 outcomes, one for each person. Let M be the event that a man is chosen. M has 27 possible outcomes. Since the drawing for the radio is done at random, each of the 61 people has an equal chance to be chosen. We say that there is a $\frac{27}{61}$ chance that a man is chosen, or $P(M) = \frac{27}{61}$.

This is trickier to interpret, since we normally would not continue choosing prize winners the way we can keep flipping a coin. But we can imagine that if this choice were made 1 million times that a man would be chosen about $\frac{27}{61}$ of the time, or $\frac{27}{61} \cdot 1,000,000 = 442,623$ times.

We can generalize this procedure for computing probabilities. Suppose that U is the sample space of our experiment, and the number of outcomes in U is written $N(U)$. If each outcome has as much chance of occurring as any other outcome, we say that the outcomes are **equally likely** or **equiprobable**. This democratic assumption is natural to make with coin tossing, dice rolling, card drawing, random people choosing, and the like, for which all outcomes are equally likely.

Suppose that we are interested in an event A with $N(A)$ number of possible outcomes. Then the formula for the probability of A is

Formula 6–2–1

$$P(A) = \frac{N(A)}{N(U)} = \frac{\text{number of outcomes in } A}{\text{total number of outcomes}}$$

This says that the probability of A is the fraction we get by dividing the number of outcomes in A by the total number of outcomes. We call this **countable probability**, since we can count up all the possibilities before we even do the experiment.

Problem   What is the probability of drawing an ace from a shuffled deck of cards?

ANSWER   Let $U = \{\text{deck of cards}\}$; then $N(U) = 52$. Let $A = \{\text{aces}\}$; then $N(A) = 4$. By Formula 6–2–1, we get

$$P(A) = \frac{N(A)}{N(U)} = \frac{4}{52} = \frac{1}{13}$$

This says that there is a $\frac{1}{13}$ chance of pulling an ace from a deck of cards. This means that, if one card each was drawn from 13 million decks, about 1 million would be aces.

Problem   A letter is drawn at random from the alphabet. What is the probability that the letter is a vowel?

ANSWER   If $U = \{\text{alphabet}\}$, then $N(U) = 26$. And if $V = \{a, e, i, o, u\}$, then $N(V) = 5$. So

$$P(V) = \frac{N(V)}{N(U)} = \frac{5}{26} \quad \text{or} \quad 0.192 \quad \text{or} \quad 19.2\%$$

Notice that this is not the same as asking for a vowel chosen from a page of print. Why? Because the letters aren't used in print equally often. You will be asked about this again in one of the problems of the next section.

Problem   A brown-eyed man (with a blue recessive gene) marries a brown-eyed woman (with a blue recessive gene). What is the probability that their first child will have brown eyes?

ANSWER   In genetics, B = brown (dominant) and b = blue (recessive) are the labels for the genes. The possible combinations (genotypes) are BB (brown), Bb (brown), and bb (blue). Here our couple both have Bb. In genetics, the following table is used to examine the possibilities.

		Man	
		B	b
Woman	B	BB	Bb
	b	bB	bb

The experiment is having a child. The sample space (for eye color) is $U = \{BB, Bb, bB, bb\}$. The event $A = \{BB, Bb, bB\}$ is the child having brown eyes. Thus

$$P(\text{brown eyes}) = P(A) = \frac{N(A)}{N(U)} = \frac{3}{4}$$

1. Define or discuss:
   (a) Equally likely events.
   (b) Countable probability.

Use Formula 6–2–1 to determine the following probabilities:

	N(A)	N(U)	P(A)	P(not A)	Answer form
2.	3	8			Write as fraction
3.	23	129			Write as fraction
4.	1	367			Write as fraction
5.	56	57			Write as decimal
6.	3	1,245			Write as decimal
7.	689	689			Write as decimal
8.	123	5,561			Write as percent
9.	334	1,123			Write as percent
10.	4378	12,345			Write as percent

11. A single die is rolled. What is the probability of rolling a 1 or 2?

A card is drawn from a 52-card deck. Find the probability of getting

12. A king.   13. The king of hearts.

14. A red card.   15. A black, even-numbered card.

16. A church sells 10,000 lottery tickets for a new car. Janet buys 3 tickets. What is the probability of her winning the car?

Professor Peters has a rigid grading scale. For a class of 30 he will give 1 A, 5 B's, 16 C's, 7 D's, and 1 F. Ron is one of his students. Find the probability of Ron

17. Passing the course.

18. Getting an A.

19. Getting a C or better.

20. Paula must take Mathematics 152, but she doesn't know who will teach it. Of 19 mathematics professors, 7 are good. What is the probability that she will get a good teacher?

21. George is given a key ring with 12 keys on it. What is the probability (as a percent) that he will pick the right key on the first try?

22. Howie's wife is expecting a baby soon. Howie works 8 hours a day, 5 days a week. What is the probability that the baby will be born while Howie is at work?

23. Tim is playing five-card draw poker. He has 4, 5, 6, 7, K. He discards the king, hoping to get a 3 or an 8 and make a straight. What is the probability of Tim filling the straight?

24. Len has a girlfriend, Ellen, on the east side of town, and Wendy, on the west side. Every Saturday, he walks to the bus stop. If the eastbound bus comes first, he sees Ellen; if the westbound bus comes first, he sees Wendy. The eastbound busses run at 9:00, 9:30, 10:00, 10:30, and so on. The westbound busses run at 9:21, 9:51, 10:21, 10:51, and so on. What is the probability that he will see Ellen?

Counting Principles

Problem Paul has nine shirts and seven pairs of pants. How many total combinations can he wear?

ANSWER We can draw a picture of all the shirt–pants pairs.

We can see that this is $9 \times 7 = 63$ total combinations.

In the last section we saw that the key to many probability problems is being able to count all the possible outcomes. We will now give a very useful formula for counting, the **fundamental principle of counting**. Let A be a task with $N(A)$ number of ways to do A; and let B be a task after A with $N(B)$ ways to do B. Let $N(A, B)$ be the total number of ways to be A and then B. Then we have the formula

Formula 6–2–2

$$N(A, B) = N(A) \cdot N(B)$$

To find the number of ways to do tasks A, then B, and so on:

1. Find the number of ways to do A; this is $N(A)$.

2. Find the number of ways to do B once A is done; this is $N(B)$.

3. Continue finding $N(C)$, $N(D)$, and so on.

4. Multiply $N(A) \cdot N(B) \cdot N(C) \ldots$.

Formula 6–2–3

$$N(A, B, C, \ldots) = N(A) \cdot N(B) \cdot N(C) \ldots$$

This formula tells us to multiply all the ways to do each of the different tasks.

Problem Art walks into an automobile showroom. There are 6 models from which to choose. Each model has 10 colors, 3 types of transmission, 4 interior packages, 3 exterior packages, is air conditioned or not, has power brakes or not, and 5 types of radio/stereo. How many different cars could Art design?

ANSWER Using Formula 6–2–3, we multiply all the different ways to make each choice. We have 2 choices for air conditioning: take it or not. The same is true for the power brakes. So the number of different combinations is $6 \cdot 10 \cdot 3 \cdot 4 \cdot 3 \cdot 2 \cdot 2 \cdot 5 = 43,200$.

Formula 6–2–3

$$\underset{\text{model}}{6} \cdot \underset{\text{color}}{10} \cdot \underset{\text{transmission}}{3} \cdot \underset{\text{interior packages}}{4} \cdot \underset{\text{exterior packages}}{3}$$

$$\cdot \underset{\text{a/c or not}}{2} \cdot \underset{\text{power brakes or not}}{2} \cdot \underset{\text{radio}}{5} = 43,200$$

Problem Coach Phillips of the baseball team has chosen his 9 starting players. How many different batting orders are there?

ANSWER There are 9 players who can bat first. Once he is chosen there are 8 players who can bat second; then there are 7 who can bat third; and so on. We can use the following chart to picture the possibilities.

Formula 6–2–3

$$\underset{\text{first}}{9} \cdot \underset{\text{second}}{8} \cdot \underset{\text{third}}{7} \cdot \underset{\text{fourth}}{6} \cdot \underset{\text{fifth}}{5}$$

$$\cdot \underset{\text{sixth}}{4} \cdot \underset{\text{seventh}}{3} \cdot \underset{\text{eighth}}{2} \cdot \underset{\text{ninth}}{1}$$

Using Formula 6–2–3, the number of ways to choose a batting order is $9 \cdot 8 \cdot 7 \cdot 6 \cdot 5 \cdot 4 \cdot 3 \cdot 2 \cdot 1 = 362,880$.

Problem In a new game on the market, one player, called the codemaker, must fill 4 hidden holes with 4 pegs. There are 6 different colors to choose from, and repeats are allowed. How many possible combinations are there? What is the probability of the other player guessing the code?

ANSWER There are 4 holes to be filled. In each hole we can place any one of 6 different colors. Thus we have

| Formula 6–2–3 | ● ● ● ●
$6 \cdot 6 \cdot 6 \cdot 6 = 1296$ combinations |

Since we were allowed to repeat colors, we used 6 for each hole. Since there are 1296 possibilities, and only 1 is correct, the probability of guessing is $\frac{1}{1296}$.

Problem Suppose that we are not allowed to repeat a color. How many possibilities are there now?

ANSWER We still have 6 choices for the first hole. After that, there are only 5 choices for the second hole (since we cannot repeat). There are 4 for the next hole, and 3 for the last hole.

| Formula 6–2–3 | ● ● ● ●
$6 \cdot 5 \cdot 4 \cdot 3 = 360$ combinations |

PROBLEM SET 6–2–2

1. What is the fundamental principle of counting?

2. A dating service has 312 men and 259 women on file. How many possible man–woman dates can they set up?

3. Fred is rolling two dice. Each die, of course, has six possible outcomes. How many total outcomes are there for the two dice?

4. In a certain state the license plates are made as follows. There are six characters. The first is either a letter of the alphabet or a digit between 0 and 9. The next five are digits between 0 and 9. How many total license plates are possible?

5. A fraternity has 32 members. They must choose a president, vice-president, treasurer, and secretary. How many different ways are there to choose the officers?

6. Sherri has 3 pairs of shoes, 9 pairs of pants, and 12 sweaters. How many different outfits does she have?

7. Helen the home economist has four types of soup, two types of salad, eight types of meat, six types of vegetable, three types of potato, and seven types of dessert. How many total meals can she prepare?

Suppose, in the game discussed, that there were seven colors and four holes.

8. How many combinations are there if repeated colors are allowed?

9. How many combinations are there if no repeats are allowed?

There is also a super version with eight colors and five holes.

10. How many combinations are there if repeated colors are allowed?

11. How many combinations are there if no repeats are allowed?

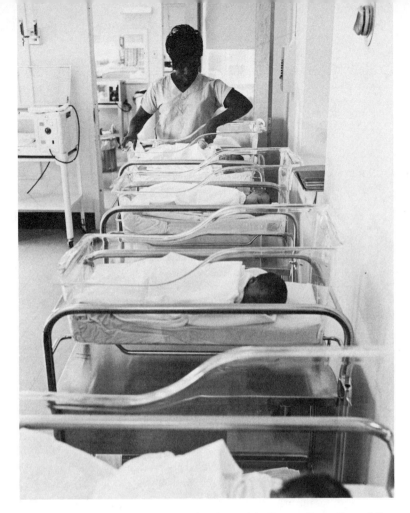

The life of each of these babies is charted in the mathematics of the mortality table. (© Degast from Rapho/Photo Researchers)

Scientific Probability

Question The probability of a person living to be 50 years old is 87.6%. What does this mean and how was it calculated?

ANSWER On a person's fiftieth birthday, he or she is either dead or alive, but *not* 87.6% alive. Unlike coin tossing or dice rolling, we cannot simply count some things to determine the probability of being alive (unless the person is playing Russian roulette). We have to rely on data already gathered. In this case, insurance companies have determined that in the past, for every 10,000,000 babies born alive, 8,762,306 lived to be 50. This is about 87.6%. This is a *historical* figure, since it says that in the past 87.6% of all babies lived to be 50. Since the basis of science is that the future behaves like the past, we assume that there is an 87.6% chance of a person living to be 50.

Question   Jim buys a new car, an Aardvark. He reads that there is an 8% chance that his car will be recalled to correct a factory error. What does this mean, and how was it calculated?

ANSWER   Again, either Jim's car will be recalled or it won't. This estimate is based on past records. Perhaps, in the past, 400,000 Aardvarks were manufactured, and 32,129 of them had to be recalled. This is 8.032% or about 8%. So it is natural to assume that, if things are the same at Aardvark, 8% of the new cars will be recalled. So we say Jim's car has an 8% chance of being recalled.

These probabilities are different from countable probabilities, for which all the events are equally likely. Here the outcomes are *not* necessarily equally likely. For example, in the case above, P(car recalled) = 0.08, and P(car not recalled) = 0.92. These numbers cannot be gotten by counting possibilities on your fingers. We must study historical records, or conduct experiments to gather the needed data. For this reason we call this **scientific** or **historical probability**. It is based on the past and assumes that the future will act the same. In fact, most of science and our day-to-day living is based on this assumption.

If we want the scientific probability that event A will occur, we have to repeat the experiment U many times and observe how many times A occurs. We use the following formula:

Formula 6–2–4

$$P(A) = \frac{N(A)}{N(U)} = \frac{\textbf{number of times } A \textbf{ occurred}}{\textbf{number of repetitions of experiment}}$$

This formula should look very similar to Formula 6–2–1 for countable probabilities. In both we are measuring the relative frequency with which A occurs. But they are different in some ways. Formula 6–2–1 is used by counting up the possible outcomes in *one* experiment. Formula 6–2–4 is used by tallying the occurrence of A in *many* experiments.

Problem   In a certain 55-year period (1913–1968) there were 1075 presidential vetoes, 37 of which were overridden by Congress. What is the probability that a presidential veto will be overridden?

ANSWER   We can think of U as a presidential veto; so $N(U) = 1075$. The event A is a congressional override; so $N(A) = 37$. If we use Formula 6–2–4, we get

$$P(A) = \frac{37}{1075} = 0.0344 = 3.4\%.$$ So, historically, 3.4% of all vetoes were overridden. We then assume that any veto has a 3.4% chance of being overridden.

PROBLEM SET 6–2–3

1. Define or discuss scientific probability and how it is calculated. What are some of its drawbacks?

2. One day, 254 are asked if they are happy; 196 responded "yes." Use this to calculate the probability that a person is happy.

3. Casey Bigbat has 156 hits in 471 times at bat. What is the probability that he will get a hit the next time up? (Write the answer as a three-place decimal.)

4. In 23 years, 1953–1976, the Academy Award for best picture and best director have gone to the same movie 20 times. What is the probability of it happening this year?

TABLE OF MORTALITY
(Commissioners' Standard Ordinary 1958)

Age	Number Living	Deaths Each Year	Deaths Per 1,000	Expectation of Life	Age	Number Living	Deaths Each Year	Deaths Per 1,000	Expectation of Life
0	10,000,000	70,800	7.08	68.30	50	8,762,306	72,902	8.32	23.63
1	9,929,200	17,475	1.76	67.78	51	8,689,404	79,160	9.11	22.82
2	9,911,725	15,066	1.52	66.90	52	8,610,244	85,758	9.96	22.03
3	9,896,659	14,449	1.46	66.00	53	8,524,486	92,832	10.89	21.25
4	9,882,210	13,835	1.40	65.10	54	8,431,654	100,337	11.90	20.47
5	9,868,375	13,322	1.35	64.19	55	8,331,317	108,307	13.00	19.71
6	9,855,053	12,812	1.30	63.27	56	8,223,010	116,849	14.21	18.97
7	9,842,241	12,401	1.26	62.35	57	8,106,161	125,970	15.54	18.23
8	9,829,840	12,091	1.23	61.43	58	7,980,191	135,663	17.00	17.51
9	9,817,749	11,879	1.21	60.51	59	7,844,528	145,830	18.59	16.81
10	9,805,870	11,865	1.21	59.58	60	7,698,698	156,592	20.34	16.12
11	9,794,005	12,047	1.23	58.65	61	7,542,106	167,736	22.24	15.44
12	9,781,958	12,325	1.26	57.72	62	7,374,370	179,271	24.31	14.78
13	9,769,633	12,896	1.32	56.80	63	7,195,099	191,174	26.57	14.14
14	9,756,737	13,562	1.39	55.87	64	7,003,925	203,394	29.04	13.51
15	9,743,175	14,225	1.46	54.95	65	6,800,531	215,917	31.75	12.90
16	9,728,950	14,983	1.54	54.03	66	6,584,614	228,749	34.74	12.31
17	9,713,967	15,737	1.62	53.11	67	6,355,865	241,777	38.04	11.73
18	9,698,230	16,390	1.69	52.19	68	6,114,088	254,835	41.68	11.17
19	9,681,840	16,846	1.74	51.28	69	5,859,253	267,241	45.61	10.64
20	9,664,994	17,300	1.79	50.37	70	5,592,012	278,426	49.79	10.12
21	9,647,694	17,655	1.83	49.46	71	5,313,586	287,731	54.15	9.63
22	9,630,039	17,912	1.86	48.55	72	5,025,855	294,766	58.65	9.15
23	9,612,127	18,167	1.89	47.64	73	4,731,089	299,289	63.26	8.69
24	9,593,960	18,324	1.91	46.73	74	4,431,800	301,894	68.12	8.24
25	9,575,636	18,481	1.93	45.82	75	4,129,906	303,011	73.37	7.81
26	9,557,155	18,732	1.96	44.90	76	3,826,895	303,014	79.18	7.39
27	9,538,423	18,981	1.99	43.99	77	3,523,881	301,997	85.70	6.98
28	9,519,442	19,324	2.03	43.08	78	3,221,884	299,829	93.06	6.59
29	9,500,118	19,760	2.08	42.16	79	2,922,055	295,683	101.19	6.21
30	9,480,358	20,193	2.13	41.25	80	2,626,372	288,848	109.98	5.85
31	9,460,165	20,718	2.19	40.34	81	2,337,524	278,983	119.35	5.51
32	9,439,447	21,239	2.25	39.43	82	2,058,541	265,902	129.17	5.19
33	9,418,208	21,850	2.32	38.51	83	1,792,639	249,858	139.38	4.89
34	9,396,358	22,551	2.40	37.60	84	1,542,781	231,433	150.01	4.60
35	9,373,807	23,528	2.51	36.69	85	1,311,348	211,311	161.14	4.32
36	9,350,279	24,685	2.64	35.78	86	1,100,037	190,108	172.82	4.06
37	9,325,594	26,112	2.80	34.88	87	909,929	168,455	185.13	3.80
38	9,299,482	27,991	3.01	33.97	88	741,474	146,997	198.25	3.55
39	9,271,491	30,132	3.25	33.07	89	594,477	126,303	212.46	3.31
40	9,241,359	32,622	3.53	32.18	90	468,174	106,809	228.14	3.06
41	9,208,737	35,362	3.84	31.29	91	361,365	88,813	245.77	2.82
42	9,173,375	38,253	4.17	30.41	92	272,552	72,480	265.93	2.58
43	9,135,122	41,382	4.53	29.54	93	200,072	57,881	289.30	2.33
44	9,093,740	44,741	4.92	28.67	94	142,191	45,026	316.66	2.07
45	9,048,999	48,412	5.35	27.81	95	97,165	34,128	351.24	1.80
46	9,000,587	52,473	5.83	26.95	96	63,037	25,250	400.56	1.51
47	8,948,114	56,910	6.36	26.11	97	37,787	18,456	488.42	1.18
48	8,891,204	61,794	6.95	25.27	98	19,331	12,916	668.15	.83
49	8,829,410	67,104	7.60	24.45	99	6,415	6,415	1,000.00	.50

5. For the last 84 Christmases, it has snowed on 47. What is the probability of a white Christmas?

6. In one year 2,290,000 students graduated high school. Four years later 666,710 students graduated from college. Use this to approximate the probability that a high school senior will finish college 4 years later.

7. There are about 220,000,000 people in the United States. Every year about 50,000 of them are killed in automobile accidents. What is the probability of a person dying in a car crash next year?

The table on page 342 is a standard mortality table used by insurance companies for computing premium rates. Notice that it starts with 10,000,000 babies (age 0) and follows them through 100 years. For example, at age 25, of the original 10,000,000 babies, 9,575,636 are still alive; 18,481 will die before they reach 26. The *Death Per 1,000* column means that of 1000 25 years olds, 1.93 will die that year. Use the table to answer the following questions:

8. What is the probability of living to be 40? 50? 60? 70? 80? 90?

9. At about what age are half the people still living and half dead?

10. At what age do more people die than any other?

11. The *Deaths Each Year* column represents the length of people's lives. For instance, 72,902 out of 10,000,000 live to be exactly 50. Make a frequency graph of these columns versus age. Use ages 0, 10, 20, 30, and so on.

12. If a person lives to be 40, what is the probability that he or she will live to be 80? (How many are alive at 40? How many of these make it to 80?)

13. Notice that as many 25-year-olds will die as will 97-year-olds (both about 18,450). Why are the deaths per 1,000 so much higher for the 97 year old?

14. What is the probability of a person dying between 65 and 75?

15. In a given year, the airlines flew 86,900,000,000 passenger miles. There were 258 passenger deaths due to crashes. What is the probability of being killed in a given mile of flight?

16. What is the probability that a letter chosen at random from a page of print will be a vowel? (Take any page of print and count.)

17. Shake four coins from your hand. Count the heads. Do this 50 times. Complete the table.

Number of heads	Tally	Total	Probability
0			
1			
2			
3			
4			

18. Shake 10 coins from your hand. Count the heads. Do this 50 times. Of the 500 total coins, how many heads were there?

Number of heads	Tally	Total	Probability
0			
1			
2			
3			
4			
5			
6			
7			
8			
9			
10			

19. Since there are six faces on a die, it might seem reasonable that rolling three dice would give a 50–50 chance of getting at least one 6. Design and perform an experiment to check this.

20. With two dice, the sample space is $\{2, 3, 4, \ldots, 11, 12\}$. Design and perform an experiment to find the probabilities for this sample space.

Design and perform an experiment to determine the probabilities of the following events among your fellow students.

21. A male student likes football.

22. A female student likes football.

23. A male student likes beer.

24. A female student likes beer.

25. A student expects to go to graduate or professional school.

26. A student is confused about life.

27. A student enjoys cafeteria food.

28. A student is understood by his parents.

29. A student likes mathematics.

30. A student . . . (invent your own question).

Guesstimate Probability and Odds

Question   Teams A and B are playing in the Super Bowl. Timmy says the probability of team A winning is 40%. What does this mean, and where did he get it?

ANSWER At the end of the game, team *A* will either be the winner or the loser, not a 40% winner. Unlike coin tossing or life expectancy, there are no rules or history that Timmy can use. This is the first and last time that these two teams will be exactly the way they are now. So it makes no sense to say that if this game were played 1 million times, team *A* would win about 400,000 times. So what does the 40% mean? This number is purely an educated guess by a shrewd gambler.

All we can say is that, if Timmy were actually betting on the game, he would consider a $40 bet on team *A* against a $60 bet on team *B* to be fair. In other words, he could take either side of the bet and not feel cheated.

Question Wendy the Weatherwoman predicts a 70% chance of rain tomorrow. How did she get this, and what does it mean?

ANSWER Again, there is only one tomorrow, and it will either rain or not. Wendy must look at all the meteorological reports, look at the records for this day, look out the window, and check her Aunt Matilda's corns, and then come up with an intelligent estimate of the likelihood of rain.

The 70% means that if Wendy were a gambler, she would bet $70 on rain versus $30 on no rain as a fair bet. It also means that it would be wise to carry an umbrella.

This is our last type of probability. The outcomes are not necessarily equally likely, so counting won't work. The experiment has no history and cannot be

Playing craps. (Michael Abramson/Black Star)

repeated, so a scientific measurement won't work. All that's left is a "seat-of-the-pants" guess based on one's knowledge and instincts. We call this **guesstimate probability**. This type of probability is most common in sports, weather, political elections, wars, and other one-shot events. We cannot give any standard procedures for determining such probabilities.

Question The odds against Pammy winning the fourth race at Balmoral are 5 to 2. Is this a probability?

ANSWER Yes and no. Yes, it is a numerical measure of the likelihood of Pammy winning the race. No, it is not a number between 0 and 1. Technically, 5 to 2 are called **odds** and not probability, but obviously they are somehow related.

(Courtesy of Carol Olson)

Problem What is the probability that Pammy will win the race?

ANSWER Since the odds are 5 to 2 against Pammy, this is like 5 chances that she will lose and 2 chances that she will win the race. Now we think of this as a counting probability. There are 2 chances to win out of 7 total chances; so, $P \text{ (winning)} = \frac{2}{7}$.

Problem There is a 70% chance for rain tomorrow. What are odds for it raining?

ANSWER There is a 70% chance of rain, so there is a 30% chance that it won't rain. We then say that the odds are 70 to 30 that it will rain. We can reduce this to 7 to 3 for it to rain.

The *odds for* an event are a ratio of the chances that the event will happen to the chances that the event won't happen. We can convert from odds to probability, and vice versa. Suppose that the odds *for* an event A are x to y. Then there are x + y total chances, and x of them are *for* event A. So

Formula 6–2–5

$$P(A) = \frac{x}{x + y}, \text{ if odds are } x \text{ to } y \text{ for}$$

Suppose that the odds *against* A are m to n. Now there are m + n total chances, and n of them are for event A. Thus

Formula 6–2–6

$$P(A) = \frac{n}{m + n}, \text{ if odds are } m \text{ to } n \text{ against}$$

Suppose that P(A) is the probability of the event A. Then P(*not* A) = 1 − P(A) is the probability that A won't happen. Therefore,

Formula 6–2–7

$$\textbf{Odds } for \text{ } A = \frac{P(A)}{1 - P(A)} = P(A) \text{ to } \left[1 - P(A)\right]$$

Formula 6–2–8

$$\textbf{Odds } against \text{ } A = \frac{1 - P(A)}{P(A)} = \left[1 - P(A)\right] \text{ to } P(A)$$

Since odds are a ratio, they can be reduced and simplified just like fractions.

Problem   The odds against the Bently Beavers winning their softball league championship are 7 to 4. What is the probability of them winning?

ANSWER   Since we have odds *against*, we use Formula 6–2–6. There are 4 favorable chances out of a total of 11. Hence

Formula 6–2–6 $P(\text{Beavers winning}) = \dfrac{4}{7 + 4} = \dfrac{4}{11} = 36\%.$

Problem   The odds for Senator Calhoun winning reelection are 8 to 5. What is the probability of his winning reelection?

ANSWER   Here we have odds *for*, so we use Formula 6–2–5. We have 8 favorable chances out of 13 total. Thus

Formula 6–2–5 $P(\text{Calhoun winning}) = \dfrac{8}{8 + 5} = \dfrac{8}{13} = 0.615 = 61.5\%$

Problem The probability of snow on March 30 is 5%. What are the odds for and against snow?

ANSWER Since $P(\text{snow}) = 0.05$, then $P(\text{no snow}) = 0.95$. Using Formulas 6–2–7 and 6–2–8, we get

| Formula 6–2–7 |

$$\text{Odds } \textit{for} \text{ snow} = \frac{0.05}{0.95} = \frac{5}{95} = 1 \text{ to } 19$$

| Formula 6–2–8 |

$$\text{Odds } \textit{against} \text{ snow} = \frac{0.95}{0.05} = \frac{95}{5} = 19 \text{ to } 1$$

Most bets are in the form of odds. If the odds for event A are p to q, we put up p dollars on A versus q dollars on A not happening.

Problem The odds for the Peoria Peepers winning a game are 3 to 5. What is a fair bet?

ANSWER There are 3 chances of them winning to 5 chances of them losing. To make the bet fair, we should win $5 if they win and lose $3 if they lose. This way, there are 3 chances for us to win $5, and there are 5 chances for us to lose $3. This is fair.

PROBLEM SET 6–2–4

1. Define or discuss:
   (a) Guesstimate probability.
   (b) Odds.

For the following odds, determine the probability of the event happening.

2. 4 to 1 against

3. 3 to 2 for

4. 7 to 5 for

5. even (1 to 1)

6. 2 to 9 for

7. 20 to 1 against

8. 3 to 5 against

9. 100 to 1 for

For the following probabilities, determine the odds for and against the event. (Write in simplified form.)

10. $P(A) = 0.3$

11. $P(B) = 60\%$

12. $P(C) = \dfrac{2}{3}$

13. $P(D) = 55\%$

14. $P(E) = \dfrac{1}{5}$

15. $P(F) = 0.02$

16. $P(G) = 2\frac{1}{2}\%$

17–23. For the events in Problems 10–16, determine a fair bet to make on the event.

Make an educated guess to determine the guesstimate probabilities of the following events.

24. It will snow on Christmas.

25. Ohio State will win next year's Rose Bowl.

26. You will make the dean's list this term.

27. You will have a great time on Saturday night.

28. You will get a Ph.D.

29. You will earn $50,000 a year within 10 years.

30. The Republicans will win the next presidential election.

31. There will be another Middle East war.

32–39. Determine the odds against the events in Problems 24–31 happening (using the probabilities found in Problems 24–31).

40–47. Determine a fair bet to make on the events in Problems 24–31 (using the odds found in Problems 32–39).

6–3   INDEPENDENT EVENTS AND CHAINS OF EVENTS

Question   What if we have more than one event to consider?

ANSWER   Then we have to consider the effect the events have on each other. It is easier if the events have no effect on each other.

Independence

Question   What is the probability of flipping a head with a fair coin?

ANSWER   Easy, $\dfrac{1}{2}$.

Question Suppose thet I tell you today is Tuesday. Now what is the probability of flipping a head?

ANSWER It is still $\frac{1}{2}$.

Question Suppose that I have just flipped 10 straight heads. Now what is the probability of flipping a head?

ANSWER Sorry, it is still $\frac{1}{2}$. The coin has no memory. It does not say, "I ought to come up tails this time."

Quite often we will deal with two events that have no effect on each other. For instance, flipping a head and today being Tuesday have no effect on each other. We say events A and B are **independent** if they have no influence or effect on each other.

Question If A is the event living to be 65, and B is the event being named John, are A and B independent?

ANSWER Yes. There is no known relationship between one's name and one's age.

Question If C is the event being a college graduate, and D is the event earning over $15,000 a year, are C and D independent?

ANSWER No, since having a college degree generally means one can earn a higher income.

Question If T is the event being a Taurus, and S is the event being stubborn, are S and T independent?

ANSWER That depends. An astrology enthusiast would say "no," since he or she feels that there is a relationship. A skeptic would say "yes," they are independent.

Question Is independence the same as mutually exclusive?

ANSWER No! Mutually exclusive events *cannot* happen at the same time. For instance, if you flip one coin, you cannot get a head and a tail at the same time.
On the other hand, independent events may or may not happen at the same time. For example, if today is Tuesday, I may or may not flip a head on a coin toss. A good way to visualize the difference is with the following pictures.
Mutually exclusive events are out of the *same* sample space:

Notice that A and B cannot happen at the same time.

Independent events are from *two* different experiments:

	A = heads	tails
Sunday		
Monday		
B = Tuesday	×	
Wednesday		
Thursday		
Friday		
Saturday		

Notice that here A and B may or may not happen together.

PROBLEM SET 6–3–1

1. Define or discuss:
   (a) Independence.
   (b) Mutually exclusive.

Tell whether the following pairs of events are independent, mutually exclusive, neither, or both.

2. A = rolling a 2 on a die
   B = flipping a head

3. C = living to be 90
   D = smoking three packs of cigarettes a day

4. E = driving a blue car
   F = winning the state lottery

5. G = driving a sedan
   H = driving a station wagon

6. I = flipping 10 heads in a row
   J = flipping another head

7. K = your mother being born on May 28
   L = your father being born on December 23

8. M = the last digit of your phone number is 3
   N = being a woman

9. O = it will snow at noon, Christmas 1985
   P = it will snow at noon, Christmas 1986

10. Q = it will snow at noon, Christmas 1985
    R = it will be over 60° at noon, Christmas 1985

11. S = a black man carrying a sickle-cell-anemia gene
    T = a black woman carrying a sickle-cell-anemia gene

Probability for Independent Events

Problem   Mike has a $\frac{2}{7}$ chance of getting a promotion in his accounting job. Judy, his wife, has a $\frac{3}{5}$ chance of getting a promotion in her teaching job. Are these events independent? What is the probability of their *both* getting a promotion?

ANSWER   Since Mike is an accountant and Judy is a teacher, it is reasonable to assume that one promotion has no effect on the other. So their promotions are independent events. We can picture their promotion chances as shown.

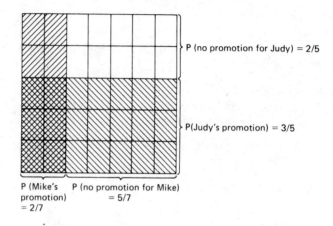

We see that this gives 35 total chances, and 6 of them are for both of them being promoted. Notice that this is just $\frac{2}{7} \times \frac{3}{5} = \frac{6}{35}$. So the probability of both of them being promoted is $\frac{6}{35}$.

We can do this for any two independent sets A and B with probabilities $P(A)$ and $P(B)$. We again make a square 1×1. On one edge we put A and *not A*; on the other edge we put B and *not B*.

Since we want the area of $P(A \text{ and } B)$, we look at the area of the A *and* B

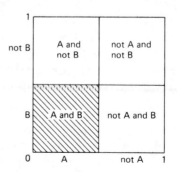

rectangle. The area of the whole square is 1, and the area of the *A and B* rectangle is its length times width, or $P(A) \cdot P(B)$. So if *A* and *B* are independent, we get

Formula 6–3–1

$$\boxed{P(A \text{ and } B) = P(A) \cdot P(B), \text{ if } A \text{ and } B \text{ are independent}}$$

Furthermore, if $A_1, A_2, A_3, \ldots, A_n$ are all independent events (no influence on each other), then

Formula 6–3–2

$$\boxed{\begin{array}{l} P(A_1 \text{ and } A_2 \text{ and } \ldots \text{ and } A_n) = P(A_1) \cdot P(A_2) \ldots P(A_n), \\ \text{if } A_1, A_2, A_3, \ldots, A_n \text{ are independent} \end{array}}$$

To compute the probability of independent events *A, B, C, D,* and so on, all happening:

1. Determine all the probabilities, $P(A)$, $P(B)$, and so on.
2. Multiply $P(A) \cdot P(B) \cdot P(C) \cdot P(D) \ldots$.

Problem   What is the probability of flipping a head twice in a row?

ANSWER   Let *A* be the event head on the first flip, and let *B* be the event head on the second flip. Since a coin does not have a memory, we can say that *A* and *B* are independent, each with probability $\frac{1}{2}$. So, By Formula 6–3–1,

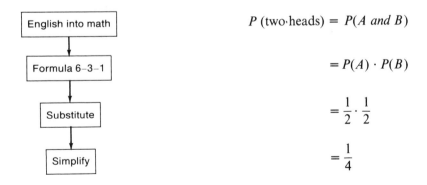

$$P(\text{two-heads}) = P(A \text{ and } B)$$

$$= P(A) \cdot P(B)$$

$$= \frac{1}{2} \cdot \frac{1}{2}$$

$$= \frac{1}{4}$$

Problem   What is the probability of flipping 10 straight heads?

ANSWER   Let A_1 be a head on the first flip, A_2 be a head on the second, $\ldots, A_{10}$ be a head on the tenth flip. All these events are independent, and all have probability $\frac{1}{2}$. Now we can use Formula 6–3–2.

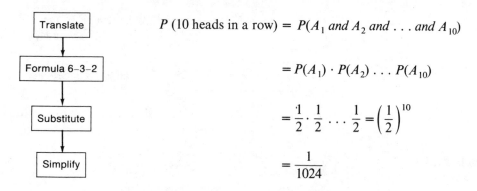

$$P \text{ (10 heads in a row)} = P(A_1 \text{ and } A_2 \text{ and } \ldots \text{ and } A_{10})$$

$$= P(A_1) \cdot P(A_2) \ldots P(A_{10})$$

$$= \frac{1}{2} \cdot \frac{1}{2} \cdots \frac{1}{2} = \left(\frac{1}{2}\right)^{10}$$

$$= \frac{1}{1024}$$

So there is one chance in 1024 of flipping 10 heads in a row.

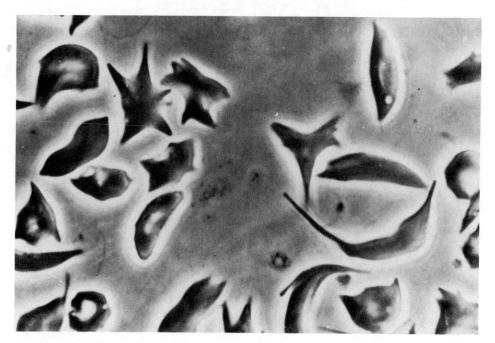

Photomicrograph of sickle cell anemia. (Courtesy of the Sickle Cell Disease Foundation of Greater New York)

Problem There is a 10% chance that a black man or woman is a sickle-cell-anemia carrier. (Not a victim, but just a carrier of the genetic trait.) What is the probability that in a black family both the mother and father are sickle-cell carriers?

ANSWER Let M be the event that the mother is a carrier, and F be the event that the father is a carrier. These are independent since there is no connection between the mother's genes and the father's. Since $P(M) = 0.10$ and $P(F) = 0.10$, we get

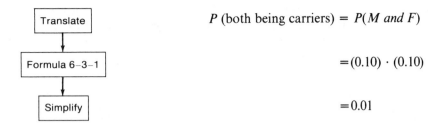

$$P \text{ (both being carriers)} = P(M \text{ and } F)$$

$$=(0.10) \cdot (0.10)$$

$$=0.01$$

So there is a $\dfrac{1}{100}$ chance of a black mother and father both being a sickle cell carrier.

Problem Bob Arbuckle is a baseball player with a .300 batting average; in other words, $P(\text{hit}) = 0.300$. Bob bats four times in a game.
  (a) What is the probability of Bob getting four hits?
  (b) What is the probability of Bob getting no hits?
  (c) What is the probability of Bob getting at least one hit?

ANSWER Let H mean that Bob gets a hit; so $P(H) = 0.3$. Then $P(\textit{not } H) = 0.7$. There is a 30% chance that he will get a hit, and a 70% chance that he won't. We assume that each time at bat is independent of the others.
  (a) To find the probability of four hits, we use Formula 6–3–2 for the four events of getting a hit.

Formula 6–3–2 → Substitute → Simplify

$$P(4 \text{ hits}) = P(H) \cdot P(H) \cdot P(H) \cdot P(H)$$

$$=(0.3) \cdot (0.3) \cdot (0.3) \cdot (0.3)$$

$$=0.0081$$

  (b) To find the probability of getting no hits, we have to look at the event that he doesn't get a hit each time at bat. This is the event *not H and not H and not H and not H*. We can use Formula 6–3–2 to get

Translate → Formula 6–3–2 → Simplify

$$P(\text{no hits}) = P(\textit{not } H) \cdot P(\textit{not } H) \cdot P(\textit{not } H) \cdot P(\textit{not } H)$$

$$=(0.7) \cdot (0.7) \cdot (0.7) \cdot (0.7)$$

$$=0.2401$$

There is less than a 1% chance of him getting four hits, and a 24% chance of him getting no hits.

(c) It is tricky to compute the probability of getting at least one hit. We could compute the probability of one hit or two hits or three hits or four hits. But this is very hard, so we do something easier. We say that Bob will either get no hits or at least one hit.

No hits	At least one hit
0	1, 2, 3, 4
Prob. = 24%	Prob. = 76%

We see from the table that getting at least one hit is the complement of getting no hits. Since $P(\text{no hits}) = 24\%$, $P(\text{at least one hit}) = 76\%$.

Problem Ten coins are flipped. What is the probability of at least one tail coming up?

ANSWER Again, it would be very hard to compute the probability of 1 tail *or* 2 tails *or . . . or* 10 tails. So we use the same trick as in the last question. We say that getting at least 1 tail is the same as *not* getting all heads. This is true, since the only way not to get all heads is to get at least 1 tail.

We already know the probability of getting 10 heads. That was $P(10 \text{ heads}) = \left(\dfrac{1}{2}\right)^{10} = \dfrac{1}{1024}$. Now we use Formula 6–1–4.

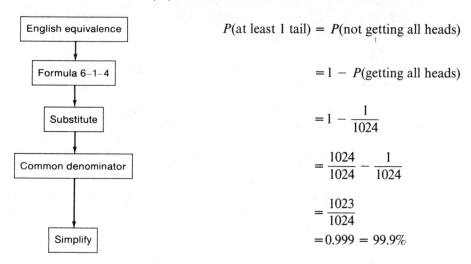

| English equivalence | $P(\text{at least 1 tail}) = P(\text{not getting all heads})$ |

$$= 1 - P(\text{getting all heads})$$

Formula 6–1–4

$$= 1 - \frac{1}{1024}$$

Substitute

$$= \frac{1024}{1024} - \frac{1}{1024}$$

Common denominator

$$= \frac{1023}{1024}$$

Simplify

$$= 0.999 = 99.9\%$$

If we flip 10 coins, there is a 99.9% chance of getting at least one tail.

None	At least one
0	1, 2, 3, 4, 5, etc.

We can see from the table that at least one is always the complement of none. So instead of finding the probability of at least one directly, we find the probability of none first; then we subtract this from 1. Thus we get

Formula 6–3–3

$$P(\text{none}) = P(not\ A) \cdot P(not\ B) \cdot P(not\ C) \ldots$$

Formula 6–3–4

$$P(\text{at least one}) = 1 - P(\text{none})$$

To calculate the probability of at least one of A, B, C, and so on, happening, when they are all independent,

1. Calculate the probabilities $P(not\ A)$, $P(not\ B)$, and so on.
2. Multiply these together to get $P(\text{none})$, using Formula 6–3–3.
3. Subtract this from 1, using Formula 6–3–4.

Problem Jack has a 10% chance of getting an A in mathematics, a 25% chance of an A in English, a 40% chance of an A in political science, and a 5% chance of an A in chemistry. What is the probability (a) of Jack getting all A's? (b) of getting no A's? (c) of Jack getting at least one A?

ANSWER To sort this problem out, let us fill in the following table:

Event	P(event)	P(not event)
M = A in mathematics	0.10	0.90
E = A in English	0.25	0.75
P = A in political science	0.40	0.60
C = A in chemistry	0.05	0.95

(a) To find the probability of all A's we use Formula 6–3–2 for the probabilities in column P(event).

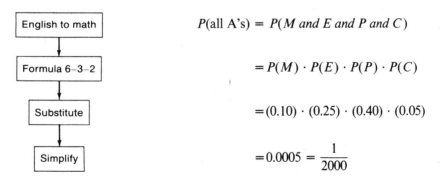

$$P(\text{all A's}) = P(M\ and\ E\ and\ P\ and\ C)$$

$$= P(M) \cdot P(E) \cdot P(P) \cdot P(C)$$

$$= (0.10) \cdot (0.25) \cdot (0.40) \cdot (0.05)$$

$$= 0.0005 = \frac{1}{2000}$$

(b) To find the probability of getting no A's, we use Formula 6–3–3 for the probabilities in column P(not event).

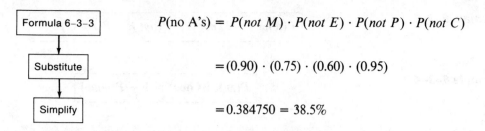

$$P(\text{no A's}) = P(not\ M) \cdot P(not\ E) \cdot P(not\ P) \cdot P(not\ C)$$

$$=(0.90) \cdot (0.75) \cdot (0.60) \cdot (0.95)$$

$$=0.384750 = 38.5\%$$

(c) Jack has a 38.5% chance of getting no A's at all. To find the probability of his getting at least one A, we use this and Formula 6–3–4.

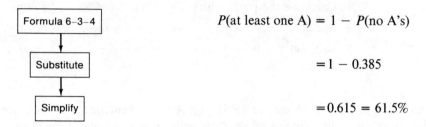

$$P(\text{at least one A}) = 1 - P(\text{no A's})$$

$$= 1 - 0.385$$

$$= 0.615 = 61.5\%$$

This makes sense. If there is a 38.5% chance of getting no A's, then there is a 61.5% chance of getting at least one A, which is the same as not getting no A's.

PROBLEM SET 6–3–2

1. Define or discuss the term *at least one*.

Complete the following table for coin flipping.

	Event	Probability
2.	Flip 1 head	
3.	Flip 2 straight heads	
4.	Flip 3 straight heads	
5.	Flip 4 straight heads	
6.	Flip 5 straight heads	
7.	Flip 10 straight heads	
8.	Flip 20 straight heads	
9.	Flip 100 straight heads	

10. If 9 straight heads have been flipped, what is the probability of the tenth flip being a head? How is this different from asking for the probability of 10 straight heads?

11. Mary has an 80% chance of passing history, and a 90% chance of passing mathematics. What is the probability of her passing both courses?

12. The CIA is sending a secret message in three parts: A, B, and C. The Russians are trying to intercept the message. There is a 5% chance to intercept message A, a 10% chance to intercept B, and a 20% chance to intercept C. If the Russians need all three parts to understand the message, what is the probability of their getting the whole message?

Alice and Paula are both fans of Terry Charisma, the actor. There is a 50% chance that Alice saw him on the local talk show, and a 30% chance that Paula did.

13. What is the probability that both saw Terry?

14. What is the probability that neither saw Terry?

15. What is the probability that at least one of them saw Terry?

Lisa is taking four courses, and has a certain probability of passing each of the courses. Fill in the following table.

Subject	P (passing)	P (not passing)
English	0.90	
History	0.80	
Mathematics	0.40	
Psychology	0.95	

16. What is the probability of passing English and psychology?

17. What is the probability of passing all four courses?

18. What is the probability of her failing at least one course?

19. What is the probability of failing all four courses?

Frannie is a waitress at a restaurant. There is an 80% chance she can get a plate of spaghetti from the kitchen to the customer's table. Five people have just ordered spaghetti from Frannie.

20. What is the probability that all five will get their spaghetti without an accident?

21. What is the probability that Frannie will spill at least one plate?

22. What is the probability that Frannie will spill all five plates?

Danny rolls three dice.

23. What is the probability of getting three 6's?

24. What is the probability of not getting any 6's?

25. What is the probability of at least one 6?

Al and Nancy are the same age. According to insurance tables, Al has a 32% chance of living to be 75, and Nancy has a 47% chance of living to be 75.

26. What is the probability that they will both live to be 75?

27. What is the probability that Nancy will live to 75 and Al won't?

28. What is the probability that Al will live to 75 and Nancy won't?

29. What is the probability that neither of them will live to be 75?

The odds against Neil's Nebish winning the first race are 3 to 1. The odds against Scoop Susan winning the second race are 4 to 1.

30. What is the probability of Neil's Nebish winning the first race?

31. What is the probability of Scoop Susan winning the second race?

32. What is the probability of them both winning?

33. What are the odds against them both winning?

34. What is the probability that at least one of them will win?

Chains of Events and Trees

Problem   What is the probability of pulling two straight aces out of a deck of cards?

ANSWER   The probability of pulling the first ace is $\frac{4}{52}$ or $\frac{1}{13}$. If we do pull this first ace, there are 3 aces left in the remaining 51 cards. So, there is a $\frac{3}{51}$ or $\frac{1}{17}$ probability. This says that $\frac{1}{13}$ of the time we will pull the first ace; then $\frac{1}{17}$ of those times, we will draw the second ace. So

$$P(\text{two straight aces}) = \frac{1}{13} \cdot \frac{1}{17} = \frac{1}{221}$$

Problem   There is a 90% chance that a person will sometime in their life get married. Once married, there is a 35% chance that he or she will ultimately get a divorce. What is the probability that a person will marry and then get a divorce?

ANSWER   Suppose that we start out with 100 people. Ninety of these will someday get married. Of this 90, 35% will get a divorce. This is 35% of 90, or 31.5. Of the original 100 people, 31.5 will get a divorce; this is 31.5%. This is also

$$P(\text{marriage, then divorce}) = (0.90) \cdot (0.35) = 0.315 = 31.5\%$$

In both of these cases, we had a chain of events. A **chain of events** is a sequence of events: A, then B, then C, and so on. First A happened, then B

happened, and so on. Suppose that we have two events, A and then B. Suppose that we know $P(A)$, which is the probability of A happening, and $P(B$ after $A)$, which is the probability of B happening after event A has already happened. Then we have

Formula 6-3-5

$$\boxed{P(A, \text{ then } B) = P(A) \cdot P(B \text{ after } A)}$$

In this formula, A and B do not even have to be independent. To make up for this we have to know what effect A has on B. That is why we must use $P(B$ after $A)$ in Formula 6-3-5.

Why is this formula true? Suppose that $P(A) = x\% = \dfrac{x}{100}$, and $P(B$ after $A) = y\% = \dfrac{y}{100}$. Suppose that we have 10,000 trials of this experiment. (We use 10,000 because it is a convenient number.) Now event A will happen about $x\%$ of 10,000, which is $\dfrac{x}{100} \cdot (10,000) = 100x$. Of these $100x$ times that A happens, B will follow $y\%$ of the time. This is $\dfrac{y}{100} \cdot (100x) = xy$. So out of 10,000 trials A and then B will occur xy times. Then the probability of A followed by B is $\dfrac{xy}{10,000}$. This is the same as

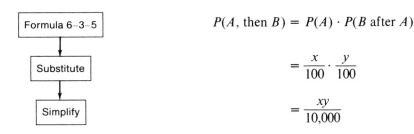

$$P(A, \text{ then } B) = P(A) \cdot P(B \text{ after } A)$$

$$= \frac{x}{100} \cdot \frac{y}{100}$$

$$= \frac{xy}{10,000}$$

Problem There is a $\dfrac{1}{100}$ chance that a black mother and father are both carriers of a sickle-cell gene. If they are both carriers, there is a $\dfrac{1}{4}$ chance that a child of theirs will actually have sickle-cell anemia. What is the probability of a black man and woman having a child with sickle-cell anemia?

ANSWER Let C be the event that both parents are carriers. Let S be the event that the child gets sickle-cell anemia. We know that $P(C) = \dfrac{1}{100}$, and $P(S$ after $C) = \dfrac{1}{4}$. We can use Formula 6-3-5.

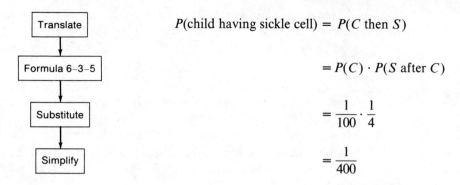

$P(\text{child having sickle cell}) = P(C \text{ then } S)$

$= P(C) \cdot P(S \text{ after } C)$

$= \dfrac{1}{100} \cdot \dfrac{1}{4}$

$= \dfrac{1}{400}$

So every black child has a $\dfrac{1}{400}$ chance of having sickle-cell anemia.

Problem There is a 70% chance that Becky will be asked to the big dance. If she is asked, there is an 80% chance that she will have a good time. What is the probability that she will be asked and have a good time?

ANSWER We again use Formula 6–3–5 to get

$$P(\text{asked, then a good time}) = P(\text{asked}) \cdot P(\text{good time after asked})$$

$$= (0.70) \cdot (0.80) = 0.56$$

We can also use a tree diagram to picture the possibilities. First, she may be asked or not asked. We draw this as shown.

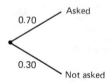

Notice that we have put the probabilities on each of the paths. Now, if she is asked, she may have good time or a bad time. We draw this as shown.

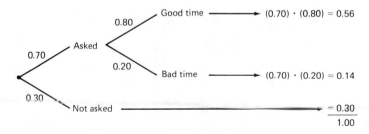

Notice that we multiplied all the probabilities along the paths. So there is a 56% chance that she will be asked and have a good time, a 14% chance that she will be asked and have a bad time, and a 30% chance that she will not be asked at all. This, of course, totals 100% since these are all the possibilities, and something has to happen.

The picture used in the preceding is called a tree diagram. **A tree diagram** is a picture that shows all the possible outcomes of two or more consecutive experiments. If an event B follows an event A, we draw a *branch* from A to B. If events B and C can both follow A, we draw branches from A to B and A to C as shown.

Then, on each branch, we can put the probabilities of B after A, and C after A. For example, if $P(B$ after $A) = 0.6$, and $P(C$ after $A) = 0.4$, we get the tree as sketched.

To find the probability of a chain, we multiply the probabilities on the branches of the chain.

Problem Two coins are flipped. What are possible outcomes and their probabilities?

ANSWER We use a tree diagram. The first toss can come up heads or tails. Regardless of the first toss, the second toss can also come up heads or tails. Everything has probability $\frac{1}{2}$. We draw this as shown.

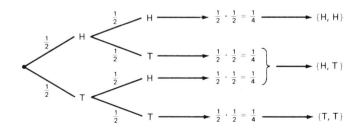

So the events and their probabilities are

Event	Probability
{H, H}	$\frac{1}{4}$
{H, T}	$\frac{1}{2}$
{T, T}	$\frac{1}{4}$

Problem The Beavers and the Raccoons are playing a two out of three series for the world's 16-inch softball title. The Beavers are a 3 to 2 favorite to win any single game. What is the probability that they will win the two out of three series?

ANSWER Since the odds for the Beavers winning a game are 3 to 2, this says that P(Beavers winning a game) $= \dfrac{3}{3+2} = 0.6$. For simplicity, let B be the Beavers winning a game, and R be the Raccoons winning a game. So $P(B) = 0.6$, and $P(R) = 0.4$.

We will draw the tree diagram. Remember that a second win for either team ends the series. We will circle a game that wins the series. We will then follow the paths that lead to the Beavers winning the series.

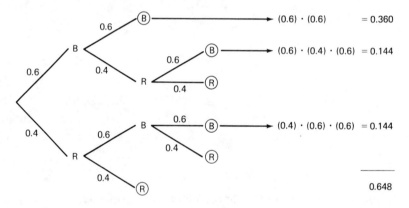

There are three chains that end in a Beavers' victory in the series. The total probability is $0.648 = 64.8\%$. It is surprising that the Beavers have a 60% chance to win each separate game, but a 64.8% chance to win the series. Why do you think this is so?

Problem Ron is graduating from college in June. There is a 50% chance that he will get a job with the government (federal, state, or local.) There is a 30% chance that he will get a business job, and a 20% chance that he will drive a cab. If he works for the government, there is a 20% chance that he will advance; but if he works in business, there is a 40% chance that he will advance. If he drives a cab, there is no chance for advancement. What are Ron's chances for advancement?

ANSWER To sort this problem out, let us label all the events. Let

The job interview. (Sybil Shelton/Monkmeyer Press Photo Service)

G = work for the government

B = work for business

C = drive a cab

A = advance on the job

At the first stage is getting the job: G, B, or C. This gives the diagram shown.

Now, for each one of these we have two possibilities: A or *not A*. Since he has a 20% chance of advancing with a government job, the branch from G to A will have a 0.2 on it; the branch from G to *not A* will have a 0.8 on it. The others are the same. If we follow out the advancement branches, we see that Ron has a 22% chance of advancing when he graduates.

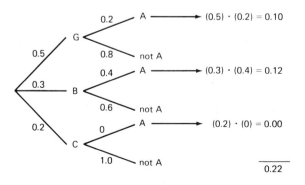

Problem Jenifer has two tennis serves, a hard serve (first serve) and a soft serve (second serve.) She knows the following:

> *Hard serve*: It has a 40% chance of getting in (being good). If it is good, she has a 70% chance of winning the point.
>
> *Soft serve*: It has a 90% chance of getting in (being good). If it is good, she has a 50% chance of winning the point.

Draw a tree diagram for her service possibilities. What is the probability of her winning the point?

ANSWER Like most tennis players, Jenifer serves hard on the first serve. It is either good or bad. If it is good, she can then win or lose the point. If it is bad, she serves a soft second serve. This can either be good or bad. If it is good, she can win or lose the point. If it is bad, she loses the point by double fault. In the tree shown, we see that there are two chains that lead to winning the point. We multiply the branches and get $(0.4)(0.7) = 0.28$ for one of them, and $(0.6)(0.9)(0.5) = 0.27$ for the other. Together, this says that Jenifer has a 55% chance to win the point on her serve.

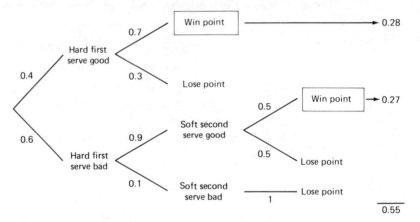

Problem Suppose that Jenifer were to hit both serves hard. Draw the tree diagram and find the probability of her winning the point.

ANSWER

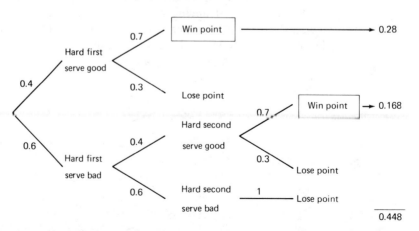

The tree will look similar to the last one. Notice that there are also two ways to win the point. The probabilities are $(0.4)(0.7) = 0.28$, and $(0.6)(0.4)(0.7) = 0.168$. Together, this says her chance of winning the point is $0.448 = 44.8\%$. This is probably why most tennis players use a soft second serve.

PROBLEM SET 6-3-3

1. Define or discuss:
   (a) Chain of events.
   (b) Tree diagram.

For a normal deck of 52 cards:

2. What is the probability of pulling a club?

3. What is the probability of pulling another club (after the first is drawn)?

4. What is the probability of pulling a third straight club? A fourth straight club? A fifth straight club?

5. What is the probability of pulling five straight clubs?

6. Mary has registered for English 252. There is a 20% chance that she will get Professor Anglophiletti. If she does, there is a 40% chance that she will fail the course. What is the probability that she get Anglophiletti and then fail?

7. The Chicago Cubs have a $\frac{1}{50}$ chance of winning their division. If they win their division, there is a $\frac{2}{5}$ chance of them winning the pennant. If they win the pennant, there is a $\frac{1}{3}$ chance of them winning the World Series. What is the probability of the Cubs winning the World Series?

A jar contains 10 white balls, 7 red balls, and 5 green balls. What is the probability of

8. Pulling out a white ball?

9. Pulling out three straight white balls?

10. Pulling a white ball, then a green ball, then a red ball?

11. Not pulling a white ball on three straight pulls?

12. For the jar in Problems 8–11, draw the tree for pulling out two balls in succession. Use the tree to compute the probability that both balls are the same color.

Use the following tree to answer the questions.

13. Find $P(B) =$

14. Find $P(B, \text{then } H) =$

15. Find $P(E) =$

16. Find $P(G \text{ or } J) =$

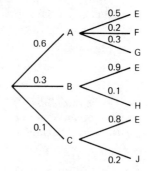

There is a 70% chance that Art will get married before he graduates from college. If he gets married then, he has only a 20% chance of going to law school. If he does not get married, he has a 65% chance of going to law school.

17. Draw a tree diagram for this situation.

18. Use the diagram to compute the probability of Art going to law school.

Consider again the tennis problem in which Jenifer has the two serves. Draw the tree and find the probability of winning the point in the following cases:

19. Soft serve, soft serve.

20. Soft serve, hard serve.

6-4   EXPECTED VALUE

Question   Tom and Lee are flipping coins. If a head comes up, Tom wins $1. If a tail comes up, Tom loses $1. Is this a fair game?

ANSWER   Yes. On any flip, Tom has a 50% chance of winning $1 and a 50% chance of losing $1. This cancels out to give $0. So the game is fair.

Meaning of Expectation

Problem   Sherri has won an $800 TV on "Let's Make a Deal." Should she trade the TV for a choice of three doors, where one door has a $2700 vacation, another has a $1200 bedroom set, and the third has $60 worth of candy bars?

ANSWER   Here we have a sample space of three outcomes: $U = \{$vacation, bedroom set, candy$\}$. The probability of choosing any one of these is $\frac{1}{3}$. Also, in addition to a probability, each outcome has some payoff. Consider the following chart:

Event	Probability	Payoff	Product
Vacation	$\frac{1}{3}$	$2700	$900
Bedroom set	$\frac{1}{3}$	1200	400
Candy	$\frac{1}{3}$	60	20
			$E = \overline{\$1320}$

In the last column, we multiplied the probability of each event by its payoff. Sherri has a $\frac{1}{3}$ chance of winning $2700, which is $900; she has a $\frac{1}{3}$ chance of winning $1200, which is $400; and she has a $\frac{1}{3}$ chance of winning $60, which is $20. If we add $900 + $400 + $20, we get $1320, which we call the expected value, E.

Question   What does the $1320 mean?

ANSWER   Obviously, Sherri will win either $2700, $1200, or $60. But the $1320 represents a form of average winnings. We can give a reasonable interpretation of this. If Sherri were allowed to make the choice 1000 times, her total winnings would be about $1,320,000, or about $1320 per choice on the average. Since $1320 is greater than the $800 TV, she should probably take her chances with the doors.

We have seen that every outcome in an experiment has a certain probability of occurring. Also, in a great many experiments, each outcome has some payoff. This **payoff** is a set of numbers that goes with each outcome of the experiment. This payoff is usually in dollars, but it doesn't have to be.

Suppose that the outcomes of an experiment are $O_1, O_2, \ldots, O_n$, each with probability $p_1, p_2, \ldots, p_n$ and payoff $m_1, m_2, \ldots, m_n$. Then we define the **expected value** as

Formula 6–4–1

$$E = p_1m_1 + p_2m_2 + \cdots + p_nm_n$$

In other words, for each outcome we multiply the payoff by the probability of getting that payoff. The expected value is essentially an average payoff. We can summarize the computation in the following table:

Outcome	Probability	Payoff	Product
O_1	p_1	m_1	p_1m_1
O_2	p_2	m_2	p_2m_2
.	.	.	
.	.	.	.
.	.	.	.
O_n	p_n	m_n	p_nm_n
			$E = \overline{p_1m_1 + p_2m_2 + \cdots + p_nm_n}$

> To compute expected value:
>
> 1. Determine all outcomes.
> 2. Determine the probability for each outcome.
> 3. Determine the payoff for each outcome.
> 4. For each outcome, multiply the probability by the payoff.
> 5. Add up all the products.

Problem Jerry Narman plays $10 on red at the roulette wheel. What are his expected winnings? What does this mean?

ANSWER On a roulette wheel there are 18 reds, 18 blacks, and 2 greens. If the balls lands on red, Jerry wins $20; if it lands on black or green, he gets nothing. So we have

Outcome	Probability	Payoff	Product
Red	$\dfrac{18}{38}$	$20	$\dfrac{\$360}{38}$
Black	$\dfrac{18}{38}$	0	0
Green	$\dfrac{2}{38}$	0	0
			$E = \dfrac{\$360}{38}$

So Jerry's expected winnings are $\dfrac{360}{38} = \$9.47$. But Jerry made a $10 bet, so his net winnings are -53 cents; in other words, he averages a 53 cents loss every time he plays. This means that if he played $10 on red 1000 times, he could expect to lose about $530, or an average of 53 cents a game.

Question But Jerry has a system. How are his chances now?

ANSWER Las Vegas loves people with a system. An old joke exposes the myth about systems.
Merchant: Mister, I lose 50 cents on every radio I sell.
Customer: How do you make any money?
Merchant: Volume!
The joke may not be terribly funny, but its absurdity is clear. Now consider Jerry's system in the light of that joke.
Jerry: My expected loss on every roulette game is 50 cents.
Friend: How do you win any money?
Jerry: I play a lot with a system.
The mathematics of Las Vegas is inescapable. If you play long enough, you will probably be a big loser. Of course, there will be a few winners, but the house is still averaging about 53 cents on every $10 bet. The games are unfair. We say that a

game is **fair** if the net winnings are $0, or the price of playing the game equals the expected payoff. This means that there are no expected winnings or losings, so the game is fair to both players. Otherwise, the game is **unfair**.

More Examples

Problem   Marv is 30 years old. He wants to buy $50,000 worth of life insurance. An insurance agent quotes him a rate of $5 per thousand, or $250 for a year. Is this fair to Marv?

ANSWER   Insurance can be viewed as a game. You are making a bet that you will die. Unfortunately, you have to die to win the bet. According to the mortality tables (see Section 6–2), the chances of a 30 year old dying are 2.13 per 1000 or 0.00213. The probability of Marv living is then 0.99787. If Marv dies, the payoff to his widow is $50,000. If he lives, he gets nothing. So we have

Outcome	Probability	Payoff	Product
Lives	0.99787	$ 0	$ 0.00
Dies	0.00213	50,000	106.50
			$E = \overline{\$106.50}$

    Marv's expected payoff is $106.50, and since his premium is $250, this means his expected *loss* is $250 − 106.50 = \$143.50$. (As bad a bet as this is, most of us still must buy insurance.) This loss is easy to interpret. If the insurance company sells 1000 policies just like this one to people just like Marv, the company will make about $143,500 profit, or an average of $143.50 per customer.

(Courtesy of Illinois State Lottery)

Problem   Ellen buys a lottery ticket for $1. There are 1 million tickets sold. The grand prize is $300,000; there are two second-place prizes of $50,000; there are also four

$10,000 prizes, ten $1000 prizes, one hundred $100 prizes, and one thousand $20 prizes. What are Ellen's expected winnings?

ANSWER The lottery is summed up in the following table.

Outcome	Probability	Payoff ($)	Product ($)
First prize	$\dfrac{1}{1,000,000}$	300,000	$\dfrac{300,000}{1,000,000}$
Second prize	$\dfrac{2}{1,000,000}$	50,000	$\dfrac{100,000}{1,000,000}$
Third prize	$\dfrac{4}{1,000,000}$	10,000	$\dfrac{40,000}{1,000,000}$
Fourth prize	$\dfrac{10}{1,000,000}$	1,000	$\dfrac{10,000}{1,000,000}$
Fifth prize	$\dfrac{100}{1,000,000}$	100	$\dfrac{10,000}{1,000,000}$
Sixth prize	$\dfrac{1000}{1,000,000}$	20	$\dfrac{20,000}{1,000,000}$
No prize	$\dfrac{998,883}{1,000,000}$	0	$\dfrac{0}{1,000,000}$

$$E = \frac{\$480,000}{1,000,000}$$
$$= \$0.48$$

So the expected payoff for a ticket is 48 cents. Since Ellen paid $1 for the ticket, she can expect to lose 52 cents per ticket. Thus this lottery like most lotteries is unfair to the player.

Expected value can also be used to help people decide among several alternative courses of action. We calculate the expected value for each course of action, and choose the one with the highest expected value.

Problem Glen is trying to decide what career to pursue: sales, teaching, or managing a store. Here are the data that Glen has.

1. In sales, there is a 10% chance of making $60,000, a 20% chance of making $20,000, and a 70% chance of making $7000 per year.
2. In teaching, there is an 80% chance of making $13,000 and a 20% chance of making $18,000 per year.
3. In managing, there is a 30% chance of making $17,000 and a 70% chance of making $11,000 per year.

What career should Glen choose?

ANSWER We will compute the expected earnings of each career.

Sales

Outcome	Probability	Payoff	Product
Great	0.10	$60,000	$6,000
Good	0.20	20,000	4,000
Poor	0.70	7,000	4,900

$$E_{sales} = \$14,900$$

Teaching

Outcome	Probability	Payoff	Product
Good	0.20	$18,000	$3,600
Fair	0.80	13,000	10,400

$$E_{teaching} = \$14,000$$

Managing

Outcome	Probability	Payoff	Product
Good	0.30	$17,000	$5,100
Fair	0.70	11,000	7,700

$$E_{managing} = \$12,800$$

Assuming Glen is interested purely in money, he will choose sales, since the expected earnings are the highest.

On the other hand, examples like this can trick us into thinking that expected value is always the number to use.

Question Which would you prefer:
(a) I will give you $1 million.
(b) I will flip a coin; heads you get $4 million, tails you get $0.

ANSWER All but the most pathological gambler would choose (a), a sure million. But the mathematics of (b) is

Outcome	Probability	Payoff	Product
Heads	0.5	$4,000,000	$2,000,000
Tails	0.5	0	0

$$E = \$2,000,000$$

Expected value is merely one number that somehow averages the payoffs in proportion to the likelihood that they will occur. We will leave it to the social scientists to determine why people would prefer a sure million instead of a game with an expected value of two million, why people buy life insurance, or why people play the lottery. These are not mathematics questions.

Expected Number

Problem   A die is rolled 300 times. What is the expected number of 4's?

ANSWER   The probability of a 4 is $\frac{1}{6}$. So out of 300 rolls we would expect $\frac{1}{6} \cdot (300) = 50$ fours.

This suggests one last formula, which should seem common sense. Suppose that A is an event with probability $P(A) = p$. If the experiment is repeated N times, the expected number of times A will occur is given by

Formula 6–4–2

$$\boxed{E = N \cdot p}$$

Why should this formula be true? Remember that most of our probabilities were defined by a formula

$$p = \frac{E}{N} = \frac{\text{number of times } A \text{ occurs}}{\text{total number of experiments}}$$

If we know p and N, we can multiply across to get $E = N \cdot p$.

Problem   The probability of a person being born a twin is about $\frac{1}{45}$. There are about 220,000,000 people in America. What is the expected number of Americans who are twins?

ANSWER   Here we have $p = \frac{1}{45}$ and $N = 220,000,000$. By Formula 6–4–2,

$$E = N \cdot p = (220,000,000) \cdot \frac{1}{45} = 4,888,889$$

Problem   At a certain university, there is a 0.6 chance that an entering freshman will ultimately graduate. Last year the freshman class was 2500. What is the expected number of students who will graduate?

ANSWER   Here $p = 0.6$ and $N = 2500$; so $E = (0.6) \cdot (2500) = 1500$.

PROBLEM SET 6–4–1

1. Define or discuss:
   (a) Payoff.
   (b) Expected value.
   (c) Fair game.
   (d) Unfair game.

Compute the expected value for the following sample spaces.

2.

Outcome	A	B	C	D
Probability	0.1	0.2	0.3	0.4
Payoff	10	40	100	200

3.

Outcome	x	y	z	t	w
Probability	0.1	0.2	0.4	0.2	0.1
Payoff	100	200	300	400	500

4.

Outcome	m	n	o	p	q	r
Probability (%)	20	30	40	5	3	2
Payoff	1000	2000	3000	4000	5000	6000

5. Bob is playing a dice game. He rolls one die. If the die comes up 1, 2, 3, or 4, he wins as many dollars. If it comes up 5 or 6, he loses as many dollars. What is Bob's expected payoff? Is this a fair game?

6. Ann signs up for a chemistry course. She figures she has a 10% chance to get an A, a 20% chance to get a B, a 30% chance for a C, a 30% chance for a D, and a 10% chance for an F. Using a scale with A = 4, compute Ann's expected grade.

7. Charlie bets $10 on 17 at the roulette wheel. This will pay off $360 if a 17 comes up, but nothing if any other number comes up. What are Charlie's expected winnings? (Compare this to betting $10 on Red.)

8. Lisa rates boys on a scale from 0 to 10, where 0 = disgusting and 10 = great. Lisa is going out on a blind date. She and her friends have figured the probabilities of getting the differently rated boys as follows.

Rating	0	1	2	3	4	5	6	7	8	9	10
Probability (%)	8	13	20	23	18	8	4	3	2	1	0

Find the expected rating of her blind date. (The rating is the payoff in this case.)

9. In a certain raffle, 1000 tickets are sold. There is one $250 prize, three $100, ten $10, and twenty $5 prizes.
(a) What is the expected payoff for a ticket?
(b) If the ticket is $1, is this lottery fair?

Fred is 38 years old. He needs $40,000 of life insurance. Safe Form Insurance charges him a $290 premium.

10. What is the probability of Fred dying while he's 38? (Use Table 6-1.)

11. What is the expected payoff of this policy?

12. What is the expected profit for Safe Form?

13. Ajax TV estimates that the probability of producing a defective TV is 0.02. A big chain of stores has just bought 5000 Ajax TV's. What is the expected number of defective TV's?

14. Ken Klutch is a .310 hitter in baseball. If he bats 550 times in a season, what is the expected number of hits he will get?

A scientist has accumulated the following data for a certain species of rat:

Life span (weeks)	1	2	3	4	5	6
Probability (of a rat living that long)	0.1	0.1	0.4	0.3	0.05	0.05

15. Are these probabilities countable, scientific, or guesstimate?

16. What is the probability that a rat will live exactly 4 weeks?

17. What is the probability that a rat will live 4 weeks or more?

18. What is the probability of a given rat living to be 6 weeks?

19. What is the probability of two given rats both living to be 6 weeks?

20. What is the probability of a rat dying before it reaches 6 weeks?

23. If the scientist has 300 rats, how many can be expected to live 5 or 6 weeks?

REVIEW EXERCISES Give an appropriate English word or phrase for the likelihood that the following situations will happen.

1. Richard Nixon will re-enter politics.

2. A rock group's next album will sell 2,000,000 copies.

For the following experiments, write out the sample space of all possible outcomes. Give an appropriate phrase for their likelihoods. Then assign each outcome a probability.

3. The situation in the Middle East.

4. Asking someone for a date.

5. Asking your father for $20.

Let U = {standard deck of cards}, F = {face cards}, and C = {clubs}.

6. Find *not C*.

7. Find *F and C*.

8. Find *F or C*.

For the set $\{a, b, c, d, e\}$, which of the following are possible probabilities for the set? Tell why or why not.

9. $P(a) = 15\%$ 10. $P(a) = \dfrac{1}{2}$

   $P(b) = 27\%$ $P(b) = \dfrac{1}{4}$

   $P(c) = 36\%$ $P(c) = \dfrac{1}{8}$

   $P(d) = 20\%$ $P(d) = \dfrac{1}{16}$

   $P(e) = 4\%$ $P(e) = \dfrac{1}{16}$

11. If $P(A) = 0.43$, what is $P(not\ A)$?

12. If $P(B) = \dfrac{2}{15}$, what is $P(not\ B)$?

13. What is the probability of pulling a red king from a shuffled 52-card deck?

14. A radio station has 10 new cars to give away. If there are 3 million listeners who are eligible for the car, what is the probability of a listener winning a car?

15. Denise is buying a new car. There are 7 models, 12 colors, 3 transmissions, 5 interiors, 2 exteriors, 3 types of tires, 4 types of radio/stereo, power windows or not, power steering or not, air conditioning or not, and 3 types of doors. How many combinations is this?

16. In a recent year 394 out of every 717 high school graduates entered college. What is the probability that a high school graduate will enter college?

17. If $P(A) = 0.3$, find the odds against A happening.

18. If the odds for A are 6 to 5, find $P(A)$.

19. If $P(A) = 0.15$, what is a fair bet to make on the event A?

Which of the following pairs of events are independent?

20. A = your mother is a Pisces.
   B = flipping a head on a coin.

21. C = this is April.
   D = it is raining today.

22. What is the probability of flipping four straight heads?

Al, Bonnie, Chuck, and Donna are applying to medical school. Their chances of getting in are given in the following table:

Name	Probability of getting into medical school
Al	0.4
Bonnie	0.6
Chuck	0.2
Donna	0.3

23. What is the probability that all of them will get in?

24. What is the probability that none of them will get in?

25. What is the probability that at least one of them will get in?

There are 20 people in a room. Three of them are women. Two door prizes are to be chosen randomly.

26. What is the probability that a woman will win the first prize?

27. If a woman wins the first prize, what is the probability that another woman will win the second prize?

28. What is the probability that two women will win the two prizes?

The Sioux City Squirrels are 7 to 3 favorites over the Council Bluff Bluffers in each game of a best of three-game series.

29. What is the probability that the Squirrels will win any game?

30. Draw the tree diagram to show the possibilities. (Remember that the first team to win two games wins the series.)

31. What is the probability of the Squirrels winning the series?

32. Find the expected value of the following sample space.

Outcome	A	B	C	D
Probability	0.25	0.30	0.35	0.10
Payoff	100	200	300	400

33. In a certain raffle there are 10,000 tickets. There is one $3000 prize, one $1000 prize, ten $150 prizes, 100 $10 prizes, and 100 $2 prizes. What are the expected winnings of a ticket? If a ticket is $1, is this a fair lottery?

34. There are about 2500 problems in this textbook. The probability that the author will make a mistake in answering any one of the problems is 0.01. How many errors can the author expect to make in working out all the problems for an answer book?

Row houses, San Francisco. (© Fred Lyon from Rapho/Photo Researchers)

7

STATISTICS

Question How would we make heads or tails out of a group of numbers such as:

78	83	117	27	815	431	552	418	107
607	20	315	871	661	310	293	458	263

ANSWER These days, everyone collects data: governments, schools, baseball teams, businesses, farmers, weathermen, scientists, and even astrologers. The only way to make sense out of billions and billions of numbers is with statistics.

Statistics is a set of methods used to study and simplify groups of numbers and other data. In Chapter 4, we saw how to use graphs to display numbers. In this chapter, we will study other ways to study mountains of numbers.

7–1 CENTRAL TENDENCY

Problem Seven men are sitting at a table. Their weights are 163, 172, 179, 185, 191, 192, and 304. What is the typical weight?

ANSWER One of the goals of statistics is to take a pile of numbers and represent it by one typical number. Here are two popular ways to find one typical number to represent all the weights.

1. The first method is to take a simple average. We add up all the weights and divide by 7.

$$\text{Average} = \frac{163 + 172 + 179 + 185 + 191 + 192 + 304}{7}$$

$$= \frac{1386}{7} = 198 \text{ lb}$$

The average weight is 198 pounds.

2. Another method to find the typical weight is to find the weight that cuts the group in half.

$$163 \quad 172 \quad 179 \quad \boxed{185} \quad 191 \quad 192 \quad 304$$

When written in order, the weight 185 cuts the group in half: three are above, and three are below.

Question Which is right, 198 or 185?

ANSWER Both methods are common methods of finding one typical number to represent a whole group of numbers. (The term for this is **measure of central tendency**.) We could debate the case for each number:

1. We get the number 198 by adding up all the weights and dividing by 7. This number came from all the men's weights. There were 1386 total pounds at the table. So the average weight is $\frac{1386}{7} = 198$ pounds for each man. This is called the *mean* weight.

2. Someone else might say that 198 is not typical since six of the seven men are below this weight. The 304 pounder forced up the average and distorted the figure. The number 185 is more typical since it is right in the middle. This is called the *median* weight.

It still isn't clear which number is more typical. Both the mean and median have their uses. In this section, we will look at both of them.

Mean

Question   What is the mean?

ANSWER   One way to find a typical number for a group of N numbers is simply to add up all the numbers and divide by N. This is called the **mean**.

Formula 7–1–1

$$\text{mean} = \frac{x_1 + x_2 + x_3 + \cdots + x_N}{N}$$

where $x_1, x_2, \ldots, x_N$ are the numbers in the group, and N = total number in the group.

> To find the *mean* of a group:
>
> 1. Add up all the numbers.
> 2. Divide by the total number, N.

In the previous weight problem, the mean was 198 pounds. We also use the mean when we find average test scores, average income, average age, or average number of televisions in a house.

Problem   Find the mean of the following numbers: 17, 29, 35, 49, 26, 51, 33, 41, 68, 72.

ANSWER   The mean is just the average of these numbers. There are 10 numbers, so $N = 10$. Formula 7–1–1 tells us to add up all these numbers and divide by 10.

Formula 7–1–1

$$\text{mean} = \frac{x_1 + x_2 + \ldots + x_N}{N}$$

Substitute

$$\text{mean} = \frac{17 + 29 + \ldots + 68 + 72}{10}$$

Simplify

$$= \frac{421}{10} = 42.1$$

Thus the mean is 42.1.

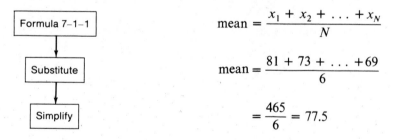

MATH FLASH
`1 2 3 4 5 6 7 8`

Hand Calculator Instant Replay

PUNCH	DISPLAY	MEANING
C	0.	Clear
1 7	17.	
+ 2 9	29.	
+ 3 5	35.	
+ 4 9	49.	
+ 2 6	26.	Add up numbers
+ 5 1	51.	
+ 3 3	33.	
+ 4 1	41.	
+ 6 8	68.	
+ 7 2	72.	
=	421.	Total sum
÷ 1 0	10.	Divided by N
=	42.1	Mean (answer)

Problem Ann takes six math exams and gets the following scores: 81, 73, 79, 88, 75, 69. What is the mean?

ANSWER Here, again, the mean is the usual average.

Formula 7–1–1	$\text{mean} = \dfrac{x_1 + x_2 + \ldots + x_N}{N}$
Substitute	$\text{mean} = \dfrac{81 + 73 + \ldots + 69}{6}$
Simplify	$= \dfrac{465}{6} = 77.5$

Problem The following are the temperatures (Fahrenheit) taken in Iceberg, Idaho, every day at noon for a week in February: $+12$, $+3$, -6, -8, -10, -2, $+4$. What was the mean temperature for the week?

ANSWER To find the mean, we again add up all the numbers and divide by 7. We must be careful here since we have both positive and negative temperatures.

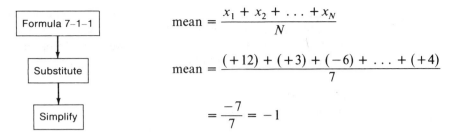

Formula 7-1-1

$$\text{mean} = \frac{x_1 + x_2 + \ldots + x_N}{N}$$

Substitute

$$\text{mean} = \frac{(+12) + (+3) + (-6) + \ldots + (+4)}{7}$$

Simplify

$$= \frac{-7}{7} = -1$$

This tells us that the mean temperature for the week was $-1°F$. Notice that it is perfectly legitimate to have a negative mean.

Problem The Humperts are a family of five: Dick, 34; Pat, 32; Jason, 7; Scott, 4; Amy, 2. What is the mean age of the household (a) now? (b) in 5 years? (c) in 10 years? (d) in 15 years? (e) in 20 years?

ANSWER   (a)   The mean age now is simply

$$\text{mean} = \frac{34 + 32 + 7 + 4 + 2}{5} = \frac{79}{5} = 15.8 \text{ yr}$$

(b)   In 5 years, everyone is 5 years older, so

$$\text{mean} = \frac{39 + 37 + 12 + 9 + 7}{5} = \frac{104}{5} = 20.8 \text{ yr}$$

Notice that the mean age in 5 years is exactly 5 more than it is now.
(c)   In 10 years, everyone will be 10 years older, so

$$\text{mean} = \frac{44 + 42 + 17 + 14 + 12}{5} = \frac{129}{5} = 25.8 \text{ yr}$$

Notice that the mean is exactly 10 years more than it is now since everyone is 10 years older.
(d)   By now, you are probably tricked into thinking that in 15 years the mean household age will be 15 years more, or 30.8. Wrong! Jason and Scott, now 22 and 19, have moved out of the house, leaving only Dick, 49; Pat, 47; and Amy, 17. So

$$\text{mean} = \frac{49 + 47 + 17}{3} = \frac{113}{3} = 37.7 \text{ yr}$$

Notice that the mean age jumped by almost 12 years in just 5 years' time. This was because the group size changed.
(e)   In 20 years, Amy will also have left home, leaving only Dick, 54; and Pat, 52. Now

$$\text{mean} = \frac{54 + 52}{2} = \frac{106}{2} = 53 \text{ yr}$$

Notice how the mean jumped again. This can happen when someone is computing the age of a town. Sometimes the mean can jump way up as many young adults leave home. The mean might drop many years if a lot of young couples have babies.

PROBLEM SET 7–1–1 **1.** Define or discuss:
   (a) Measure of central tendency.
   (b) Mean.

Find the mean for each of the following sets of numbers.

2. 17, 25, 29, 45, 31, 56, 62, 44, 51, 83, 42, 51, 68

3. 125, 223, 317, 419, 1275

4. 17,000, 22,500, 19,400, 22,900, 25,100, 32,000, 27,100, 24,600, 18,600, 14,100, 25,500, 27,600, 29,800, 30,700

5. 27.1, 8.2, 5.7, 16.1, 15.8, 13.7, 18.7, 16.8, 18.0, 14.2, 21.7, 23.3, 20.1, 17.5, 18.8, 19.9, 21.6, 22.0, 14.7

6. 79, 82, 53, 93, 77, 54, 75, 68, 81, 87, 73, 65, 62, 71, 82, 72, 61, 91, 100, 49, 83, 84, 75, 82, 89, 75, 79

7. 605, 776, 832, 651, 481, 503, 442, 517, 609, 667, 723, 801, 889, 635, 552, 617, 844, 655, 765, 572

8. The following are the number of home runs hit by the National League Home Run Champions for 30 years (1947–1976): 51, 40, 54, 47, 42, 37, 47, 49, 51, 43, 44, 47, 46, 41, 46, 49, 44, 47, 52, 44, 39, 36, 45, 45, 48, 40, 44, 36, 38, 38. Find the mean.

Below is a table of average temperatures in certain U.S. cities for each month. Find the mean temperature for each city for the entire year. (All temperatures are Fahrenheit.)

City	Jan.	Feb.	Mar.	Apr.	May	June	July	Aug.	Sep.	Oct.	Nov.	Dec.	Mean
9. Atlanta	42	45	51	61	69	76	78	78	72	62	51	44	?
10. Boston	29	30	38	49	59	68	73	71	65	55	45	33	?
11. Chicago	24	27	37	50	60	71	75	74	66	55	40	29	?
12. Dallas	45	49	56	66	74	82	86	86	78	68	56	48	?
13. Fairbanks	−12	−3	10	29	47	59	61	55	44	25	3	−10	?
14. Honolulu	72	72	73	75	77	79	80	81	80	79	77	74	?
15. Los Angeles	57	48	59	62	65	68	73	74	73	68	63	58	?
16. Miami	67	68	71	75	78	81	82	83	82	78	72	68	?
17. Nashville	38	41	49	60	69	77	80	79	72	61	48	40	?
18. New York	32	33	41	52	62	72	77	75	68	58	47	35	?
19. Phoenix	51	58	57	67	81	88	94	93	85	74	61	55	?
20. San Francisco	48	51	53	55	58	62	63	63	64	61	55	50	?

21. The following are the ages of the U.S. presidents on the day that they first became president: 57, 61, 57, 57, 58, 57, 61, 54, 68, 51, 49, 64, 50, 48, 65, 52, 56, 46, 54, 49, 51, 47, 55, 54, 42, 51, 56, 55, 51, 54, 51, 60, 62, 43, 55, 56, 61, 52. Find the mean age of a new president.

22. The following are the heights of the Sioux City Tigers basketball team: 74, 79, 73, 78, 80, 82, 75, 74, 77, 76, 79, 78. Find the mean height (in inches).

23. Fran belongs to a weight-loss club. The following are the number of pounds each member is overweight: 10, 26, 35, 42, 18, 22, 47, 73, 52, 24, 18, 33, 41, 15, 28, 32, 48, 91, 53, 41, 32, 39, 20, 32, 29, 36, 13. Find the mean amount of overweight.

24. Professor Gaines has the following test scores: 78, 85, 72, 91, 98, 63, 75, 84, 89, 99, 73, 79, 69, 88, 84, 87, 92, 76, 74, 83, 82, 71, 66, 96, 83, 91, 75, 70. Find the mean score.

25. Professor Smith has the following test scores: 53, 47, 63, 21, 38, 57, 45, 61, 41, 32, 42, 12, 71, 52, 61, 40, 37, 27, 44, 54, 62, 42, 19, 33. Find the mean score.

26. Gary is an entomologist (a scientist who studies insects). He has measured the wingspan (in millimeters) of 13 different butterflies as follows: 10, 23, 41, 32, 17, 28, 50, 37, 32, 41, 29, 25, 35. Find the mean wingspan.

27. On a certain street, Joan recorded the number of televisions each family had: 2, 4, 3, 2, 2, 1, 2, 1, 1, 0, 1, 2, 1, 1, 2, 3, 0, 3, 4, 3, 0, 0, 1, 2, 2, 1, 2, 3, 2, 2, 1, 3, 2. Find the mean.

28. Lillian wants to know the mean number of pages in a library book. She pulls 20 books at random off the shelves and records their lengths as follows: 196, 310, 248, 272, 451, 118, 177, 507, 475, 356, 331, 513, 688, 521, 433, 277, 378, 292, 487, 963. Find the mean.

29. Ten alumni of a state college are chosen at random and asked their annual incomes, which are: $17,100, $22,500, $11,000, $52,000, $33,000, $27,300, $19,900, $30,100, $25,400, $14,200. Find the mean income.

Median

Question   What is the median?

ANSWER   A second way to find a typical number for a group is to find the number that cuts the group in half. This number is called the **median**.

Problem   Find the median of the numbers 16, 29, 23, 35, 18, 24, 31, 27, 29.

ANSWER   To find the median or the middle number, we must rearrange them in order.

16 18 23 24 (27) 29 29 31 35
↑
median

When we arrange the numbers in order, we see that 27 is right in the middle. Four are below 27, and four are above 27. Thus 27 is the median.

Problem Find the median of the following scores: 89, 116, 73, 105, 123, 174, 132, 120.

ANSWER We first arrange these scores in order, smallest to largest.

$$73 \quad 89 \quad 105 \quad 116 \quad 120 \quad 123 \quad 132 \quad 174$$
$$\uparrow$$
$$\text{median} = 118$$

Notice that we have a little problem here. We have 8 (an even number) scores, so it is impossible to find one score right in the middle. Instead, we take the number that is halfway between the central two numbers. Here, 116 and 120 are the two middle numbers; so we let the median be 118, the number halfway between 116 and 120.

> To find the *median* of a group of N numbers:
>
> 1. Arrange them in order, smallest to largest.
> 2. (a) If N is odd, the median is the middle number.
>   (b) If N is even, the median is halfway between the two middle numbers.

Problem Find the median temperature for the Iceberg, Idaho, problem of the last section.

ANSWER Recall that we added the seven temperatures and divided by 7 to get a mean of $-1°$. To find the median, we first arrange them in order:

$$-10 \quad -8 \quad -6 \quad \boxed{-2} \quad +3 \quad +4 \quad +12$$
$$\uparrow$$
$$\text{median}$$

Since we have an odd number of days (7), we see that the median is simply the middle number, $-2°$. Notice that the mean ($-1°$) and the median ($-2°$) are not the same.

Question Is there another way to find the median without rearranging the whole group of numbers?

ANSWER Yes. Notice that when we had nine numbers, the median was the fifth number from the bottom. When we had seven numbers, the median was the fourth number from the bottom. When we had eight numbers, the median was between the fourth and fifth numbers from the bottom. We can generalize this to the following rule:

To find the *median* of N numbers:

1. Compute the number $\dfrac{N + 1}{2}$. (Add 1 to N; then divide by 2.)

2. Start counting (1, 2, 3, and so on) from the smallest number of the group until you get to $\dfrac{N + 1}{2}$.

3a. If N is odd, this is the median.

3b. If N is even, the median is between two numbers.

Problem The following are Janet's test scores for her sixth-grade class. Find the median.

79	63	91	38	75
82	88	73	66	92
52	74	81	75	69
59	87	70	81	72
68	77	72	83	70

ANSWER There are $N = 25$ scores here. We could rearrange them in order, but it would be painful. Instead, we use the new rule and find $\dfrac{N + 1}{2} = \dfrac{25 + 1}{2} = \dfrac{26}{2} = 13$.

So, we count up to the thirteenth number from the smallest. We circle the numbers as we count.

79	63④	91	38①	75
82	88	73⑫	66⑤	92
52②	74⑬*	81	75	69⑦
59③	87	70⑨	81	72⑪
68⑥	77	72⑩	83	70⑧

* = median

We start with 38, the smallest number. We continue until we get to the thirteenth biggest number, which is 74. Thus 74 is the median. Incidentally, the mean happens to be 73.48, very close to the median.

Question Are the mean and median usually close?

ANSWER Usually. Both are supposed to be typical numbers, and usually are close. It is a coincidence if they are equal. The main difference is that the mean can be affected by a few very small or large numbers, and the median cannot.

The mean can be affected by an odd man out. (© Jim Jowers/Nancy Palmer Photo Agency)

Problem There are 20 employees at Chuck's Discount House. Their incomes are given below. Find the mean and median.

$8100	$8800	$10,200	$11,200	$20,000
$8100	$9300	$10,200	$11,700	$30,000
$8300	$9400	$10,300	$12,000	$50,000
$8400	$9700	$10,700	$13,000	$100,000

ANSWER First we will find the mean. This is done by simply adding up all the numbers and dividing by 20.

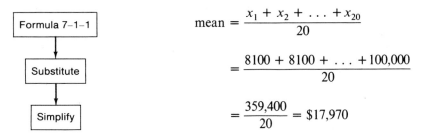

$$\text{mean} = \frac{x_1 + x_2 + \ldots + x_{20}}{20}$$

$$= \frac{8100 + 8100 + \ldots + 100,000}{20}$$

$$= \frac{359,400}{20} = \$17,970$$

To find the median, we compute $\dfrac{N + 1}{2} = \dfrac{20 + 1}{2} = \dfrac{21}{2} = 10\frac{1}{2}$. What is the $10\frac{1}{2}$th number? That sounds odd, but it means take the number halfway between the tenth and eleventh. The tenth number up from the smallest is $10,200. The eleventh number is $10,300. The median is then halfway in between, or $10,250.

Notice that the mean, $17,970, is much higher than the median, $10,250. Why? This is because there are three or four big salaries; all the others are smaller. The big salaries affect the mean, but do not affect the median.

PROBLEM SET 7–1–2

1. Define or discuss *median*.

2–29. Find the median for Problems 2–29 in Problem Set 7–1–1.

7–2 MEASURES OF SPREAD OR VARIABILITY

Problem Mitch and Connie are both sales representatives for a certain company. For the year, their monthly sales were as follows:

Month	Jan.	Feb.	Mar.	Apr.	May	June	Jul.	Aug.	Sep.	Oct.	Nov.	Dec.
Mitch	159	61	147	112	187	101	41	124	154	85	93	152
Connie	121	113	119	116	113	111	125	122	119	114	117	126

Find both of their mean and median monthly sales.

ANSWER First, let's find their mean sales. In both cases, we add up the sales and divide by 12.

Formula 7–1–1
↓
Simplify

$$\text{Mitch's mean} = \frac{159 + 61 + \ldots + 93 + 152}{12}$$

$$= \frac{1416}{12} = 118$$

$$\text{Connie's mean} = \frac{121 + 113 + \ldots + 117 + 126}{12}$$

Simplify

$$= \frac{1416}{12} = 118$$

Notice that their means are the same, 118.

Now we find the median sales by putting them in order, smallest to largest.

Mitch: 41 61 85 93 101 112   124 147 152 154 159 187

↑

median = 118

↓

Connie: 111 113 113 114 116 117   119 119 121 122 125 126

Since there are 12 months (an even number), we must go halfway between the sixth and seventh numbers to find the medians.

Notice again that the two medians are the same, 118. In fact, all the means and medians are the same, 118, which is a coincidence.

Question But something is obviously different between Mitch's and Connie's sales records. What is it?

ANSWER The mean or median can be used as a typical number for a group of numbers. As this problem shows, the mean and median cannot tell the whole story.

Mitch's sales jumped all over the place (high, low, medium). Connie's sales were fairly steady (all about medium). We would like to measure how much their sales jumped around. (The term for this is **measure of variability** or **measure of spread**.)

Question How do we measure the spread of the data?

ANSWER There are several ways. We will give two of the more common ones. The simplest way is to subtract the biggest number minus the smallest number. This difference is called the **range**. Put in a simple formula,

Formula 7–2–1

> **Range = biggest − smallest**

Problem Find the range for Mitch's and Connie's sales.

ANSWER We simply use Formula 7–2–1.

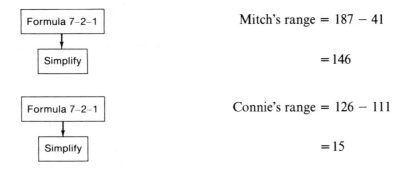

| Formula 7–2–1 | Mitch's range $= 187 - 41$ |
| Simplify | $= 146$ |

| Formula 7–2–1 | Connie's range $= 126 - 111$ |
| Simplify | $= 15$ |

Now we can see the difference between them that the mean and the median didn't tell us. The illustration shows how much more spread out Mitch's sales were than Connie's.

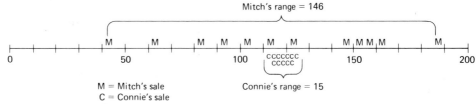

Mitch's range = 146

M = Mitch's sale
C = Connie's sale

Connie's range = 15

Problem Jeff gives an English test and a mathematics test to the 11 students. The scores are

English: 95, 48, 64, 62, 84, 68, 75, 65, 73, 77, 37
Mathematics: 42, 68, 91, 93, 88, 52, 48, 97, 39, 45, 85

Find the mean, median, and range for each test.

ANSWER (a) To find the means, we add up the scores and divide by 11.

$$\text{English mean} = \frac{\text{sum of scores}}{11} = \frac{748}{11} = 68$$

$$\text{Mathematics mean} = \frac{\text{sum of scores}}{11} = \frac{748}{11} = 68$$

(b) To find the medians, we put the scores in order.

English: 37 48 62 64 65 ⑥⑧ 73 75 77 84 95
median = 68

Mathematics: 39 42 45 48 52 ⑥⑧ 85 88 91 93 97

Notice, again, that by coincidence all the means and medians are the same, 68.

(c) To find the ranges, we subtract the smallest from the biggest score.

$$\text{English range} = 95 - 37 = 58$$

$$\text{Mathematics range} = 97 - 39 = 58$$

Notice that the ranges are the same, 58. But even though the means, medians, and range are the same, the data are still different. The illustration shows how the mathematics scores are more spread out than the English.

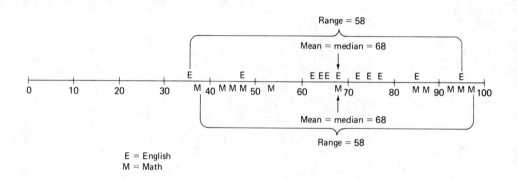

E = English
M = Math

People in statistics like to study how close all the data stay to the mean. The name for this is **standard deviation**. The formula for standard deviation is involved, but it is not as difficult as many formulas in statistics. We usually find the standard deviation in two steps. First we find the variance.

Formula 7–2–2

$$\text{Var} = \frac{x_1^2 + x_2^2 + \cdots + x_N^2}{N} - m^2$$

where m = the mean, N = total number in group, $x_1, x_2, \ldots, x_N$ are the scores, and Var = variance.

Once we have the variance, we find the standard deviation.

Formula 7–2–3

$$\sigma = \sqrt{\text{Var}}$$

where σ = standard deviation.

To find the *standard deviation*:

1. Find the mean.
2. Square all the scores.
3. Add up the squares.
4. Divide by N.
5. Subtract the (mean)^2 to get the variance.
6. Take the square root of the variance to get the standard deviation, σ.

Question   What does the standard deviation mean?

ANSWER   Essentially, the standard deviation is a measure of how much the data tend to spread away from the mean. We will see this in the next examples.

Problem   Find the standard deviation of Jeff's mathematics scores.

ANSWER   We use Formulas 7–2–2 and 7–2–3 together. We already know that the mean $m = 68$.

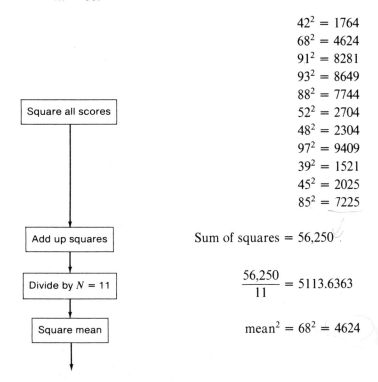

$$42^2 = 1764$$
$$68^2 = 4624$$
$$91^2 = 8281$$
$$93^2 = 8649$$
$$88^2 = 7744$$

Square all scores

$$52^2 = 2704$$
$$48^2 = 2304$$
$$97^2 = 9409$$
$$39^2 = 1521$$
$$45^2 = 2025$$
$$85^2 = 7225$$

Add up squares

Sum of squares = 56,250

Divide by $N = 11$

$$\frac{56,250}{11} = 5113.6363$$

Square mean

$$\text{mean}^2 = 68^2 = 4624$$

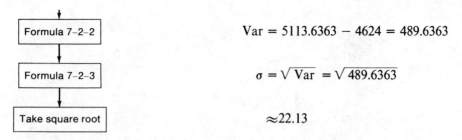

Formula 7–2–2	$\text{Var} = 5113.6363 - 4624 = 489.6363$
Formula 7–2–3	$\sigma = \sqrt{\text{Var}} = \sqrt{489.6363}$
Take square root	≈ 22.13

What does this mean? It tells us that on the mathematics test a typical spread from the mean 68 was about 22.13 points.

Hand Calculator Instant Replay

Find the standard deviation of:

$$42 \quad 68 \quad 91 \quad 93 \quad 88 \quad 52 \quad 48 \quad 97 \quad 39 \quad 45 \quad 85$$

(a) Find the mean first.

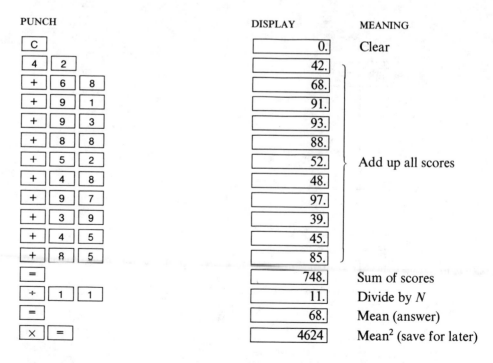

PUNCH	DISPLAY	MEANING
C	0.	Clear
4 2	42.	⎫
+ 6 8	68.	
+ 9 1	91.	
+ 9 3	93.	
+ 8 8	88.	
+ 5 2	52.	⎬ Add up all scores
+ 4 8	48.	
+ 9 7	97.	
+ 3 9	39.	
+ 4 5	45.	
+ 8 5	85.	⎭
=	748.	Sum of scores
÷ 1 1	11.	Divide by N
=	68.	Mean (answer)
× =	4624	Mean^2 (save for later)

(b) Find the variance and standard deviation.

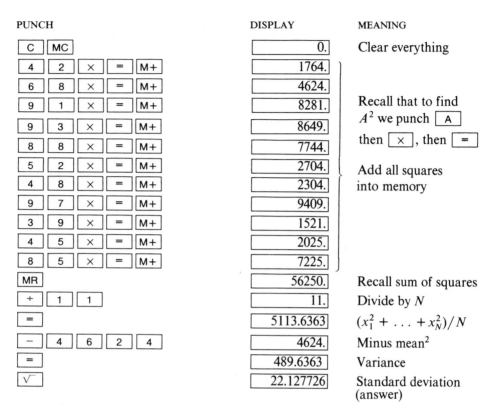

PUNCH	DISPLAY	MEANING
C MC	0.	Clear everything
4 2 × = M+	1764.	
6 8 × = M+	4624.	
9 1 × = M+	8281.	Recall that to find
9 3 × = M+	8649.	A^2 we punch A
8 8 × = M+	7744.	then ×, then =
5 2 × = M+	2704.	Add all squares
4 8 × = M+	2304.	into memory
9 7 × = M+	9409.	
3 9 × = M+	1521.	
4 5 × = M+	2025.	
8 5 × = M+	7225.	
MR	56250.	Recall sum of squares
+ 1 1	11.	Divide by N
=	5113.6363	$(x_1^2 + \ldots + x_N^2)/N$
− 4 6 2 4	4624.	Minus mean^2
=	489.6363	Variance
√	22.127726	Standard deviation (answer)

Problem   Find the standard deviation of the English scores.

ANSWER   We again know that the mean is 68. Recall that the data are

95   48   64   62   84   68   75   65   73   77   37.

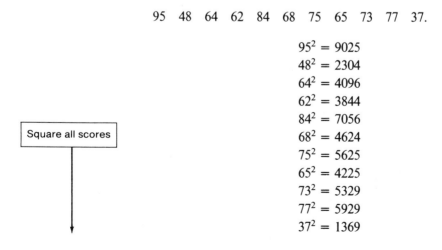

$$95^2 = 9025$$
$$48^2 = 2304$$
$$64^2 = 4096$$
$$62^2 = 3844$$
$$84^2 = 7056$$
$$68^2 = 4624$$
$$75^2 = 5625$$
$$65^2 = 4225$$
$$73^2 = 5329$$
$$77^2 = 5929$$
$$37^2 = 1369$$

Square all scores

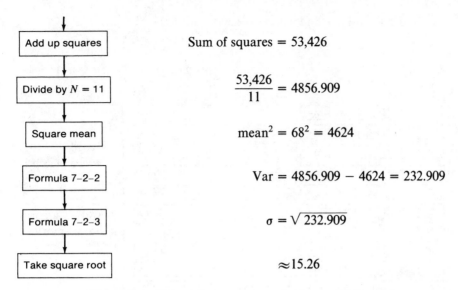

Add up squares	Sum of squares = 53,426
Divide by $N = 11$	$\dfrac{53,426}{11} = 4856.909$
Square mean	$\text{mean}^2 = 68^2 = 4624$
Formula 7–2–2	$\text{Var} = 4856.909 - 4624 = 232.909$
Formula 7–2–3	$\sigma = \sqrt{232.909}$
Take square root	≈ 15.26

This tells us that the average spread on the English test was about 15 points. Remember that the spread on the mathematics test was about 22. This was seen in the illustration, where the mathematics scores were more spread out than the English scores, even though they had the same range.

Problem The following are the premiums for $1000 of whole life insurance for a 25-year-old male. These are from 10 different insurance companies.
  (a) Find the mean premium.
  (b) Find the median premium.
  (c) Find the range.
  (d) Find the standard deviation.

$16.95	$17.82	$17.66	$18.19	$17.37
$15.66	$17.16	$18.34	$17.31	$17.72

ANSWER These are 10 typical premiums. It should be noted that each company's policy is slightly different.
  (a) To find the mean, we add up the premiums and divide by 10.

Formula 7–1–1	$\text{mean} = \dfrac{(16.95 + \ldots + 17.72)}{10}$
Simplify	$= \dfrac{174.18}{10} = \$17.418$
Round off	$= \$17.42$

(b) To find the median, we compute $\dfrac{10 + 1}{2} = \dfrac{11}{2} = 5\dfrac{1}{2}$. We want to find the number between the fifth and sixth from the smallest. This number is halfway between $17.37 and $17.66. So the median is $17.52. Notice that this is a little more than the mean of $17.42.

(c) The range is simply the largest premium minus the smallest. This is

$$\text{Range} = \$18.34 - \$15.66 = \$2.68$$

(d) To find the standard deviation, we first find the variance. To do this, we square all the premiums and add the squares. We then divide by 10 and subtract the square of the mean.

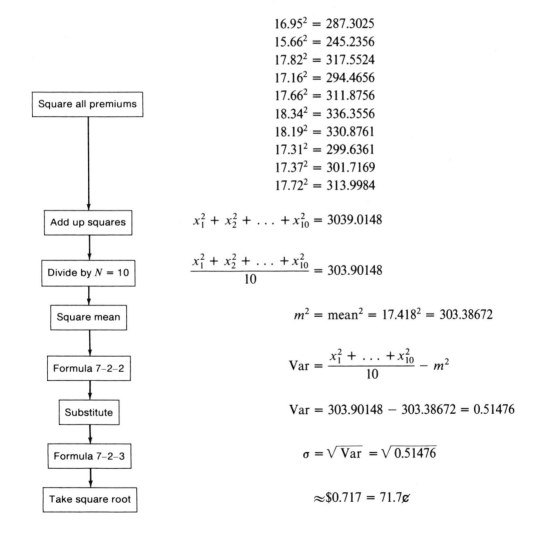

$$16.95^2 = 287.3025$$
$$15.66^2 = 245.2356$$
$$17.82^2 = 317.5524$$
$$17.16^2 = 294.4656$$
$$17.66^2 = 311.8756$$
$$18.34^2 = 336.3556$$
$$18.19^2 = 330.8761$$
$$17.31^2 = 299.6361$$
$$17.37^2 = 301.7169$$
$$17.72^2 = 313.9984$$

Square all premiums

Add up squares

$$x_1^2 + x_2^2 + \ldots + x_{10}^2 = 3039.0148$$

Divide by $N = 10$

$$\frac{x_1^2 + x_2^2 + \ldots + x_{10}^2}{10} = 303.90148$$

Square mean

$$m^2 = \text{mean}^2 = 17.418^2 = 303.38672$$

Formula 7-2-2

$$\text{Var} = \frac{x_1^2 + \ldots + x_{10}^2}{10} - m^2$$

Substitute

$$\text{Var} = 303.90148 - 303.38672 = 0.51476$$

Formula 7-2-3

$$\sigma = \sqrt{\text{Var}} = \sqrt{0.51476}$$

Take square root

$$\approx \$0.717 = 71.7\text{¢}$$

MATH FLASH

1 2 3 4 5 6 7 8

√	%	+/-	÷	MC
7	8	9	X	MR
4	5	6	−	M-
1	2	3	+	M+
CE	0	·	=	C

Hand Calculator Instant Replay

	PUNCH	DISPLAY	MEANING
(a)	C	0.	Clear
	1 6 · 9 5	16.95	
	+ 1 5 · 6 6	15.66	
	+ 1 7 · 8 2	17.82	
	+ 1 7 · 1 6	17.16	
	+ 1 7 · 6 6	17.66	Add all numbers
	+ 1 8 · 3 4	18.34	
	+ 1 8 · 1 9	18.19	
	+ 1 7 · 3 1	17.31	
	+ 1 7 · 3 7	17.37	
	+ 1 7 · 7 2	17.72	
	=	174.18	Total sum
	÷ 1 0	10	Divide by N
	=	17.418	Mean (answer)
	× =	303.38672	Mean^2 (Save for later)

	PUNCH	DISPLAY	MEANING
(b)	C	0.	Clear
	1 7 · 3 7	17.37	Fifth premium
	+ 1 7 · 6 6	17.66	Sixth premium
	=	35.03	Total
	÷ 2	2.	Divide by 2
	=	17.515	Median (answer)

	PUNCH	DISPLAY	MEANING
(c)	C	0.	Clear
	1 8 · 3 4	18.34	Biggest
	− 1 5 · 6 6	15.66	Minus smallest
	=	2.68	Range (answer)

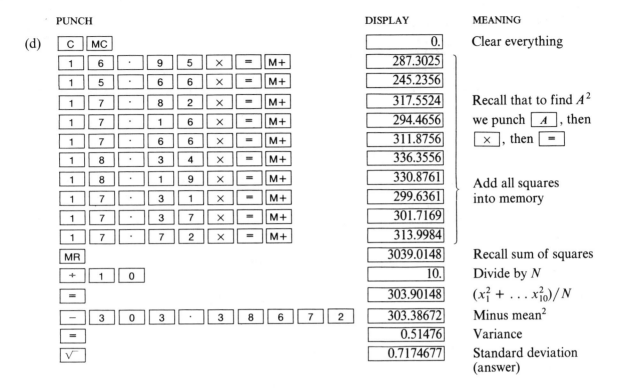

PUNCH	DISPLAY	MEANING
(d) C MC	0.	Clear everything
1 6 . 9 5 × = M+	287.3025	
1 5 . 6 6 × = M+	245.2356	
1 7 . 8 2 × = M+	317.5524	Recall that to find A^2
1 7 . 1 6 × = M+	294.4656	we punch A, then
1 7 . 6 6 × = M+	311.8756	$\times$, then $=$
1 8 . 3 4 × = M+	336.3556	
1 8 . 1 9 × = M+	330.8761	Add all squares
1 7 . 3 1 × = M+	299.6361	into memory
1 7 . 3 7 × = M+	301.7169	
1 7 . 7 2 × = M+	313.9984	
MR	3039.0148	Recall sum of squares
÷ 1 0	10.	Divide by N
=	303.90148	$(x_1^2 + \ldots x_{10}^2)/N$
− 3 0 3 . 3 8 6 7 2	303.38672	Minus mean^2
=	0.51476	Variance
$\sqrt{}$	0.7174677	Standard deviation (answer)

Question What do these numbers mean?

ANSWER We have 10 different premiums for basically the same insurance policy: $1000 whole life for a 25-year-old male. The typical premium is $17.42 if we use the mean, and $17.52 if we use the median. This is about what we can expect to pay for the insurance if we choose an insurance company randomly from the Yellow Pages.

The range is $2.68. This tells us that there is a difference of $2.68 between the cheapest and most expensive insurance in this survey.

The standard deviation is about 72 cents. This tells us that the average spread of the premiums is about 72 cents either way. Usually, about two thirds of the data are within this standard deviation. The figure summarizes this.

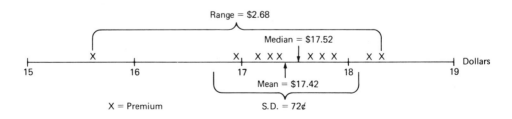

PROBLEM SET 7–2–1

PROBLEM SET 7–2–1

1. Define or discuss:
   (a) Range.
   (b) Measure of spread.
   (c) Standard deviation.

2–29. Find the range and the standard deviation of the data in Problems 2–29 of Problem Set 7–1–1.

7–3 NORMAL CURVE

Question What is the normal curve?

ANSWER The **normal curve** or **normal distribution** is a special curve in mathematics and statistics. It is sometimes called the *bell-shaped curve* because it looks somewhat like a bell.

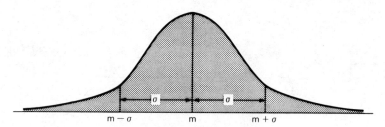

The figure shows a typical normal (or bell-shaped) curve. Notice that the peak is exactly at the mean, m. It is symmetric about the mean. (This means it looks exactly the same to the right and left of the mean.)

Not all normal curves look exactly like this figure. In fact, each normal curve depends on the mean, m, and the standard deviation, σ. In the next figure we have drawn two more normal curves. The one on the left has its mean at 30 and a standard deviation of about 10. The normal curve on the right has a mean of 60 and a standard deviation of about 30. These two curves also show how the standard deviation affects the curve. When the standard deviation is small (as on the left), the curve is compact with a tall peak. When the standard deviation is large (as on the right), the curve is wide with a lower peak.

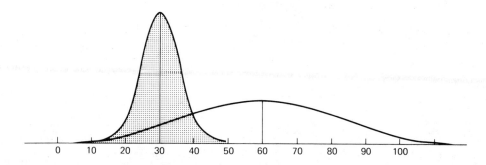

Question Why is the normal curve important?

ANSWER Recall in Chapter 4 that we discussed frequency graphs, which show how often a quantity occurs. The normal curve is very important since many things in real life have a frequency graph like this curve.

IQ scores, heights, weights, wages, and the like tend to have a normal curve when graphed. They have a peak at the mean, and most of the data are fairly close to the mean (depending on the standard deviation). The accompanying figure shows how the standard deviation is related to the amount of data close to the mean.

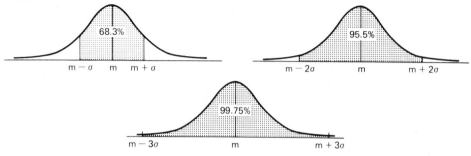

1. Within one standard deviation (1σ) is about 68.3% of the data.
2. Within two standard deviations (2σ) is about 95.5% of the data.
3. Within three standard deviations (3σ) is about 99.75% of the data.

Problem At Park Forest State College, the average height of the 1500 women is 5 feet, 4 inches (64 inches). The standard deviation is 2 inches. The heights have a normal distribution.
  (a) How many women are between 62 and 66 inches?
  (b) How many women are between 60 and 68 inches?
  (c) How many women are between 58 and 70 inches?

ANSWER Here is a problem where we can use the normal curve rule, since we are told that the heights have a normal distribution. The figure shows what this means graphically.

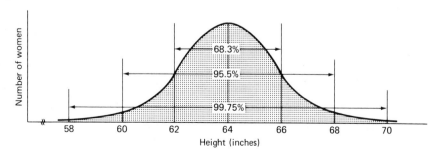

(a) The mean is 64 inches. Since the standard deviation σ is 2, 1σ either side of the mean is from 62 to 66. The rule tells us that this is 68.3% of 1500, or $(0.683)(1500) = 1025$ women.

(b) Two σ ($= 4$) either side of the mean is from 60 to 68. The rule tells us that this is 95.5% of 1500, or $(0.955)(1500) = 1433$ women.

(c) Three σ ($= 6$) either side of the mean is from 58 to 70. The rule tells us that this is 99.75% of 1500, or $(0.9975)(1500) = 1496$ women.

Problem Jane takes a standardized test that is scored with a mean of 500 and a standard deviation of 100. Suppose that 10,000 students take this test.

(a) How many score between 400 and 600?

(b) How many score between 300 and 700?

(c) How many score between 200 and 800?

ANSWER It is known that scores on this test have a normal (or bell-shaped) distribution. The figure shows what this means graphically.

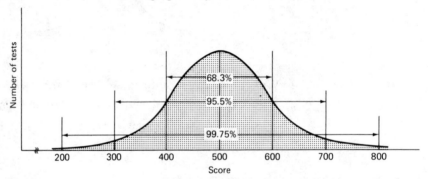

The mean is 500, and the standard deviation is 100. We can use the normal curve rule.

(a) One σ ($= 100$) either side of the mean 500 is from 400 to 600. The rule tells us that this is about 68.3% of 10,000, or $(0.683)(10,000) = 6830$ students.

(b) Two σ ($= 200$) either side of the mean 500 is from 300 to 700. The rule tells us that this is about 95.5% of 10,000, or $(0.955)(10,000) = 9550$ students.

(c) Three σ ($= 300$) either side of the mean 500 is from 200 to 800. The rule tells us that this is about 99.75% of 10,000, or $(0.9975)(10,000) = 9975$ students.

Notice that there are still about 25 students $(10,000 - 9975)$ who are not between 200 and 800. About 12 or 13 of them are very smart and are above 800. The other 12 or 13 are very dumb and are below 200.

PROBLEM SET 7–3–1 1. Define or discuss normal curve.

A math placement test is given to 960 entering freshmen. The mean score is 18. The standard deviation is 5. The scores have a normal curve.

2. How many students scored between 13 and 23?

3. How many students scored between 8 and 28?

4. How many students scored between 3 and 33?

Sugar-Junx is a box of cereal with a mean weight of 16 ounces. The standard deviation is 0.1 ounce. Suppose that the manufacturer packages 400,000 boxes of Sugar-Junx whose weights have a normal curve.

5. How many boxes weigh between 15.9 and 16.1 ounces?

6. How many boxes weigh between 15.8 and 16.2 ounces?

7. How many boxes weigh between 15.7 and 16.3 ounces?

8. How many boxes weigh under 15.7 or over 16.3 ounces?

9. How many boxes weigh under 15.7 ounces (half of the answer to Problem 8)?

The mean IQ score is 100 with a standard deviation of 15. Eight hundred children in a township take an IQ test, and their scores have a normal curve.

10. How many children will have an IQ between 85 and 115?

11. How many children will have an IQ between 70 and 130?

12. How many children will have an IQ between 55 and 145?

13. How many children will have an IQ under 55 or over 145?

14. How many children will have an IQ over 145?

The mean IQ of these children is 100 with a standard deviation of 15.
(Sam Falk/Monkmeyer Press Photo Service)

In a certain town the mean income is $10,500 with a standard deviation of $2000. The distribution of incomes has a normal curve. There are 10,000 workers in the town.

15. What is the range of incomes one standard deviation either side of the mean?

16. How many workers are in this range?

17. What is the range of incomes two standard deviations either side of the mean?

18. How many workers are in this range?

19. What is the range of incomes three standard deviations either side of the mean?

20. How many workers are in this range?

z-Scores

Question   Suppose that we want a percent of scores that is not exactly one, two, or three standard deviations from the mean?

ANSWER   We can use a special table to help us. But first we have to convert the scores to a z-score. A **z-score** is a standard score that tells us how many standard deviations a score is from the mean.

Formula 7–3–1

$$z = \frac{x - m}{\sigma}$$

where z = z-score, x = regular score, m = mean, and σ = standard deviation.

Problem   On a test, 75 is the mean and 8 is the standard deviation. What is the z-score of a score of 86?

ANSWER   Here 86 is the regular score. We use Formula 7–3–1 to find the z-score.

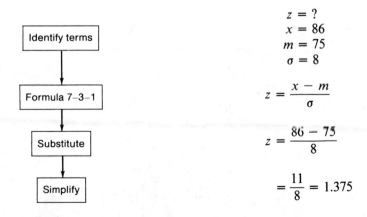

Identify terms

$$z = ?$$
$$x = 86$$
$$m = 75$$
$$\sigma = 8$$

Formula 7–3–1

$$z = \frac{x - m}{\sigma}$$

Substitute

$$z = \frac{86 - 75}{8}$$

Simplify

$$= \frac{11}{8} = 1.375$$

What does the number 1.375 mean? It tells us that a score of 86 is 1.375 times σ *above* the mean. We can see this in the figure.

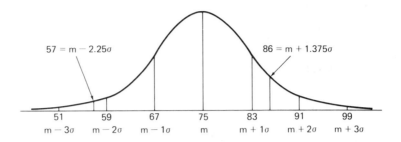

$$57 = m - 2.25\sigma \qquad\qquad 86 = m + 1.375\sigma$$

51	59	67	75	83	91	99
$m - 3\sigma$	$m - 2\sigma$	$m - 1\sigma$	m	$m + 1\sigma$	$m + 2\sigma$	$m + 3\sigma$

Problem On the same test, what is the *z*-score of a 57?

ANSWER Here 57 is the regular score. Again, the mean is 75 and standard deviation is 8.

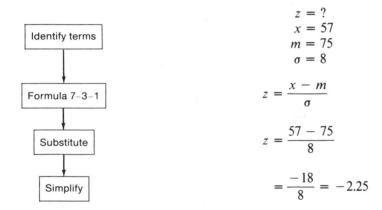

$$z = ?$$
$$x = 57$$
$$m = 75$$
$$\sigma = 8$$

$$z = \frac{x - m}{\sigma}$$

$$z = \frac{57 - 75}{8}$$

$$= \frac{-18}{8} = -2.25$$

What does -2.25 mean? It tells us that a score of 57 is 2.25 times σ *below* the mean. We can see this in the figure. A negative *z*-score always means that the score is below the mean.

Problem For a certain group of men, the mean weight is 182 pounds, with a standard deviation of 11 pounds. Bob's weight has a *z*-score of 1.4. What is his weight?

ANSWER We again use Formula 7–3–1. Now the unknown is *x*, but we are given $z = 1.4$.

$$z = 1.4$$
$$x = ?$$
$$m = 182$$
$$\sigma = 11$$

$$z = \frac{x - m}{\sigma}$$

$$1.4 = \frac{x - 182}{11}$$

$$15.4 = x - 182$$

$$197.4 = x$$

This means that Bob weighs about 197.4 lbs.

Question   How do we use z-scores?

ANSWER   We use z-scores to help us determine the percent of scores that are below or above a certain score, or the percent of scores that are between any two scores. To do this we have a special function, $F(z)$. If we have a normal curve, $F(z)$ tells us the percent of the scores below z. (Notice that we have to have a z-score to use this.) The values for $F(z)$ are given in Table 7–1.

Table 7–1   Values of $F(z)$

z	$F(z)$ (%)	z	$F(z)$ (%)	z	$F(z)$ (%)
− 3.0	0.1	− 1.0	15.9	+ 1.0	84.1
− 2.9	0.2	− 0.9	18.4	+ 1.1	86.4
− 2.8	0.3	− 0.8	21.2	+ 1.2	88.5
− 2.7	0.4	− 0.7	24.2	+ 1.3	90.3
− 2.6	0.5	− 0.6	27.4	+ 1.4	91.9
− 2.5	0.6	− 0.5	30.9	+ 1.5	93.3
− 2.4	0.8	− 0.4	34.5	+ 1.6	94.5
− 2.3	1.1	− 0.3	38.2	+ 1.7	95.5
− 2.2	1.4	− 0.2	42.1	+ 1.8	96.4
− 2.1	1.8	− 0.1	46.0	+ 1.9	97.1
− 2.0	2.3	0.0	50.0	+ 2.0	97.7
− 1.9	2.9	+ 0.1	54.0	+ 2.1	98.2
− 1.8	3.6	+ 0.2	57.9	+ 2.2	98.6
− 1.7	4.5	+ 0.3	61.8	+ 2.3	98.9
− 1.6	5.5	+ 0.4	65.5	+ 2.4	99.2
− 1.5	6.7	+ 0.5	69.1	+ 2.5	99.4
− 1.4	8.1	+ 0.6	72.6	+ 2.6	99.5
− 1.3	9.7	+ 0.7	75.8	+ 2.7	99.6
− 1.2	11.5	+ 0.8	78.8	+ 2.8	99.7
− 1.1	13.6	+ 0.9	81.6	+ 2.9	99.8
				+ 3.0	99.9

Question   How do we use Table 7–1?

ANSWER   We use the following rule.

To find $F(z)$ for any score on a normal curve:

1. Use Formula 7–3–1 to make it a z-score.
2. Find this z-score in Table 7–1.
3. $F(z)$ is next to it on the right.

For scores on a *normal curve*, we have the following formulas that use Table 7–1. The accompanying figure illustrates these formulas.

Formula 7–3–2

$$\% \text{ below } z = F(z)$$

Formula 7–3–3

$$\% \text{ above } z = 100\% - F(z)$$

Formula 7–3–4

$$\% \text{ between } z_1 \text{ and } z_2 = F(z_2) - F(z_1)$$

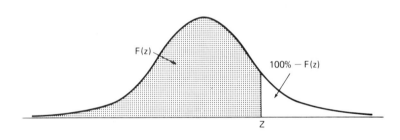

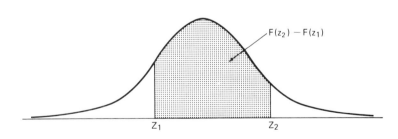

Problem On a test, 79 is the mean and 7 is the standard deviation. Find $F(z)$ for a score of 84.

ANSWER We first convert 84 to a z-score; then we use Table 7–1 to find $F(z)$.

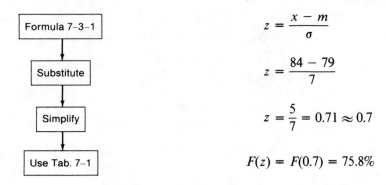

Formula 7–3–1	$z = \dfrac{x - m}{\sigma}$
Substitute	$z = \dfrac{84 - 79}{7}$
Simplify	$z = \dfrac{5}{7} = 0.71 \approx 0.7$
Use Tab. 7–1	$F(z) = F(0.7) = 75.8\%$

For this score of 84, the z-score is 0.71. In Table 7–1, we look up a z-score of 0.7 (the closest one) and find that $F(0.7) = 75.8\%$. This means that on a normal curve 75.8% scored below 84.

Problem   Kathy takes a standardized exam where 500 is the mean and 100 is the standard deviation. She scores 621. What percent of people scored below her? What percent scored above her? (The scores have a normal curve.)

ANSWER   We must first find the z-score for 621. Then we find $F(z)$ in Table 7–1. This is the percent below her score.

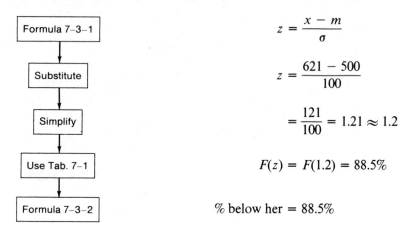

Formula 7–3–1	$z = \dfrac{x - m}{\sigma}$
Substitute	$z = \dfrac{621 - 500}{100}$
Simplify	$= \dfrac{121}{100} = 1.21 \approx 1.2$
Use Tab. 7–1	$F(z) = F(1.2) = 88.5\%$
Formula 7–3–2	% below her = 88.5%

Thus about 88.5% of the people taking the exam scored below Kathy. If 88.5% scored below her, then $100 - 88.5 = 11.5\%$ scored above her. This is what Formula 7–3–3 says, and the figure shows this result.

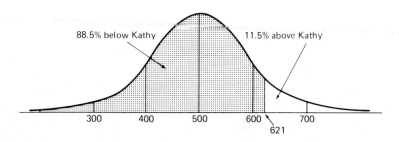

88.5% below Kathy   11.5% above Kathy

300   400   500   600   700

621

410

Problem The mean height of a group of men is 69 inches, with a standard deviation of 2.5 inches. The heights have a normal curve. What percent of these men are between 65 and 74 inches tall?

ANSWER Since this is a percent between two values, we will use Formula 7–3–4. We must first find the z-scores of both 65 and 74 inches. We will do this at the same time.

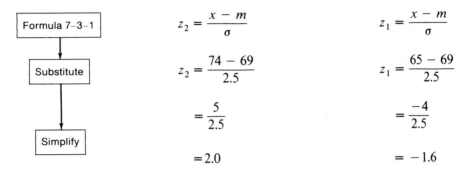

Formula 7–3–1	$z_2 = \dfrac{x - m}{\sigma}$	$z_1 = \dfrac{x - m}{\sigma}$
Substitute	$z_2 = \dfrac{74 - 69}{2.5}$	$z_1 = \dfrac{65 - 69}{2.5}$
	$= \dfrac{5}{2.5}$	$= \dfrac{-4}{2.5}$
Simplify	$= 2.0$	$= -1.6$

The z-scores are -1.6 and 2.0. Now we use Formula 7–3–4.

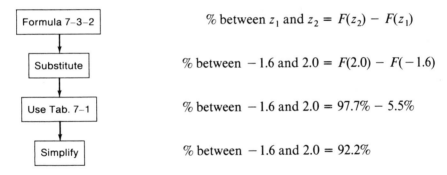

Formula 7–3–2	% between z_1 and $z_2 = F(z_2) - F(z_1)$
Substitute	% between -1.6 and $2.0 = F(2.0) - F(-1.6)$
Use Tab. 7–1	% between -1.6 and $2.0 = 97.7\% - 5.5\%$
Simplify	% between -1.6 and $2.0 = 92.2\%$

This tells us that 92.2% of the men are between 65 and 74 inches tall? This is shown in the figure.

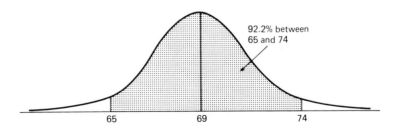

Problem For an adult IQ score, the mean is 100 with a standard deviation of 15.
(a) Mary has an IQ of 124. What percent of people are below her?
(b) What percent of people have IQ's between 105 and 120?

ANSWER (a) To find the percent of people below Mary, we first convert her 124 to a z-score. Then we use Table 7-1 to find $F(z)$, the percent below her.

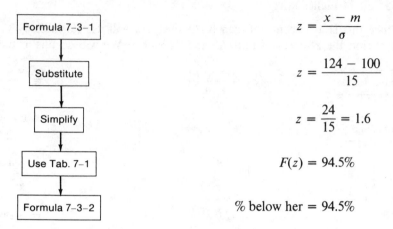

Formula 7–3–1

$$z = \frac{x - m}{\sigma}$$

Substitute

$$z = \frac{124 - 100}{15}$$

Simplify

$$z = \frac{24}{15} = 1.6$$

Use Tab. 7–1

$$F(z) = 94.5\%$$

Formula 7–3–2

$$\% \text{ below her} = 94.5\%$$

About 94.5% of the people have an IQ below Mary.
(b) To find the percent between 105 and 120, we find both their z-scores; then we use Formula 7–3–4.

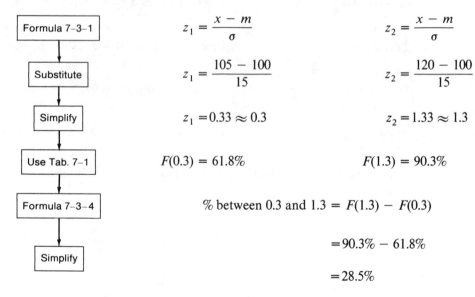

Formula 7–3–1

$$z_1 = \frac{x - m}{\sigma} \qquad\qquad z_2 = \frac{x - m}{\sigma}$$

Substitute

$$z_1 = \frac{105 - 100}{15} \qquad\qquad z_2 = \frac{120 - 100}{15}$$

Simplify

$$z_1 = 0.33 \approx 0.3 \qquad\qquad z_2 = 1.33 \approx 1.3$$

Use Tab. 7–1

$$F(0.3) = 61.8\% \qquad\qquad F(1.3) = 90.3\%$$

Formula 7–3–4

$$\% \text{ between 0.3 and 1.3} = F(1.3) - F(0.3)$$

Simplify

$$= 90.3\% - 61.8\%$$

$$= 28.5\%$$

Thus about 28.5% of the people have IQ's between 105 and 120.

PROBLEM   1. Define or discuss z-score.
SET 7–3–2

Complete the following table using Formula 7–3–1.

	x	m	σ	z
2.	67	60	5	?
3.	955	1,000	25	?
4.	610	550	40	?
5.	423	500	100	?
6.	350	250	60	?
7.	13,500	14,000	1,000	?
8.	?	600	75	1.8
9.	?	400	50	−0.7
10.	?	100	15	−0.3

Use Table 7-1 to complete the following table.

	z	F(z)			z	F(z)
11.	+ 1.9	?		16.	?	78.8%
12.	0.0	?		17.	?	90.3%
13.	− 0.7	?		18.	?	18.4%
14.	− 2.8	?		19.	?	1.1%
15.	+ 1.1	?		20.	?	50%

Complete the following table using Formulas 7–3–2 and 7–3–3.

	x	m	σ	z	F(z)	% below x	% above x
21.	108	100	15	?	?	?	?
22.	539	500	100	?	?	?	?
23.	431	500	100	?	?	?	?
24.	87	100	15	?	?	?	?
25.	1,045	1,000	150	?	?	?	?
26.	880	1,000	200	?	?	?	?
27.	660	600	70	?	?	?	?
28.	22,000	20,000	1,000	?	?	?	?
29.	72	69	2.5	?	?	?	?
30.	159	180	15	?	?	?	?

Complete the following table to find the percent between A and B. Use Formula 7–3–4. (z_A = z-score for A; z_B = z-score for B.)

	A	B	m	σ	z_A	z_B	$F(z_A)$	$F(z_B)$	% between A and B
31.	590	660	600	90	?	?	?	?	?
32.	700	1,000	1,200	200	?	?	?	?	?
33.	90	115	100	15	?	?	?	?	?
34.	110	135	100	15	?	?	?	?	?
35.	450	650	500	100	?	?	?	?	?
36.	66	74	70	3	?	?	?	?	?
37.	171	205	180	15	?	?	?	?	?

38. Art takes a standardized test where 500 is mean and 100 is the standard deviation. He scores 529. What percent scored above him? (The test has a normal curve.)

39. Angela gives a test to her English class. The test has a normal curve with a mean of 72 and standard deviation of 9. After class, a student asks where a score of 65 is. What percent of scores are below 65?

40. Sugar-Junx advertises that their box holds 16 ounces. To make sure, the manufacturer fills the box to a mean weight of 16.1 ounces with a standard deviation of 0.05 ounce.

(Mimi Forsyth/Monkmeyer Press Photo Service)

(No two boxes are ever the same, but their weights do have a normal curve: $m = 16.1$, $\sigma = 0.05$.) What percent of the boxes are actually 16.0 ounces or more?

7-4 SAMPLING

Question   How do poll takers, such as Gallup, Harris, and Nielson work?

Market research interview. (Sybil Shelton/Monkmeyer Press Photo Service)

ANSWER   Generally, these poll takers are trying to find how certain percents (or proportions) of a population think and act. For example, they may want to find out: what proportion of people go to church regularly; what percent of the voters in the state will vote for a senator; what percent of a student body feel that mathematics is a fun subject; or what percent of smokers have tried to quit within a year?

The surest way to answer these questions is to ask every single person. Unfortunately, this can be very expensive. So instead, people ask a small group. This is called **sampling**. If we sample properly, we can get a lot of information about a big group from a small group.

There will be two proportions (or percents): (1) the **sample proportion**, which we get from our small sample group; and (2) the **true proportion** of the whole population, which we probably never know.

In this section, we will not discuss the way that the sample is chosen. Rather, we will look at the mathematical formula used to understand the sample results.

Problem   A TV network wants to see what percent of the households are watching their show, "A Salute to Mathematics." They call 512 families, and 28 say that they are watching this show. What percent of the whole country is watching the show?

ANSWER   We will never know the exact answer unless we ask every single family. Since this is almost impossible and very expensive, the network asks 512 people chosen at random. We first find what percent of the sample watched the show. We then assume that the entire country has the same proportion.

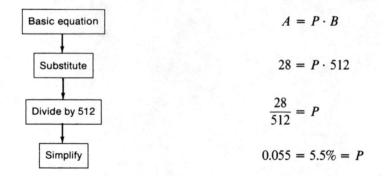

Basic equation	$A = P \cdot B$
Substitute	$28 = P \cdot 512$
Divide by 512	$\dfrac{28}{512} = P$
Simplify	$0.055 = 5.5\% = P$

Since 5.5% of the sample watched the show, we assume that 5.5% of the nation watched the show. This is what we assume for all samples.

When we have a sample, we always assume that the proportions are the same for the entire group. The answer might not be 100% accurate, but it is faster and cheaper than asking everyone.

Formula 7–4–1

$$p = \frac{E}{N}$$

where p = percent or proportion, N = total number in sample, and E = number in

HOW HAS GOVERNMENT CHANGED YOUR LIFE?

	IMPROVED IT	MADE IT WORSE	NO CHANGE
Federal government	23%	37%	34%
State government	27%	14%	52%
Local government	28%	11%	54%

Undecided omitted

(Fenga & Freyer, Inc.)

sample answering a certain way. We use Formula 7–4–1 to find the sample percent, and then assume that this is the same for the big group.

Problem A pollster asks 723 people on the street how they will vote in the next Senate election. The results are

$$
\begin{array}{lr}
\text{Senator Johnson} & 362 \\
\text{Nick Livingston} & 305 \\
\text{Others} & 26 \\
\text{Undecided} & \underline{30} \\
\text{Total} = & \overline{723}
\end{array}
$$

What are the proportions for the whole state?

ANSWER Using Formula 7–4–1, we divide all the answers by the total $N = 723$. This gives us the proportions for the sample. We convert these to percentages.

Senator Johnson	362	50.1%
Nick Livingston	305	42.2%
Others	26	3.6%
Undecided	30	4.1%
Total =	723	100.0%

Divide by 723

Convert to percent

We now assume that Senator Johnson will get about 50.1% of the vote, Mr. Livingston will get about 42.4%, other candidates will get about 3.6%, and 4.1% of the votes are undecided.

Problem Margie is taking a survey. She asks 231 people the question, "Do you believe there is life on other planets in the total universe?" To this, 129 people say yes. What percent of all people feel that there is life on other planets?

ANSWER　We use Formula 7–4–1, with $E = 129$ and $N = 231$.

$$p = \frac{129}{231}$$

$$= 0.558 = 55.8\%$$

This says 55.8% of the sample feel that there is life on other planets. So we assume that 55.8% of the whole population feels this way.

PROBLEM SET 7–4–1

1. Define or discuss:
   (a) Sample proportion.
   (b) True proportion.

Use Formula 7–4–1 to complete the following table.

	E (part)	N (total)	p (proportion or percent)
2.	65	95	?
3.	173	313	?
4.	14	45	?
5.	557	1,102	?
6.	302	?	12%
7.	159	?	0.65
8.	?	1,150	0.57
9.	?	442	23%

10. Jackie asks 155 people at random to name their state's two U.S. senators. The results are 73 could name both, 47 could name one, and 35 could name neither. What are the proportions for the whole state based on this sample?

11. Amy goes to a shopping center and surveys 225 passing shoppers. She asks them, "Do you think candy bars should be outlawed?" Twenty-two people said "yes" and 203 said "no." What are the proportions for the whole society?

12. Burt asks 331 people in downtown Chicago if they think the Chicago Cubs will go to the World Series this year. Twenty-one say "yes," and 310 say "no." What proportion of all Chicago thinks that the Cubs will go to the World Series?

Confidence Intervals

Question　How accurate are these percents that we get from small samples of big populations?

ANSWER　Obviously, if we ask a few people a question, it is not surprising that this sample percent might be slightly different from the true percent of the whole group. There is almost always some error.

The Gallup poll never claims to be exact, but is usually within about 3% of being correct. A. C. Nielson measures the popularity of TV shows by sampling about 1200 families. Nielson also claims to be within 3%. (Every star of a canceled show feels Nielson is way off.)

The statisticians are caught in a dilemma:

1. If they take a big sample, it will be accurate but too expensive.
2. If they take a small sample, it will be cheap but not too accurate.

So the polltaker has to compromise between cost and accuracy.

To measure the accuracy of a sample percent, statisticians have something called a confidence interval. Remember that there are two proportions in every situation: (1) the real proportion of the whole group (which we will probably never know), and (2) the sample proportion (which we can measure).

The **95% confidence interval (CI)** is shown in the accompanying figure. It is a region surrounding the sample percent into which we are about 95% sure that the true percent falls.

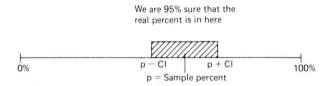

To find the confidence interval for a sample:

1. Find the sample proportion, p.

2. Find CI (using Formula 7–4–2).

3. The confidence interval goes from CI to the left of p to CI to the right of p. In other words, $p - $ CI to $p + $ CI.

4. We are 95% sure that the true percent is inside this interval.

How do we find the CI? The formula below is a very simple, yet close approximation.

Formula 7–4–2

$$CI \approx 2\sqrt{\frac{p(1 - p)}{N}}$$

where $p = $ sample proportion and $N = $ sample size.

Problem   One hundred and twenty students at Narr College are asked, "Is college life exciting or dull?" Seventy-eight students answer, "Dull."

    (a) What is the proportion who think college life is dull?

    (b) What is the confidence interval?

ANSWER   (a) To find the sample proportion, we use Formula 7–4–1, with $E = 78$ and $N = 120$. So $p = \dfrac{78}{120} = 0.65$ or 65%. We assume that 65% of the whole school feels that college life is dull.

    (b) How accurate is that 65% figure? This is where the confidence interval comes in. It will tell us how close to 65% the real percent probably is. First we compute CI with Formula 7–4–2.

(Doug Bruce/Camera 5)

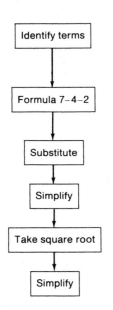

$$p = 0.65$$
$$1 - p = 0.35$$
$$N = 120$$

$$CI = 2\sqrt{\frac{p(1 - p)}{N}}$$

$$CI = 2\sqrt{\frac{(0.65)(0.35)}{120}}$$

$$= 2\sqrt{0.0018958}$$

$$= 2(0.0435)$$

$$= 0.087 = 8.7\%$$

Hand Calculator Instant Replay

PUNCH			DISPLAY	MEANING
(a) C			0.	Clear
7 8			78.	E
÷ 1 2 0			120.	Divided by N
=			0.65	p (answer)
(b) C			0.	Clear
. 6 5			0.65	p
× . 3 5			0.35	Times $(1 - p)$
÷ 1 2 0			120.	Divided by N
=			0.0018958	$p(1 - p)/N$
√			0.0435411	$\sqrt{p(1 - p)/N}$
× 2			2.	Times 2
=			0.0870822	CI (answer)

Question   What does CI = 8.7% mean?

ANSWER   We got a sample proportion that 65% of the students think college is dull. We doubt that 65% is the exact figure for the whole school, but the CI will help us to see how close it is.

The 95% confidence interval is 8.7% either way from the sample percent. So the real proportion might be up to 8.7% less (65 − 8.7 = 56.3%) or up to 8.7% more (65 + 8.7 = 73.7%). Thus we can be 95% sure that the real proportion is somewhere between 56.3 and 73.7% (see the accompanying figure).

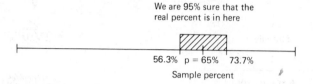

We are 95% sure that the
real percent is in here

56.3%   p = 65%   73.7%

Sample percent

Problem   The Yentex Opinion Poll surveys 1176 people on an upcoming election. The results are

$$
\begin{array}{lr}
\text{Harold Sills} & 596 \\
\text{Mike Wolff} & 580 \\
\hline
\text{Total} = & 1176
\end{array}
$$

What are the proportions for the whole population? What are the confidence intervals?

ANSWER   To find the proportions, we use Formula 7–4–1 and divide both sample counts by N = 176.

Sills	596	50.7%	= 0.507
Wolff	580	49.3%	= 0.493
Total	1176	100.0%	= 1.000

This shows Sills over Wolff by 50.7 to 49.3%. Is it safe to conclude that Sills will beat Wolff? We need to find the 95% confidence intervals to help us answer this question.

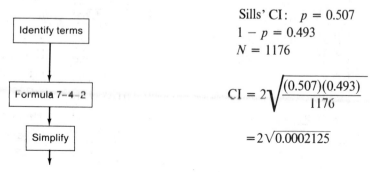

Identify terms

Formula 7–4–2

Simplify

Sills' CI: $p = 0.507$
$1 - p = 0.493$
$N = 1176$

$$CI = 2\sqrt{\frac{(0.507)(0.493)}{1176}}$$

$$= 2\sqrt{0.0002125}$$

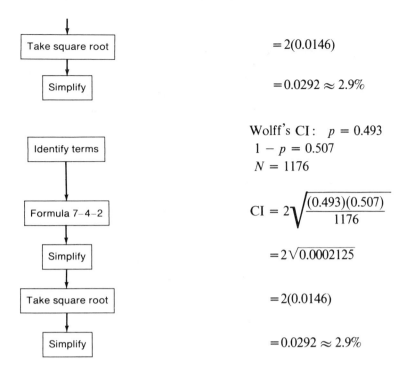

Take square root	$= 2(0.0146)$
Simplify	$= 0.0292 \approx 2.9\%$

Wolff's CI: $p = 0.493$
$1 - p = 0.507$
$N = 1176$

Formula 7–4–2

$$CI = 2\sqrt{\frac{(0.493)(0.507)}{1176}}$$

Simplify

$$= 2\sqrt{0.0002125}$$

Take square root

$$= 2(0.0146)$$

Simplify

$$= 0.0292 \approx 2.9\%$$

Notice that they are the same. This tells us that Sills' real percent might be up to 2.9% more (50.7 + 2.9 = 53.6%) or up to 2.9% less (50.7 − 2.9 = 47.8%). So Sills' true percent is probably between 47.8 and 53.6%.

At the same time, Wolff's real percent might be up to 2.9% more (49.3 + 2.9 = 52.2%) or up to 2.9% less (49.3 − 2.9 = 46.4%). So Wolff's true percent is probably between 46.4 and 52.2%.

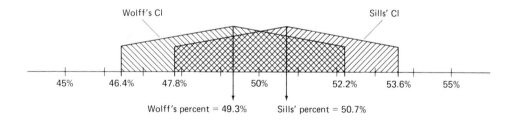

Notice in the figure that there is a large overlap of the two confidence intervals. What does this mean? It means that their percents are so close that it is possible in the overlap region that Wolff is really slightly more popular than Sills.

It is possible that Wolff might even beat Sills 52.2 to 47.8%. On the other hand, Sills might defeat Wolff 53.6 to 46.4%. These are the extremes of their 95% confidence intervals.

Sample Size

Question   How large a sample do we need to have?

ANSWER   That depends on the accuracy we want. The higher the accuracy, the higher the cost. To make a sample very accurate, we need a small confidence interval. This is because the smaller the confidence interval, the closer the true percent must be to the sample percent. Notice that part (b) of the figure, with its smaller CI, is much more accurate.

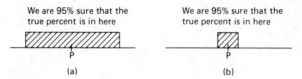

We have the following formula, which gives a rough idea of how large a sample is needed to give us our desired accuracy.

Formula 7–4–3

$$N \approx \frac{1}{CI^2}$$

where N = sample size we need, and CI = the 95% confidence interval we want.

Problem   Carol is doing a study on people's feelings about abortion. She would like the results to be within 5% of the true proportion. How large a sample must she use?

ANSWER   Since she wants to be within 5% of the true proportion, CI = 5% = 0.05. We can use Formula 7–4–3 to get the sample size N.

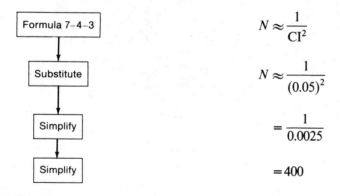

Thus Carol will be 95% sure that she is within 5% if she uses 400 people.

Problem   Yentex is doing a straw poll for an upcoming election. They would like their results to be within 2% of the true proportions. How many people must they sample?

ANSWER   Here the desired accuracy is 2%; so CI = 0.02.

Formula 7–4–3
$$N \approx \frac{1}{(0.02)^2}$$

Simplify
$$N \approx \frac{1}{0.0004}$$

Simplify
$$= 2500$$

Yentex must sample 2500 people to be 95% sure that they are within 2%.

Problem   Suppose that Yentex wants 1% accuracy.

ANSWER   Now we redo the problem with CI = 0.01.

Formula 7–4–3
$$N \approx \frac{1}{(0.01)^2}$$

Simplify
$$N \approx \frac{1}{0.0001}$$

Simplify
$$= 10,000$$

Notice that to go from 2% accuracy to 1% accuracy, Yentex has to sample four times as many people. This shows how it can get very costly to make a sampling very accurate.

The graph in the accompanying figure shows vividly how hard it is to get very

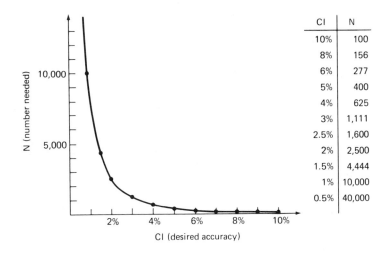

CI	N
10%	100
8%	156
6%	277
5%	400
4%	625
3%	1,111
2.5%	1,600
2%	2,500
1.5%	4,444
1%	10,000
0.5%	40,000

accurate polls. For instance, to be within $\frac{1}{2}$%, we would have to have a sample of 40,000 people. As we can see from the graph, the size N seems to "take off" when the accuracy is less than about 3%. This is probably why Gallup, Harris, and Nielson all use samples around 1200. They get reasonable accuracy without too large a sample size.

PROBLEM SET 7–4–2

1. Define or discuss:
   (a) 95% confidence interval.
   (b) Sample size.

Complete the following table for the missing proportions and 95% confidence intervals. Use Formulas 7–4–1 and 7–4–2.

	E (part)	N (total)	p	CI
2.	63	95	?	?
3.	247	610	?	?
4.	814	1,215	?	?
5.	75	918	?	?
6.	?	150	0.65	?
7.	?	420	0.21	?
8.	114	?	0.42	?
9.	217	?	0.88	?

Use Formula 7–4–3 to complete the following table:

	CI	N
10.	7%	?
11.	3.5%	?
12.	0.8%	?
13.	4.5%	?
14.	?	500
15.	?	300
16.	?	20,000
17.	?	75

The Daily News is taking a straw poll for the upcoming mayoral election; 810 people are surveyed. The results are

Dan O'Connor	421
Ken Packwood	368
Others	21
Total	$=810$

18. What proportion will Mr. O'Connor get?

19. What is the 95% confidence interval for this proportion?

20. Roger Yawnman feels a fun Saturday night is counting out-of-state license plates on the freeway. One Saturday night, he counts 82 out-of-state license plates out of 1203 passing cars. What proportion of cars are from out of state, and what is the 95% confidence interval?

"LET'S SEE NOW — YOUR ANSWERS WERE: 37 'DON'T KNOW', 42 'NO OPINION', AND ONE 'INDUSTRY SHOULD COMMIT ITSELF TO LONG-TERM PROJECTS INSTEAD OF CONCENTRATING CAPITAL EXPENDITURE IN SHORT-LIVED INVESTMENTS WHICH OFFER A QUICK CASH RETURN.'"

(Sydney Harris)

21. Kent is a quality-control inspector in a factory. He wants to know what proportion of the factory's product is defective. He can't check them all, so he checks 85 at random and finds 11 are defective. What proportion is this, and what is the 95% confidence interval?

22. To see if a coin is fair, Ed flips it 300 times. He gets 154 heads. What proportion is this? What is the 95% confidence interval? Is 50% (a fair coin) inside this confidence interval? Could the coin be fair?

23. 150 college students are asked if they know where the library is; 132 say yes. What proportion is this, and what is the 95% confidence interval?

24. In a TV survey 1125 people are asked what show they are watching. The results are as follows:

Sonny and Cher Special	215
Brady Bunch (rerun)	204
Others	706
Total	= 1125

What is the proportion watching each show? What are their 95% confidence intervals? Is it possible that more people were really watching Brady Bunch than Sonny and Cher?

A pollster samples 500 people on an upcoming election.

Ludlow Bean	275
Leonard Box	225
Total	= 500

25. What are the proportions for each candidate?

26. What is the 95% confidence interval for each candidate?

27. Is there any overlap of their confidence intervals?

28. Is it likely that Box might really defeat Bean?

7-5 B.S. MATHEMATICS

Question Is statistics ever misused?

ANSWER A Canadian humorist, Stephen Leacock, once said, "In earlier times, they had no statistics, and so they had to fall back on lies." People are afraid of mathematics and can easily be snowed by *B.S. mathematics*. In order to keep the editors and publishers of this book happy, I must tell you that B.S. means *butchered statistics*. (But you know what it really means.)

In Chapter 4, we saw how graphs could be used to fool people. In this section, we will give some more examples of B.S. mathematics.

Question   Joe asks seven families how many TV sets they own. The answers given are 2, 1, 4, 3, 2, 2, 1. The mean is $\frac{15}{7}$. Now Joe says, "The mean number of TV's in the American home is 2.14286." What is wrong with this?

ANSWER   Joe has talked to seven families, but his mean has five-decimal place accuracy. This might trick people into thinking Joe surveyed many, many families. This is called **phony accuracy.**

   Actually, the correct mean might be anywhere from 1.8 to 2.5. Joe is trying to impress people with a lot of decimal places that are not accurate.

Question   "GermBomb kills 10,000 germs in your mouth when you gargle." What is wrong with this?

ANSWER   First, we do not know for sure if those germs are bad germs. Second, we do not know how many germs are in the mouth. Suppose that there are 10,000,000 germs in the mouth. Then the percent killed is very small.

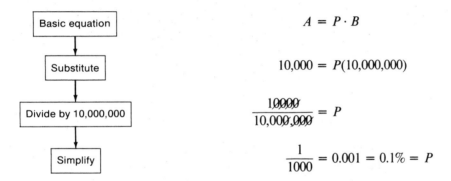

$$A = P \cdot B$$

$$10,000 = P(10,000,000)$$

$$\frac{10000}{10,000,000} = P$$

$$\frac{1}{1000} = 0.001 = 0.1\% = P$$

   Now it doesn't look as if GermBomb does too much. This trick is called the **missing base**, since we are not told the total size of the group.

Question   "In a recent study at a famous name university, students who brushed with Climax Toothpaste went out on 33% more dates." What is wrong with this?

ANSWER   There are many problems with this statement. Where was the study done? Maybe it was done at Mickey Mouse State University. (After all, Mickey Mouse is a famous name.) Also, maybe they went from school to school until they got the results they wanted, while the bad results were thrown away.

   Another serious problem is that they don't say who used Climax and who didn't. Perhaps, they let 10 cheerleaders use Climax, and had 10 other students not use Climax. Most likely, the cheerleaders will go out on more dates, but not because of their toothpaste.

   Finally, we do not know what the base is. Suppose one group had three dates and the Climax group had four dates. The Climax group did have $\frac{1}{3} = 33\%$ more dates, but this is not too convincing. This is also a case of the missing base.

Question "Four out of five dentists surveyed prefer sugarless gum for their patients who chew gum." What is wrong with this?

ANSWER The catch is that the four out of five dentists aren't saying sugarless gum is good for teeth. They only recommend it over gum with sugar. Suppose that the data from the survey of dentists looked like this.

	Number	Percent
Shouldn't chew gum at all	200	85.1
Can only chew sugarless gum	28	11.9
Can chew sugar gum	7	3.0
Total	=235	100.0

In this case, the result is clear: don't chew gum. But suppose that the advertising people throw out the 200 dentists who are against gum, and just looked at the 35 who felt gum was all right. Now the data look like this.

	Number	Percent
Can chew sugarless gum	28	80
Can chew sugar gum	7	20
Total	=35	100

Now the proportion is 80%, or four out of five dentists preferred sugarless gum for those who chew gum. This is another version of the missing base.

Question "More teenagers use Smooth Face Cream than any other. Shouldn't you?" What is wrong about this?

ANSWER The ad does not tell the actual percent of sales. Suppose that they look like this.

Product	Percent of Market
Smooth	12.6
Product 2	12.4
Product 3	12.0
Product 4	11.5
Product 5	10.3
Others	41.2
Total	= 100.0% of the market

Yes, it is true that Smooth is at the top of the list, but the top five are all pretty close. It is hard to say which one is really better. Maybe Smooth spends much more for its advertising campaign. Also, if 12.6% of people use Smooth, then 87.4% use something else.

Question "I am 31 years old. The average person lives to be 71. So I will die in 40 years." What is wrong with this statement?

ANSWER   The mean is one number that gives a typical age of death in this case. But in this case, it does not tell the whole story. Seventy-one is the average age that people live to. But this does not mean that every one will die at exactly 71 years old. In fact, only about 5% of all people die when they are exactly 71 years old. The other 95% die either younger or older than 71.

      Another fact is that only 50% of all people die between 61 and 81. The other 50% either die before 61 or after 81. This is the **fallacy of the average** in which someone thinks that everything will be exactly like the average.

Question   "Nine out of 10 pickup trucks sold in the last 10 years are still on the road." What is wrong with this statement?

ANSWER   Of course, the numbers themselves are correct. But there is a catch. This statistic gives the impression that most trucks last 10 years. Actually, the average truck in America lasts 6.9 years.

      Notice the ad did *not* say "Nine out of 10 trucks sold 10 years ago are still on the road." Their figure contains all the brand-new trucks just sold. Their figure looks good because they are averaging the few old trucks left with many new trucks.

Question   "I know a guy who smoked four packs of cigarettes a day and still lived to be 92. So, cigarettes don't affect your health." What is wrong with this statement?

ANSWER   The average nonsmoker lives about $5\frac{1}{2}$ years longer than the average smoker. This is a fact. But these are only averages, and there are bound to be exceptions.

      As we can see from the accompanying figure, the average smoker dies before the average nonsmoker. But there are rare exceptions. These don't change the average. This trick, pointing to an exception, is called the **fallacy of exceptions**.

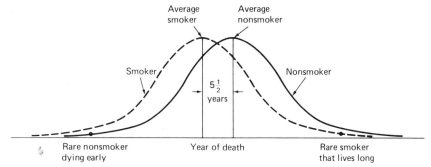

Question   "The average salary in our shop is $15,500." What could be wrong with this?

ANSWER   Perhaps nothing is wrong with this. But perhaps the salaries look like this.

$8000	$8000	$8000	$10,000	$10,000
8000	8000	8000	10,000	10,000
8000	8000	10,000	10,000	50,000
8000	8000	10,000	10,000	100,000

It is clear that 18 out of 20 people earn $8000 to $10,000. Two people (probably the owners) make $50,000 and $100,000. Let us compute the mean.

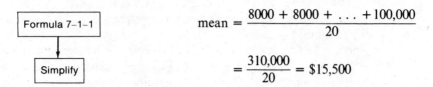

$$\text{mean} = \frac{8000 + 8000 + \ldots + 100,000}{20}$$

$$= \frac{310,000}{20} = \$15,500$$

So the mean salary is in fact $15,500. But this is not really typical. Let us find the median. Since there are 20 salaries, we look for $\frac{20 + 1}{2} = 10\frac{1}{2}$ up. In other words, we want the salary halfway between the tenth and eleventh salary. This is $9000.

The median is $9000 and this is a much more typical salary. The boss used a mean salary because it made him look better even though it wasn't typical. This is a **mean–median fallacy**.

Question "The Church of the Cheap is the fastest growing religion in the world." What is wrong with this statement?

ANSWER Like many of the other tricky statements in this section, it may be true but misleading. Consider the following table of major world religions.

Religion	Membership (1970)	Membership (1975)	Increase (number)	Increase (%)
Christianity	924,000,000	944,000,000	20,000,000	+2.2
Judaism	13,500,000	14,400,000	900,000	+6.7
Islam	493,000,000	529,000,000	36,000,000	+7.3
Hinduism	437,000,000	514,000,000	77,000,000	+17.6
Cheap	100	143	43	+43.0

Clearly, Christianity is the biggest religion in the world. Clearly, Hinduism increased by the most people in the 5 years.

But in terms of percentage, the Church of the Cheap went up 43%, the most of all. The reason is that they had so few people (small base) that even 43 more people seems like a big increase. This is also a missing base fallacy.

PROBLEM
SET 7–5–1
1. Define or discuss:
   (a) Phony accuracy.
   (b) Missing base.
   (c) Fallacy of the average.
   (d) Mean–median fallacy.
   (e) Fallacy of exceptions.

The following statements all have some trick or fallacy in them. Discuss what the possible trick is.

2. The average lawyer earns $22,000. So when I graduate law school, I will be earning $22,000.

3. The average American owns 1.517326 cars.

4. In a recent study, students using Quik-know Study Notes scored 10% higher than those students who didn't use any notes.

5. Aspireen gets in your bloodstream faster.

6. Le Boat is a cut above.

7. Extraspirin has twice as much pain reliever.

8. Easy-Clean Oven Cleaner has 33% more active ingredients than another popular spray oven cleaner.

9. Ding Dong Bread helps build healthy bodies 4 ways.

IMPORTANT WORDS

central tendency (7–1)
confidence interval (7–4)
fallacy of the average (7–5)
mean (7–1)
mean–median fallacy (7–5)
median (7–1)
missing base (7–5)
normal curve (7–3)

phony accuracy (7–5)
range (7–2)
sample proportion (7–4)
sample size (7–4)
standard deviation (7–2)
true proportion (7–4)
variability (7–2)
z-score (7–3)

REVIEW EXERCISES

Consider the following sets of data:
X: 14, 28, 41, 52, 30, 38, 47, 82, 64
Y: 82, 150, 142, 201, 183, 121, 252, 106, 144, 168, 302, 177, 193, 113
Z: 11000, 13500, 17700, 8200, 15100, 18800, 22300, 9600, 10400,
  26100, 14400, 12000, 16900, 15300, 20200, 19900, 13400

1. Find the mean of data X.

2. Find the mean of data Y.

3. Find the mean of data Z.

4. Find the median of data X.

5. Find the median of data Y.

6. Find the median of data Z.

7. Find the range of data X.

8. Find the range of data Y.

9. Find the range of data Z.

10. Find the standard deviation of data X.

11. Find the standard deviation of data Y.

12. Find the standard deviation of data Z.

Thirty-three hundred students take a test that has a normal curve. The mean is 70 and the standard deviation is 9.

13. How many students are between 61 and 79?

14. How many students are between 52 and 88?

15. How many students are between 43 and 97?

Complete the following tables.

	x	m	σ	z
16.	62	66	9	?
17.	525	800	100	?
18.	?	150	25	−0.8

	z	$F(z)$
19.	1.4	?
20.	−0.3	?
21.	?	65.5%
22.	?	13.6%

	A	B	m	σ	z_A	z_B	% above A	% between A and B
23.	2800	3500	3000	700	?	?	?	?
24.	115	122	100	15	?	?	?	?

	E (part of sample)	N (total sample)	p (proportion)	CI
25.	201	253	?	?
26.	143	?	0.23	?
27.	?	368	0.122	?

28. How large a sample size is needed to have a CI of 6%?

29. A sample size of 150 will give about what CI?

30. Linda surveys 268 people; 110 of them say that they attended a church or synagogue within the week. What is this proportion and the CI?

31. What is the trick in the following statement? "The Weasel has had the fastest growing sales of any car in America."

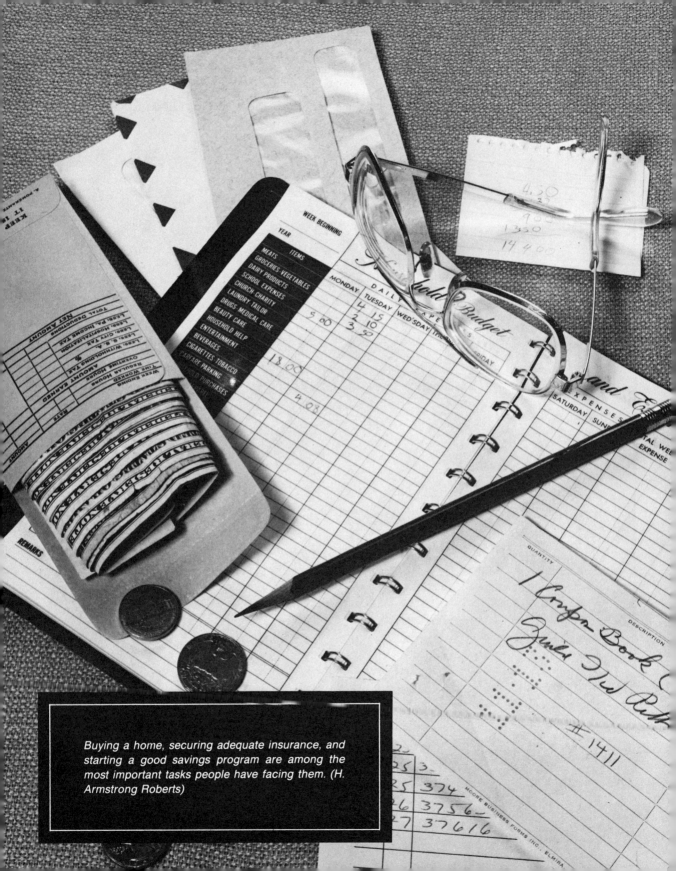

Buying a home, securing adequate insurance, and starting a good savings program are among the most important tasks people have facing them. (H. Armstrong Roberts)

8

CONSUMER
MATH II

8-1 MONEY GROWTH I: COMPOUND INTEREST

Question   What is bank interest?

ANSWER   In Section 5.2 we studied the interest we pay to someone (such as a bank) if we use their money. This is a two-way street. If they use our money, they pay us interest. This is what happens when we deposit money in a bank or savings and loan. (In this chapter, when we mention banks, we also mean savings and loan institutions, stocks, bonds, and the like).

If we put P dollars in the bank at r interest rate, we earn the following interest in one year.

Formula 8-1-1

$$I = Pr$$

where I = interest, P = bank balance, and r = interest rate.

Problem   Fred put $700 in the bank at 6% interest. How much interest does he earn in 1 year? What is his account after 1 year?

ANSWER   This is a simple problem, and we can use Formula 8–1–1.

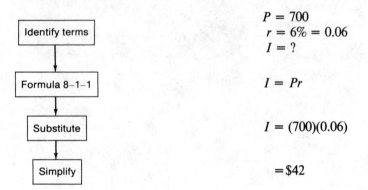

$$P = 700$$
$$r = 6\% = 0.06$$
$$I = ?$$

$$I = Pr$$

$$I = (700)(0.06)$$

$$= \$42$$

Fred now has $700 + $42 = $742 in his account at the end of 1 year.

Problem   How much does Fred have in his account after 2 years?

ANSWER   We do exactly the same thing as above, except now P = 742.

$$P = 742$$
$$r = 6\% = 0.06$$
$$I = ?$$

$$I = Pr$$

6/1/78. (Reprinted with permission of Franklin Society Federal Savings and Loan Association)

$$I = (742)(0.06)$$

$$= \$44.52$$

After 2 years, Fred has $742 + 44.52 = \$786.52$.

Notice that Fred earned a little more interest in the second year. This is because the \$42 from the first year was earning interest in the second year. When interest from one year earns interest itself in the next years, we call this **compounding**.

We can use algebra to get a simple formula to show how the money grows. We call P_0 the original balance and P_1 the balance after 1 year.

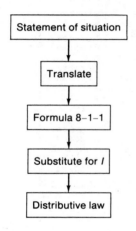

The balance after 1 year is the original balance plus interest

$$P_1 = P_0 + I$$

$$I = P_0 r$$

$$P_1 = P_0 + P_0 r$$

$$= P_0(1 + r)$$

Formula 8–1–2

$$\boxed{P_1 = P_0(1 + r)}$$

where P_1 = balance after 1 year, P_0 = original balance, and r = interest rate (as a decimal).

Problem   Diane puts \$350 in the bank at $5\frac{1}{2}\%$ interest. What is her balance after 1 year?

ANSWER   We use Formula 8–1–2. We have to be careful to write $r = 5\frac{1}{2}\% = 5.5\% = 0.055$.

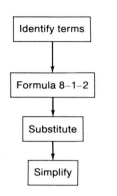

$$P_0 = 350$$
$$r = 5.5\% = 0.055$$
$$P_1 = ?$$

$$P_1 = P_0(1 + r)$$

$$P_1 = (350)(1.055)$$

$$= \$369.25$$

Hand Calculator Instant Replay

PUNCH	DISPLAY	MEANING
C	0.	Clear
3 5 0	350.	P_0
×	350.	Multiply by
1 · 0 5 5	1.055	$1 + r$
=	369.25	Answer

We can use Formulas 8–1–1 and 8–1–2 together to find what the balance is after 2 years. Notice that the interest $I = P_1 r$, since the new balance P_1 earns the interest.

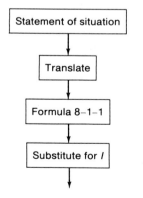

Balance after 2 years is balance
after 1 year plus interest

$$P_2 = P_1 + I$$

$$I = P_1 r$$

$$P_2 = P_1 + P_1 r$$

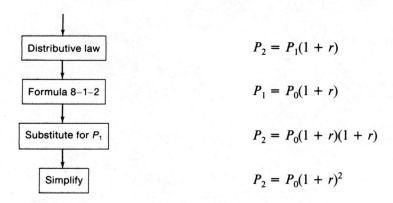

Distributive law	$P_2 = P_1(1 + r)$
Formula 8–1–2	$P_1 = P_0(1 + r)$
Substitute for P_1	$P_2 = P_0(1 + r)(1 + r)$
Simplify	$P_2 = P_0(1 + r)^2$

If we do this same thing for year 3, we will get $P_3 = P_0(1 + r)^3$. Perhaps, now you can see the pattern for compound interest.

Formula 8–1–3

$$P_n = P_0(1 + r)^n$$

when P_0 = original deposit, P_n = balance after n years, and r = rate of interest (as a decimal).

> Formula 8–1–3 tells us how to calculate the interest on money left in the bank for n years:
>
> 1. Find $1 + r$.
>
> 2. Raise $1 + r$ to the nth power.
>
> 3. Multiply this by the original deposit.

This is a fairly easy procedure if you have a calculator.

Problem Linda put $600 in the bank 7 years ago at 5% interest. What is it worth now?

ANSWER To do this, we will have to calculate $(1.05)^7$ to use in Formula 8–1–3.

$$P_0 = 600$$
$$r = 5\% = 0.05$$
$$n = 7$$
$$P_7 = ?$$

Identify terms	
Formula 8–1–3	$P_n = P_0(1 + r)^n$

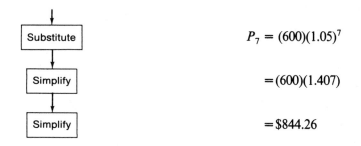

$$P_7 = (600)(1.05)^7$$

$$= (600)(1.407)$$

$$= \$844.26$$

Thus, in 7 years, Linda's account has grown from $600 to $844.26.

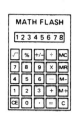

Hand Calculator Instant Replay

PUNCH	DISPLAY	MEANING
C	0.	Clear
1 . 0 5	1.05	$(1 + r)$
×	1.05	Start multiplying
=	1.1025	$(1 + r)^2$
=	1.157625	$(1 + r)^3$
=	1.2155062	$(1 + r)^4$
=	1.2762815	$(1 + r)^5$
=	1.3400955	$(1 + r)^6$
=	1.4071002	$(1 + r)^7$
×	1.4071002	Times
6 0 0	600.	P_0
=	844.26012	Answer

Recall that this is how we raise a number to an exponent, n. We punch the number, punch $\boxed{\times}$, and then punch $\boxed{=}$ $(n - 1)$ times.

Problem Andy is given $1200 in gifts when he is born. This is put into the bank at 6%. How much will this be worth when Andy is 18 and ready for college?

ANSWER Here $r = 6\% = 0.06$ and $n = 18$, so we will have to calculate $(1.06)^{18}$. Obviously, this is very boring to do by hand. But it is not too bad with your pocket calculator.

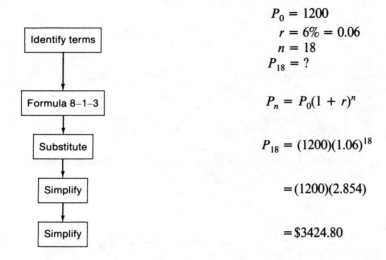

$$P_0 = 1200$$
$$r = 6\% = 0.06$$
$$n = 18$$
$$P_{18} = ?$$

Identify terms

Formula 8–1–3

$$P_n = P_0(1 + r)^n$$

Substitute

$$P_{18} = (1200)(1.06)^{18}$$

Simplify

$$= (1200)(2.854)$$

Simplify

$$= \$3424.80$$

Whenever we have money just drawing interest, we use the compound interest formula, 8–1–3.

Question Is there another way to do this without a calculator?

ANSWER There is. Instead of computing the exponent $(1 + r)^n$ each time, we can use a table. Table 8–1 is called a **compound interest table**.

Table 8–1 Compound Interest: $(1 + r)^n$

				r (rate; %)				
n (years)	$4\frac{1}{2}$	5	$5\frac{1}{2}$	6	$6\frac{1}{2}$	7	$7\frac{1}{2}$	8
1	1.045	1.050	1.055	1.060	1.065	1.070	1.075	1.080
2	1.092	1.103	1.113	1.124	1.134	1.145	1.156	1.166
3	1.141	1.158	1.174	1.191	1.208	1.225	1.242	1.260
4	1.193	1.216	1.239	1.262	1.286	1.311	1.335	1.360
5	1.246	1.276	1.307	1.338	1.370	1.403	1.436	1.469
6	1.302	1.340	1.379	1.419	1.459	1.501	1.543	1.587
7	1.361	1.407	1.455	1.504	1.554	1.606	1.659	1.714
8	1.422	1.477	1.535	1.594	1.655	1.718	1.783	1.851
9	1.486	1.551	1.619	1.689	1.763	1.838	1.917	1.999
10	1.553	1.629	1.708	1.791	1.877	1.967	2.061	2.159
11	1.623	1.710	1.802	1.898	1.999	2.105	2.216	2.332
12	1.696	1.796	1.901	2.012	2.129	2.252	2.382	2.518
13	1.772	1.886	2.006	2.133	2.267	2.410	2.560	2.720
14	1.852	1.980	2.116	2.261	2.415	2.579	2.752	2.937
15	1.935	2.079	2.232	2.397	2.572	2.759	2.959	3.172
20	2.412	2.653	2.918	3.207	3.524	3.870	4.248	4.661
25	3.005	3.386	3.813	4.292	4.828	5.427	6.098	6.848
30	3.745	4.322	4.984	5.743	6.614	7.612	8.755	10.063
35	4.667	5.516	6.514	7.686	9.062	10.677	12.569	14.785
40	5.816	7.040	8.513	10.286	12.416	14.974	18.044	21.725
45	7.248	8.985	11.127	13.765	17.011	21.002	25.905	31.920
50	9.033	11.467	14.542	18.420	23.307	29.457	37.190	46.902

Question   How do we use Table 8–1?

ANSWER   We use it as follows:

> To use Table 8–1 to find $(1 + r)^n$:
>
> 1. Find the $r\%$ column at the top.
> 2. Read down until you get to the n (year) row.
> 3. $(1 + r)^n$ is where r column meets n row.

Problem   Find $(1.07)^{12}$ in Table 8–1.

ANSWER   In the given figure we have a cut-away version of Table 8–1. Notice that we find the 7% column and then read down to the 12-year row. They meet at 2.252. So $(1.07)^{12} = 2.252$.

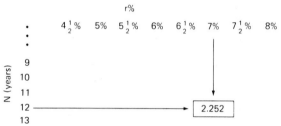

Problem   Chuck and Ann get $2100 cash at their wedding. They put it into certificates of deposit (C.D.'s) at 7% interest for 6 years. What will this be worth in 6 years?

ANSWER   We use Formula 8–1–3 and Table 8–1.

Identify terms	$P_0 = 2100$ $r = 7\% = 0.07$ $n = 6$ $P_6 = ?$
Formula 8–1–3	$P_n = P_0(1 + r)^n$
Substitute	$P_6 = (2100)(1.07)^6$
Use Tab. 8–1	$= (2100)(1.501)$
Simplify	$= \$3152.10$

Thus Chuck and Ann's C.D.'s have grown to $3152 in value.

Problem   Margie figures that she will need $3000 to take her dream trip to Europe in 8 years. She can earn 6.5% on her money. How much does she have to deposit now so that it will grow to $3000 in 8 years?

ANSWER   This problem is a little different. Here we are given the final balance after 8 years. The unknown is the original deposit, P_0. We still set up Formula 8–1–3 exactly the same.

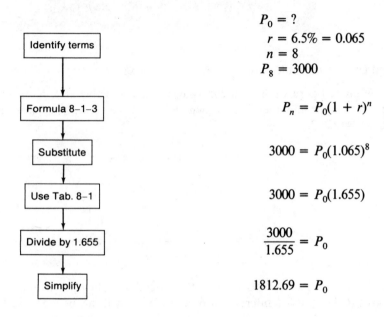

$$P_0 = ?$$
$$r = 6.5\% = 0.065$$
$$n = 8$$
$$P_8 = 3000$$

$$P_n = P_0(1 + r)^n$$

$$3000 = P_0(1.065)^8$$

$$3000 = P_0(1.655)$$

$$\frac{3000}{1.655} = P_0$$

$$1812.69 = P_0$$

Margie must put away $1812.69 now at 6.5% interest to have $3000 in 8 years.

Problem   Jerry has just received a $950 bonus from work. If he puts this into an 8% investment plan, how long will it take to grow to $1650?

ANSWER   Here we have both the original deposit, $P_0 = 950$, and the final balance $P_n = 1650$. We also have the rate $r = 8\%$. The number of years n is unknown.

$$P_0 = 950$$
$$r = 8\% = 0.08$$
$$n = ?$$
$$P_n = 1650$$

$$P_n = P_0(1 + r)^n$$

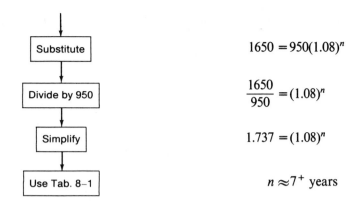

Substitute	$1650 = 950(1.08)^n$
Divide by 950	$\dfrac{1650}{950} = (1.08)^n$
Simplify	$1.737 = (1.08)^n$
Use Tab. 8–1	$n \approx 7^+$ years

Seven plus years means a little over 7 years. How did we get this? We find the 8% column in Table 8–1. We read down looking for 1.737. We cannot find it exactly, but we notice it is just a little over 1.714, which is $n = 7$ years. So we can say that n is a little over 7 years. We write this as 7^+.

$$\begin{array}{c|c} & 8\% \\ \hline 6 & 1.587 \\ 7 & 1.714 \\ 8 & 1.851 \quad \longleftarrow 1.737 \\ 9 & 1.999 \end{array}$$

When trying to find the number of years n, we find the closest numbers in the Table 8–1. Then we just approximate to the closest n. Or if the number is about halfway in between, we approximate this as a half-year.

Problem For what n is $(1.06)^n = 1.87$?

ANSWER We look at Table 8–1 under 6%. We see that our number 1.87 is between 1.791 and 1.898, but much closer to 1.898 (which is 11 years). So we say $n = 11^-$, or a little under 11 years.

$$\begin{array}{c|c} & 6\% \\ \hline 9 & 1.689 \\ 10 & 1.791 \\ 11 & 1.898 \quad \longleftarrow 1.87 \\ 12 & 2.012 \end{array}$$

Problem If $600 is invested and grows to $1125 in 9 years, what is the rate of interest?

ANSWER   Here we are given the original and final balance and the number of years. The unknown is the rate, r.

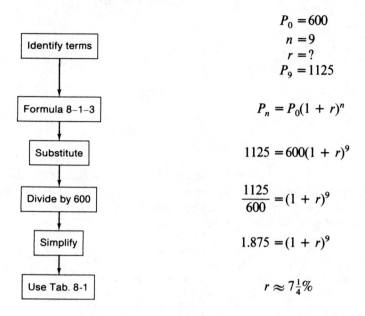

$P_0 = 600$
$n = 9$
$r = ?$
$P_9 = 1125$

| Identify terms |

| Formula 8-1-3 |

$P_n = P_0(1 + r)^n$

| Substitute |

$1125 = 600(1 + r)^9$

| Divide by 600 |

$\dfrac{1125}{600} = (1 + r)^9$

| Simplify |

$1.875 = (1 + r)^9$

| Use Tab. 8-1 |

$r \approx 7\frac{1}{4}\%$

How did we get this? We find the 9-year row in Table 8-1 and read across. Our number, 1.875, is not in the table exactly. But it is about halfway between 1.838 (for 7%) and 1.917 (for $7\frac{1}{2}\%$), so we say that the rate is about halfway between 7 and $7\frac{1}{2}\%$. Hence r is about $7\frac{1}{4}\%$ (see the illustration).

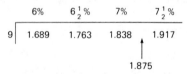

	6%	$6\frac{1}{2}\%$	7%	$7\frac{1}{2}\%$
9	1.689	1.763	1.838	1.917

1.875

1.875 corresponds to about $7\frac{1}{4}\%$

**PROBLEM
SET 8-1-1**

1. Define or discuss:
   (a) Interest.
   (b) Compound interest.

Complete the following table using Formula 8–1–1.

	I	P	r
2.	?	500	6%
3.	?	1200	6.5%

	I	P	r
4.	?	200	$4\frac{1}{2}\%$
5.	25	350	?
6.	300	3750	?
7.	12	250	?
8.	35	?	6%
9.	105	?	$5\frac{1}{2}\%$

Complete the following table using Table 8-1.

	r	n	$(1 + r)^n$
10.	5%	9	?
11.	4.5%	2	?
12.	$6\frac{1}{2}\%$	6	?
13.	$7\frac{1}{2}\%$	12	?
14.	6%	4	?
15.	5.5%	?	1.53
16.	7%	?	2.20
17.	$6\frac{1}{2}\%$	?	1.58
18.	?	8	1.60
19.	?	10	2.03

Complete the following table using Formula 8–1–3 and Table 8-1.

	P_0	r	n	P_n
20.	3100	$6\frac{1}{2}\%$	9	?
21.	500	5%	5	?
22.	?	7%	15	2,000
23.	?	$7\frac{1}{2}\%$	30	100,000

	P_0	r	n	P_n
24.	2500	6%	?	4,000
25.	420	5.5%	?	800
26.	6500	?	8	10,000
27.	400	?	6	600

28. On the day Jason was born, his grandfather put $2500 into an account for him at 7% interest. How much will this be when Jason is 20?

29. Janet and Mickey get $2900 in cash at their wedding. They plan to invest it for 6 years at $6\frac{1}{2}$% interest, and then use it for the down payment on a house. What will they have in 6 years?

30. Mary's husband has just died and left her a $20,000 insurance policy. If she invests this at 8%, how much will she have in 10 years?

31. Bob puts $300 in a bank account at 5% interest and then forgets about it for 20 years. How much is it worth when he remembers it?

32. Diane and Danny invest $1500 in mutual funds, which they are told earn 8% interest. If this is true, what will their money be worth in 6 years?

33. How much must Judy put in the bank today at $6\frac{1}{2}$% to have $1000 saved in 4 years?

34. Roger wants to have $600 in the bank in 3 years. How much must he put in today at $5\frac{1}{2}$% interest to get it?

Millard Fillmore College has a trust fund of $12,000,000. They can earn $7\frac{1}{2}$% interest on this.

35. How much interest do they earn in 1 year?

36. How much will this trust grow to in 5 years?

37. How much will this trust grow to in 10 years?

38. How much will this trust grow to in 20 years?

39. How long will it take this trust to grow to $20,000,000?

40. How long will it take this trust to grow to $30,000,000?

41. Eddie buys some stock in ABC Gold Co. at $21 a share. He sells it 8 years later for $30 a share. This is approximately what rate of interest?

42. If $1800 is left in the bank for 6 years, it grows to $2700. What rate of interest is this?

43. Take the 6% interest column of Table 8-1. Graph this, putting the years, n, on the x-axis, and the $(1.06)^n$ values on the y-axis.

Rule of 72

Question   What is the rule of 72?

ANSWER   The **rule of 72** is a simple trick for determining how quickly money doubles under compounding. The time it takes for money to double in value is called the **doubling time**. For example, if $1000 grows to be $2000 in 10 years, then 10 years is the doubling time.

(Courtesy of Suburban Federal Savings)

Formula 8–1–4

$$DT \approx \frac{72}{r}$$

where DT = doubling time and r = rate (written as a percent).

This is a very simple formula. We just divide the rate (as a percent) into 72. This is the number of years needed for money to double.

Problem Sandy put $1400 in the bank at 6% interest. How long will it take for this to double to $2800?

ANSWER Whenever doubling time is involved, we use Formula 8–1–4.

This tells us that it will take about 12 years for her money to double. Notice that for Formula 8–1–4 we take r as a whole number percent, not a decimal. This is the only place we do that.

Problem Larry can invest $1000 at 8% interest. Use the rule of 72 to get a quick idea of how his money grows.

ANSWER We use Formula 8–1–4 to find what the doubling time is.

Thus his money will double to $2000 in 9 years. In another 9 years (18 years total) his money will double again to $4000. In another 9 years (27 years total) his money will double again to $8000. The illustration shows graphically how rapidly the money grows.

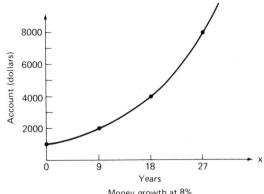

Money growth at 8%

Problem Michele's parents put $3000 in the bank at $6\frac{1}{2}\%$ interest when she is born. How long will it take to double to $6000? to $12,000? to $24,000?

ANSWER This is again a doubling-time problem. Here $DT \approx \dfrac{72}{r} = \dfrac{72}{6.5} \approx 11$ years. Thus it will take about 11 years to double to $6000, 22 years to double again to $12,000, and 33 years to double again to $24,000.

Other Uses of Compounding

Question Can Formulas 8–1–3 and 8–1–4 be used in situations other than money growing in a bank account?

ANSWER Yes. These formulas can be used in any situation in which something is growing at a steady percent rate every year. For example, inflation (price growth), population growth, and wage raises are all situations where the growth formulas 8–1–3 and 8–1–4 can be used.

Problem Suppose that the inflation rate is 5% for the next 6 years. How much will a $4500 car cost in 6 years?

ANSWER This inflation problem can be set up just like a bank interest problem. The only difference is that P_0 is the original price, and P_6 is the price 6 years later.

$$P_0 = 4500$$
$$r = 5\% = 0.05$$
$$n = 6$$
$$P_6 = ?$$

$$P_n = P_0(1 + r)^n$$

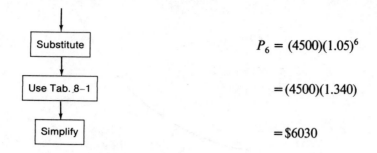

$$P_6 = (4500)(1.05)^6$$

$$= (4500)(1.340)$$

$$= \$6030$$

This tells us that a $4500 car will cost about $6030 in 6 years because of inflation.

Problem Barb puts $1000 in the bank at 6% interest, but the interest is compounded quarterly (four times a year). How much interest does she earn after 1 year with and without quarterly compounding?

ANSWER Six percent interest compounded quarterly means that the 6% is divided by 4 to give 1.5%. This 1.5% is then paid 4 times a year (every 3 months). So, instead of 6% a year, you get 1.5% four times a year.

To compute the compounding, we have to use Formula 8–1–3. Here P_4 will mean the balance after 4 quarters or 1 year. Also, $n = 4$ quarters.

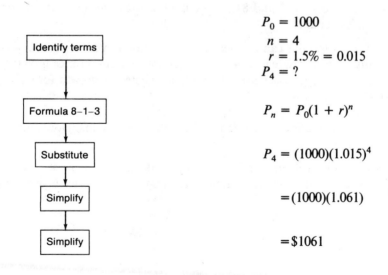

$$P_0 = 1000$$
$$n = 4$$
$$r = 1.5\% = 0.015$$
$$P_4 = ?$$

$$P_n = P_0(1 + r)^n$$

$$P_4 = (1000)(1.015)^4$$

$$= (1000)(1.061)$$

$$= \$1061$$

With quarterly compounding, Barb will have $1061 at the end of 1 year.

Hand Calculator Instant Replay

PUNCH	DISPLAY	MEANING
C	0.	Clear
1 . 0 1 5	1.015	$1 + r$
×	1.015	Start multiplying
=	1.030225	$(1 + r)^2$
=	1.0456783	$(1 + r)^3$
=	1.0613634	$(1 + r)^4$
×	1.0613634	Multiplied by
1 0 0 0	1000.	P_0
=	1061.3634	Answer

Let us now see how much she would have earned if there were no compounding; in other words, 6% for the whole year.

Formula 8–1–2 $\qquad P_1 = P_0(1 + r)$

Substitute $\qquad P_1 = (1000)(1.06)$

Simplify $\qquad = \$1060$

The difference is a grand total of about $1. As you notice, compounding quarterly is not really a big deal.

Problem The U.S. population is increasing at a rate of 0.8% every year. How long will it take the United States to double its population?

ANSWER This is a population problem, but we can treat it just like an interest growth problem for which we want the doubling time. Here $r = 0.8\%$.

$$DT \approx \frac{72}{r}$$

$$= \frac{72}{0.8}$$

$$= 90$$

This says it will take the United States about 90 years to double its population (at the current growth rate).

Sometimes, growth can be a negative rate. We saw this situation in Chapter 5 with automobile depreciation.

Problem   Jack and Roz buy an $800 color TV. This TV depreciates (loses value) at a rate of about 15% per year. What is the TV worth in 4 years?

ANSWER   We set this up just like the other growth problems, with one exception. Here the rate $r = -15\%$, since the TV is losing value. P_0 is the original price, and $(1 + r) = (1 - 0.15) = 0.85$.

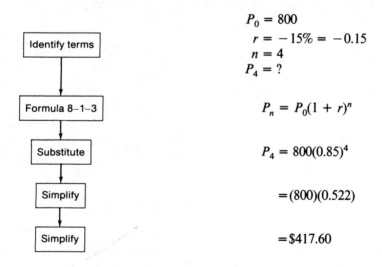

$$P_0 = 800$$
$$r = -15\% = -0.15$$
$$n = 4$$
$$P_4 = ?$$

$$P_n = P_0(1 + r)^n$$

$$P_4 = 800(0.85)^4$$

$$= (800)(0.522)$$

$$= \$417.60$$

PUNCH	DISPLAY	MEANING
C	0.	Clear
. 8 5	0.85	$(1 + r)$
×	0.85	Start multiplying
=	0.7225	$(1 + r)^2$
=	0.614125	$(1 + r)^3$
=	0.5220062	$(1 + r)^4$
×	0.5220062	Multiply by
8 0 0	800.	P_0
=	417.60496	Answer

Thus the TV is only worth about $418 four years later.

Most people don't sell TV's like they sell their cars. But in the case of a fire or theft, the insurance might only pay the depreciated value, or $418.

Problem Karyn buys a $150 antique chair. Six years later she sells it for $235. This is equivalent to what rate of interest at a bank?

Buying an antique is often a better investment than putting money in the stock market. (Zvi Lowenthal/ Editorial Photocolor Archives)

ANSWER   This is again a compound-interest problem. Here r is unknown.

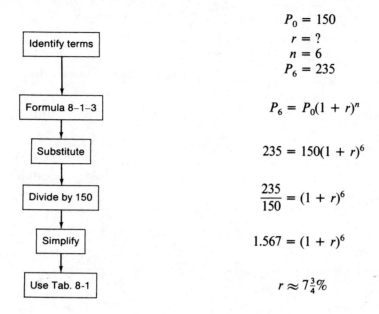

Identify terms		$P_0 = 150$
		$r = ?$
		$n = 6$
		$P_6 = 235$

Formula 8–1–3 $P_6 = P_0(1 + r)^n$

Substitute $235 = 150(1 + r)^6$

Divide by 150 $\dfrac{235}{150} = (1 + r)^6$

Simplify $1.567 = (1 + r)^6$

Use Tab. 8-1 $r \approx 7\frac{3}{4}\%$

How did we get this? We find the 6-year row and read across, as illustrated.

	$4\frac{1}{2}\%$	5%	$5\frac{1}{2}\%$	6%	$6\frac{1}{2}\%$	7%	$7\frac{1}{2}\%$	8%
6 years	1.302	1.340	1.379	1.419	1.459	1.500	1.543	1.587

1.567

We locate our number 1.567 about halfway between 1.543 ($7\frac{1}{2}\%$) and 1.587 (8%). We approximate r as halfway between, or $7\frac{3}{4}\%$.

This is a very good interest rate for Karyn, better than most banks (plus, she could sit on her investment).

PROBLEM   1. Define or discuss doubling time.
SET 8–1–2

Complete the following table using Formula 8–1–4.

	Rate of interest (%)	Doubling time (DT)	Tripling time (1.6 × DT)	Quadrupling time (QT)
2.	1			
3.	2			
4.	3			
5.	4			

	Rate of interest (%)	Doubling time (DT)	Tripling time (1.6 × DT)	Quadrupling time (QT)
6.	5			
7.	$5\frac{1}{2}$			
8.	6			
9.	$6\frac{1}{2}$			
10.	7			
11.	$7\frac{1}{2}$			
12.	8			
13.	9			
14.	10			
15.	12			
16.	15			
17.	20			

18. Pat puts $600 in the bank at $5\frac{1}{2}$% interest. How long will it take to double to $1200? quadruple to $2400?

19. A credit card company charges 18% interest in many states. How long would it take an unpaid bill to double in debt?

20. The Miser Savings and Loan advertises that "Your Money Doubles in 9.7 Years." What is the interest rate?

21. If the inflation rate holds steady at 4.5% a year, how long will it take for prices to double?

22. John and Donna speculate by buying raw land. They pay $6000 for the land. Its value goes up 9% a year. When will the land be worth $12,000? $24,000?

Bob and Carol buy a ComFee sofa for $650.

23. If inflation is 5% a year, what will the same sofa cost in 4 years?

24. If the sofa depreciates 10% a year, what will their sofa be worth in 4 years?

25. Suppose that Bob and Carol lose their sofa in a fire 4 years after they buy it. Their insurance company will only pay the depreciated value (Problem 24), but it costs more to buy a new one (Problem 23). How much do they have to pay for a new sofa out their own pockets?

26. Lisa buys an antique chest for $100. Six years later she sells it for $200. This is equivalent to what rate of bank interest?

27. Scott buys a string bass for $500. Four years later he sells it for $1000. This is equivalent to what rate of bank interest?

28. Mike and Jane buy a house for $28,000. Five years later they sell it for $40,000. This is equivalent to what rate of bank interest?

29. Sue gets a $2000 engagement ring. Seven years later, she gets divorced and sells it for $3000. What rate of interest was this?

Below is a list of many of the countries in the world with their annual population growth rates (as given in the *U.N. Demographic Yearbook, 1974*). Use Formula 8-1-4 to compute the approximate doubling time for each country's population.

30. World	1.9%	**31.** Algeria	3.2%	**32.** Argentina	1.3%
33. Australia	1.6%	**34.** Bangladesh	2.4%	**35.** Brazil	2.8%
36. Canada	1.3%	**37.** China	1.7%	**38.** Cuba	1.8%
39. Czechoslovakia	0.6%	**40.** East Germany	0.2%	**41.** Egypt	2.2%
42. England	0.3%	**43.** Finland	0.4%	**44.** France	0.8%
45. Honduras	4.0%	**46.** India	2.1%	**47.** Ireland	1.1%
48. Israel	3.2%	**49.** Italy	0.8%	**50.** Jordan	3.5%
51. Kenya	3.6%	**52.** Mexico	3.5%	**53.** North Vietnam	2.4%
54. Pakistan	3.3%	**55.** Poland	0.9%	**56.** Scotland	0.1%
57. Spain	1.1%	**58.** Syria	3.3%	**59.** Thailand	3.2%
60. U.S.A.	0.8%	**61.** U.S.S.R.	0.9%	**62.** Venezuela	3.1%
63. West Germany	0.6%	**64.** Zaire	2.8%		

8-2 MONEY GROWTH II: ANNUITIES

Question   Suppose that we keep putting money in the bank every year. What happens to it?

ANSWER   In the last section, we studied compounding in which we made *one* deposit and let it grow. In this section we will study the situation where we make regular deposits *every year* and let this grow. We call this an **annuity**.

Problem   Ron puts $200 in the bank at the end of every year. He earns 6% interest on the money. How much will he have at the end of 5 years?

ANSWER   We must first understand exactly what is happening. At the end of year 1, Ron puts $200 in the bank. At the end of year 2, he earns 6% interest on his account and he adds another $200. At the end of year 3, he earns 6% interest on his account, and he adds another $200. And so on through year 5.

The figuring goes as follows: each year we compute the new balance by multiplying by 1.06. Then we add $200. The next year we repeat the process.

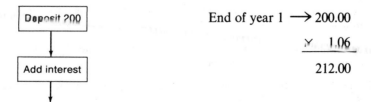

End of year 1 $\longrightarrow$ 200.00

$\times$ 1.06

212.00

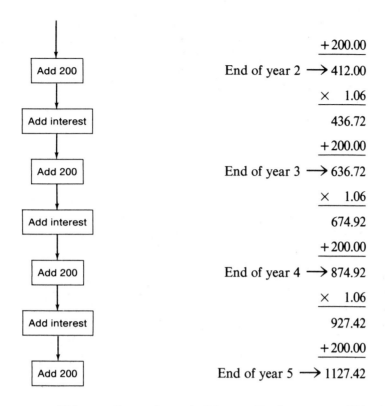

$$+\,200.00$$

End of year 2 $\longrightarrow$ 412.00

$$\times \quad 1.06$$

436.72

$$+\,200.00$$

End of year 3 $\longrightarrow$ 636.72

$$\times \quad 1.06$$

674.92

$$+\,200.00$$

End of year 4 $\longrightarrow$ 874.92

$$\times \quad 1.06$$

927.42

$$+\,200.00$$

End of year 5 $\longrightarrow$ 1127.42

This says that at the end of 5 years Ron's account will be $1127.42. Notice that every year the same thing happened: (1) he got his 6% interest (by multiplying by 1.06), and (2) he added another $200.

Problem What will Ron have after 25 years of the procedure?

ANSWER We could continue the process on through the end of year 25.

Question There must be an easier way?

ANSWER Yes, there are formulas and tables to help us. But we have to introduce a new term called the **sum of the annuity**. This is written $S_{\overline{n}|\,r}$.

This looks like an odd symbol, but don't panic. Think of it as follows:

$$S_{\overline{n}|\,r} \rightarrow \begin{cases} S \text{ stands for sum (or accumulation)} \\ n \text{ stands for number of years} \\ r \text{ stands for rate of interest} \end{cases}$$

Question How do we get $S_{\overline{n}|\,r}$, and why is it important?

ANSWER We will show why $S_{\overline{n}|\,r}$ is important very soon. First, we find $S_{\overline{n}|\,r}$ by using Table 8–2, which is very similar to Table 8–1.

To use Table 8–2 to find $S_{\overline{n}|r}$:

1. Find the $r\%$ column at the top.

2. Read down until you get to the n (year) row.

3. $S_{\overline{n}|r}$ is where the r column meets the n row.

Problem Find $S_{\overline{25}|7\%}$.

ANSWER We use Table 8–2 to find $S_{\overline{25}|7\%}$. We find the 7% column; then we read down to the 25-year row. They meet at 63.249. So $S_{\overline{25}|7\%} = 63.249$, as shown in the given figure.

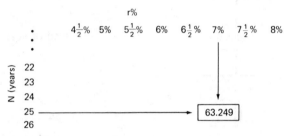

Table 8–2 Table of $S_{\overline{n}|r}$: Accumulation of $1

n (year)	5	$5\frac{1}{2}$	6	$6\frac{1}{2}$	7	$7\frac{1}{2}$	8
				r (rate; %)			
1	1.000	1.000	1.000	1.000	1.000	1.000	1.000
2	2.050	2.055	2.060	2.065	2.070	2.075	2.080
3	3.153	3.168	3.184	3.199	3.215	3.231	3.246
4	4.310	4.342	4.375	4.407	4.440	4.473	4.506
5	5.526	5.581	5.637	5.694	5.751	5.808	5.867
6	6.802	6.888	6.975	7.064	7.153	7.244	7.336
7	8.142	8.267	8.394	8.523	8.654	8.787	8.923
8	9.549	9.722	9.897	10.077	10.260	10.446	10.637
9	11.027	11.256	11.491	11.732	11.978	12.230	12.488
10	12.578	12.875	13.181	13.494	13.816	14.147	14.487
11	14.207	14.583	14.972	15.372	15.784	16.208	16.645
12	15.917	16.386	16.870	17.371	17.888	18.424	18.977
13	17.713	18.287	18.882	19.500	20.141	20.806	21.495
14	19.599	20.293	21.015	21.767	22.550	23.366	24.215
15	21.579	22.409	23.276	24.182	25.129	26.118	27.152
20	33.066	34.868	36.786	38.825	40.995	43.305	45.762
25	47.727	51.153	54.803	58.888	63.249	67.978	73.106
30	66.439	72.435	79.059	86.375	94.461	103.399	113.283
35	90.320	100.251	111.435	124.035	138.237	154.252	172.317
40	120.800	136.606	154.762	175.632	199.635	227.257	259.057
45	159.700	184.119	212.744	246.325	285.749	332.065	386.506
50	209.348	246.217	290.336	343.180	406.529	482.530	573.770

Question Why is $S_{\overline{n}|r}$ important?

ANSWER $S_{\overline{n}|r}$ is important because we can use it to compute how regular deposits grow with interest.

Formula 8–2–1

$$\boxed{A = D \cdot S_{\overline{n}|r}}$$

where A = accumulated total, D = regular yearly deposit, and $S_{\overline{n}|r}$ is found in Table 8–2.

Formula 8–2–1 tells us that if we deposit D dollars every year for n years at $r\%$, we will accumulate a total of $A = DS_{\overline{n}|r}$ dollars.

Problem Ron puts $200 in the bank every year at 6%. How much will he have after 25 years?

ANSWER This is where we use Formula 8–2–1 instead of doing the calculation for each of the 25 years. The unknown is A, total accumulated. D is the regular yearly deposit of $200. We will have to look up $S_{\overline{25}|6\%}$.

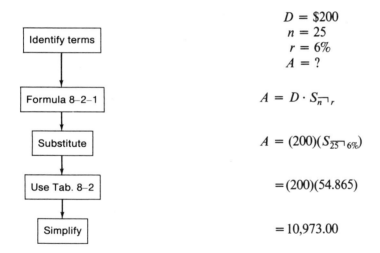

Identify terms

$D = \$200$
$n = 25$
$r = 6\%$
$A = ?$

Formula 8–2–1

$A = D \cdot S_{\overline{n}|r}$

Substitute

$A = (200)(S_{\overline{25}|6\%})$

Use Tab. 8–2

$= (200)(54.865)$

Simplify

$= 10,973.00$

This says that $200 saved every year for 25 years at 6% interest will grow to $10,973. Notice that $25 \times 200 = \$5000$ is all he put in. The rest, $10,973 - 5000 = \$5973$, is interest.

Problem Sid and Betty save $1200 per year ($100 per month) toward their retirement. They save this each year for 35 years at 7% interest. How much will this accumulate to in 35 years?

ANSWER Here again we are given the annual deposit, $D = \$1200$. We want to find the accumulation, A. We will have to look up $S_{\overline{35}|7\%}$. We use Formula 8–2–1 whenever we have a regular deposit.

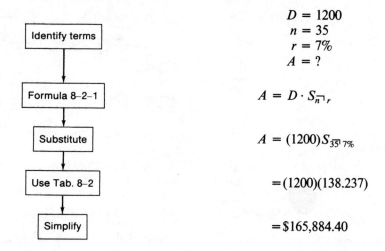

Identify terms	$D = 1200$ $n = 35$ $r = 7\%$ $A = ?$
Formula 8–2–1	$A = D \cdot S_{\overline{n}\,r}$
Substitute	$A = (1200)S_{\overline{35}\,7\%}$
Use Tab. 8–2	$= (1200)(138.237)$
Simplify	$= \$165,884.40$

So, with discipline, Sid and Betty can turn $100 a month into $165,884.40. Again, notice that only $35 \times 1200 = \$42,000$ was their money. The rest,

$$165,884.40 - 42,000 = \$123,884.40,$$

was bank interest.

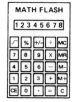

Hand Calculator Instant Replay

PUNCH	DISPLAY	MEANING
C	0.	Clear
1 2 0 0	1200.	D (deposit)
×	1200.	Multiplied by
1 3 8 . 2 3 7	138.237	$S_{\overline{n}\,r}$
=	165884.4	Answer

Problem Helen gives up smoking and figures she can save $1 per day as a result. If she puts this $1 per day (or $365 per year) into the bank at 5% interest, how much will she accumulate in 20 years?

ANSWER Since we have regular deposits, we are talking about an annuity, or Formula 8–2–1. Here the yearly deposit $D = \$365$.

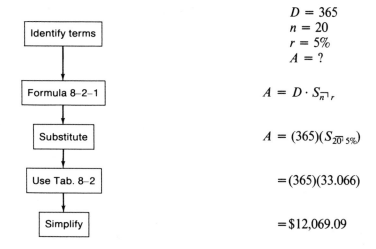

$$D = 365$$
$$n = 20$$
$$r = 5\%$$
$$A = ?$$

$$A = D \cdot S_{\overline{n}|\,r}$$

$$A = (365)(S_{\overline{20}|\,5\%})$$

$$= (365)(33.066)$$

$$= \$12,069.09$$

Strange as it may seem, Helen can accumulate \$12,069 in 20 years just by saving \$1 per day.

Problem Rich and Carol wish to retire in 30 years with a \$250,000 nest egg. How much do they have to save each year (at 7% interest) to accumulate this money?

ANSWER This is an annuity problem since the money is deposited every year. Unlike the other problems, here we are given the total accumulation $A = 250,000$. The unknown is the yearly deposit, D.

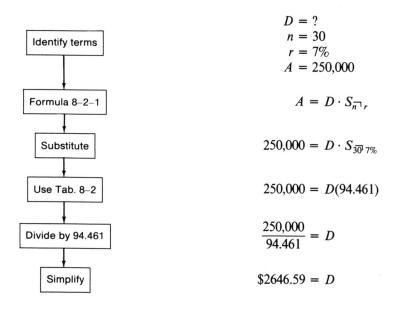

$$D = ?$$
$$n = 30$$
$$r = 7\%$$
$$A = 250,000$$

$$A = D \cdot S_{\overline{n}|\,r}$$

$$250,000 = D \cdot S_{\overline{30}|\,7\%}$$

$$250,000 = D(94.461)$$

$$\frac{250,000}{94.461} = D$$

$$\$2646.59 = D$$

Rich and Carol must save above $2647 per year (or about $221 per month) to accumulate $250,000 in 30 years.

Hand Calculator Instant Replay

PUNCH	DISPLAY	MEANING
C	0.	Clear
2 5 0 0 0 0	250000.	A (total amount)
+	250000.	Divided by
9 4 . 4 6 1	94.461	$S_{\overline{30}\,7\%}$
=	2646.5948	Yearly deposit
+ 1 2	12.	Divide by 12
=	220.54956	Monthly deposit

Problem   Lee and Harriet are saving to buy a $15,000 boat in cash. They can save about $70 a month. If they deposit this into a 6% savings account, how long will it take to grow to $15,000?

ANSWER   Here we have an annuity problem for which the yearly deposit is given, $D = 12 \times 70 = \$840$. The accumulated total is also given, $A = 15,000$. The unknown is n, the number of years.

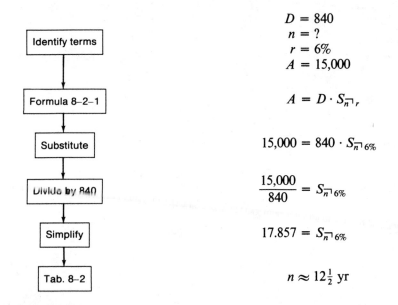

Identify terms

$$D = 840$$
$$n = ?$$
$$r = 6\%$$
$$A = 15,000$$

Formula 8–2–1

$$A = D \cdot S_{\overline{n}\,r}$$

Substitute

$$15,000 = 840 \cdot S_{\overline{n}\,6\%}$$

Divide by 840

$$\frac{15,000}{840} = S_{\overline{n}\,6\%}$$

Simplify

$$17.857 = S_{\overline{n}\,6\%}$$

Tab. 8–2

$$n \approx 12\tfrac{1}{2} \text{ yr}$$

How do we get $12\frac{1}{2}$ years? We know that $S_{\overline{n}|6\%}$ is 17.857, so we read down the 6% column of Table 8–2. We cannot find 17.857 exactly. We do find 16.870 for 12 years, and 18.882 for 13 years. Since 17.857 is about halfway in between, we say that n is about $12\frac{1}{2}$ years.

$$
\begin{array}{c|c}
 & 6\% \\
\hline
11 & 14.972 \\
12 & 16.870 \\
13 & 18.882 \quad\longleftarrow\!\!\!-\!\!\!-\!\!\!- 17.857 \\
14 & 21.015 \\
\end{array}
$$

Problem Penny joins an investment club. Each year she puts in $1500. After 10 years, her share is worth $21,000. What rate of interest is this equivalent to?

ANSWER Since she made regular payments every year, this is an annuity problem. The unknown is r.

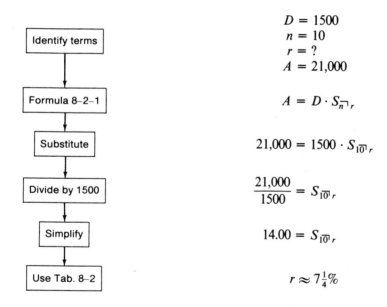

Identify terms	$D = 1500$ $n = 10$ $r = ?$ $A = 21{,}000$
Formula 8–2–1	$A = D \cdot S_{\overline{n}|r}$
Substitute	$21{,}000 = 1500 \cdot S_{\overline{10}|r}$
Divide by 1500	$\dfrac{21{,}000}{1500} = S_{\overline{10}|r}$
Simplify	$14.00 = S_{\overline{10}|r}$
Use Tab. 8–2	$r \approx 7\frac{1}{4}\%$

How do we get $7\frac{1}{4}\%$? We know $S_{\overline{10}|r} = 14$. So we read across the $n = 10$ year row. Since 14 is about halfway between 13.816 (7%) and 14.147 ($7\frac{1}{2}\%$), we say $r \approx 7\frac{1}{4}\%$ (see the given figure).

$$
\begin{array}{c|cc}
 & 7\% & 7\frac{1}{2}\% \\
\hline
10 & 13.816 & 14.147 \\
 & \underset{14}{\uparrow} & \\
\end{array}
$$

Question   How do we tell the difference between a compound interest problem and an annuity problem?

ANSWER   The following table summarizes the differences between compound interest and annuity.

Situation	Name	Formula
One deposit sits untouched	Compound interest	Formula 8–1–3
Regular amount is added yearly	Annuity	Formula 8–2–1

PROBLEM SET 8–2–1

1. What is the difference between compound interest and an annuity?

Complete the following table using Table 8–2. Approximate, if necessary.

	n	r	$S_{\overline{n}\,r}$
2.	10	5%	?
3.	15	6%	?
4.	25	7%	?
5.	13	5.5%	?
6.	?	6%	28.213
7.	?	7%	80.698
8.	10	?	13.500
9.	20	?	44.000

Complete the following table using Formula 8–2–1. Approximate, if necessary.

	D	n	r	A
10.	1000	12	6%	?
11.	1500	15	7%	?
12.	365	25	5%	?
13.	?	20	6%	20,000
14.	?	30	$6\frac{1}{2}\%$	500,000
15.	1200	?	$5\frac{1}{2}\%$	60,000

16.	500	?	5%	40,000
17.	1000	20	?	40,000
18.	500	25	?	30,000

19. Chuck saves $2000 per year at $7\frac{1}{2}\%$ interest. What will he have in 30 years?

20. Lynn saves $2 per day for 15 years at 6% interest. How much will she accumulate?

21. Frank and Bonnie plan to spend $200 every year on Christmas presents for their daughter, Jennifer. Suppose, instead, that they put the $200 in the bank at $5\frac{1}{2}\%$ interest. How much will the account be when Jennifer is 20 years old?

22. Bob saves $10 per week. He puts this into a 6% savings account. How much will this be in 25 years?

23. Judy has $70 per month taken out of her paycheck and put into a $6\frac{1}{2}\%$ savings account. What will this be in 30 years?

24. Ben and Laura want to save $6000 as a down payment on a house. They want to buy a house in 5 years. How much money must they put into a 6% account every year to do this?

25. Jack and Pam want to retire in 20 years with $150,000. How much money do they have to save each year at $6\frac{1}{2}\%$ interest to do this?

26. Sue is saving to take a trip to Europe in 7 years. The trip will cost $2500. How much must she save each week at 6% interest to save this $2500? (*Hint:* First figure how much she needs each year; then divide by 52.)

27. Wayne and Jan start a saving plan to accumulate $500,000 in 30 years. If they can get 8% interest, how much do they have to save each year? each month? each week? each day?

28. Lillian can save $10 a week at 6% interest. How long will it take her to accumulate $10,000?

29. Terry and Vicki are trying to save $8000 for a house down payment. They can save $80 a month at $6\frac{1}{2}\%$ interest. How long will it take them to get the down payment?

30. Amy and Steve invest $500 every year for 11 years. At the end of the 11 years, they have $7500. This is approximately what rate of interest?

31. Frank and Pearl spend $50 each month eating in restaurants. If they put this in the bank at 6% interest (instead of their stomachs), about how long would it take to grow to $15,000?

8–3   HOME OWNERSHIP

Question   Is owning a home better than renting?

ANSWER   It is impossible to give one answer for every person. Owning or not owning a home

(Mimi Forsyth/Monkmeyer Press Photo Service)

can be very emotional. In this section, we will just look at the dollar and cents aspects of owning a home.

Question How expensive a house can a person afford?

ANSWER The price of a house determines the costs and expenses. Our ability to pay these costs depends on our salary. Let's use the advice of experts to see what the relation is between the price of the house and our income.

> *Advice a.* The cost of owning a home (everything: mortgage, taxes, insurance, maintenance, utilities, and so on) for 1 year runs about 15% of the house's price.
>
> *Advice b.* A family should spend no more than 30% of its income toward running a house.

Let us translate these into mathematics and use algebra:

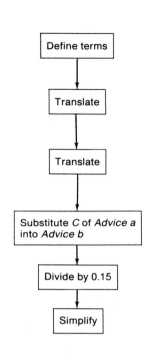

Define terms

↓

Translate

↓

Translate

↓

Substitute C of *Advice a* into *Advice b*

↓

Divide by 0.15

↓

Simplify

C = cost of running a house for a year
P = price of house
I = annual family income

Advice a. Cost is 15% of price:
$$C = 0.15 \cdot P$$

Advice b. Cost is no more than 30% of income:
$$C \leqslant 0.30 \cdot I$$

$$0.15P \leqslant 0.30I$$

$$P \leqslant \frac{0.30}{0.15} I$$

$$P \leqslant 2I$$

Formula 8–3–1

$$\boxed{P \leqslant 2I}$$

where P = price of the house and I = annual family income. Formula 8–3–1 very simply says that *the price of a house should be less than twice the family income*.

Problem Betty and Barry make $21,000 together. How expensive a house can they afford?

ANSWER Using Formula 8–3–1, we see that $I = 21,000$, so that $P \leqslant 2I = 2(21,000) = \$42,000$. Therefore, they should buy a house costing $42,000 or less.

Problem Tracy and Andy are looking at a $37,000 house. How big should their income be to afford the house?

ANSWER We use Formula 8–3–1, with $P = 37,000$.

Formula 8–3–1 → Substitute → Divide by 2

$$P \leqslant 2I$$

$$37,000 \leqslant 2I$$

$$18,500 \leqslant I$$

Hence they should make more than $18,500 to afford the house.

Down Payment

Question   What is the down payment?

ANSWER   Whenever we buy something very expensive, such as a house or car, we usually borrow most of the cost. The part of the cost that we pay right away is called the **down payment**. The other part, which we borrow, is called the **mortgage**.

For a car, the down payment is usually 25 to 35%. For a house, the down payment can vary from 5 to 30%.

Problem   Dave and Linda are interested in a used house that costs $42,000. Like many used houses, this one requires a 20% down payment. How much do Dave and Linda need to put down on this house?

ANSWER   This is a standard percentage problem.

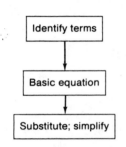

A = amount of *down payment* = ?
P = percent of *down payment* = 20% = .020
B = base = 42,000

$$A = P \cdot B$$

$$A = (0.20)(42,000) = \$8400$$

Problem   Anita and Denny have saved $6000 for the down payment on a house. They know that a used house usually requires 20% down, whereas many new houses can be bought with 10% down.
(a) How expensive a used house can they buy with $6000 down?
(b) How expensive a new house can they buy with $6000 down?

ANSWER   These are both percentage problems for which the base is unknown.
(a) For the used house, 20% down is required.

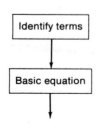

A = amount of *down payment* = 6000
P = percent of *down payment* = 20% = 0.20
B = base = ?

$$A = P \cdot B$$

$$6000 = (0.20)B$$

$$\frac{6000}{0.20} = B$$

$$30,000 = B$$

Therefore, they can buy a $30,000 used house.

Rising labor and materials costs are driving up the prices of new houses. (Sam Falk/Monkmeyer Press Photo Service)

(b) For the new house, only 10% down is required.

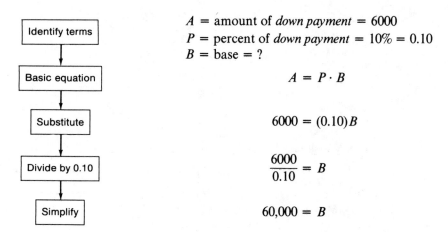

A = amount of *down payment* = 6000
P = percent of *down payment* = 10% = 0.10
B = base = ?

$$A = P \cdot B$$

$$6000 = (0.10)B$$

$$\frac{6000}{0.10} = B$$

$$60,000 = B$$

Anita and Denny could take the same $6000 and buy a $60,000 new house.

Question   If the same $6000 down payment can buy a $30,000 used house or a $60,000 new house, wouldn't everybody buy a new house?

ANSWER   Remember, the down payment is only part of owning a house. The other part is the monthly expense. (Don't forget, the price of the house should be less than twice the family income.)

PROBLEM SET 8–3–1

1. Define or discuss:
   (a) Down payment.
   (b) Mortgage.

Use Formula 8–3–1 to complete the following table.

	P (price)	I (income)
2.	$20,000	?
3.	$25,000	?
4.	?	$8,500
5.	?	$22,000
6.	$52,000	?
7.	?	$14,000
8.	$37,000	?
9.	?	$26,000
10.	?	$35,000
11.	$61,000	?

Complete the following table.

	Price of house	Down payment (%)	Down payment ($)	Mortgage = price − down payment
12.	$33,000	15	?	?
13.	$43,000	20	?	?
14.	$38,500	10	?	?
15.	$22,500	25	?	?
16.	$31,000	?	7,000	?
17.	$52,000	?	13,000	?
18.	$47,200	?	?	$40,000
19.	?	15	6,500	?
20.	?	20	8,200	?
21.	?	10	2,700	?

Mike and Connie earn $25,200 together.

22. How expensive a house can they afford?

23. If 20% down payment is needed, how much down payment must they save to buy the house?

Rob and Lorraine have saved $7200 for a down payment on a house.

24. How expensive a new house can they buy if 10% down is needed?

25. How expensive a used house can they buy if 20% down is needed?

26. If they earn $27,100 together, can they afford the new house?

27. Can they afford the used house?

Mitch and Claudia make $17,000 together.

28. How expensive a house can they afford?

29. How much do they have to save to make a 20% down payment on that house?

Ellen and Tom are interested in an older home that sells for $47,200.

30. How much must their income be to afford this house?

31. How much must they save to make a 25% down payment?

Barb and Nick make $33,100 together.

32. How expensive a house can they afford?

33. How much down payment do they need if 10% down is required?

34. If 20% down is required?

35. If 25% down is required?

Monthly Expenses

Question   What are the monthly expenses in owning a house?

ANSWER   The major expenses that occur all the time are the following:

1. Mortgage
2. Real estate taxes
3. Insurance
4. Upkeep or maintenance
5. Utilities (electricity, water, gas, phone)
6. Heating

We will discuss these different expenses next.

Question   What is a mortgage?

ANSWER   If we buy a $30,000 house, we may have to borrow from $20,000 to $27,000 from a bank (or savings and loan) to pay for it. The agreement that we sign to repay the money to the bank or savings and loan is called a **mortgage**.

The mortgage agreement will state (in very legal terms):

1. How much *money* we are borrowing, P.
2. What the *interest rate* is, r.
3. How many *years*, n, we have to repay the loan.

Each month we make the same mortgage payment to the bank. Some of it goes to repay the loan, but most of it is interest for the first few years.

Question   How do we compute the monthly mortgage payments?

ANSWER   We use another table, called the **amortization payment table** (Table 8–3). (The word amortization, comes from the Latin *mortus* and French *morte* meaning dead. We think of amortization as "killing off the loan.")

Table 8–3   Value of $L_{n,\,r}$: Monthly Payment for Each $1000

n (years)	7	$7\frac{1}{4}$	$7\frac{1}{2}$	$7\frac{3}{4}$	8	$8\frac{1}{4}$	$8\frac{1}{2}$	$8\frac{3}{4}$	9	$9\frac{1}{4}$	$9\frac{1}{2}$	$9\frac{3}{4}$	10
						r (rate, %)							
5	19.81	19.92	20.04	20.16	20.28	20.40	20.52	20.64	20.76	20.88	21.01	21.13	21.25
10	11.62	11.75	11.88	12.01	12.14	12.27	12.40	12.54	12.67	12.81	12.94	13.08	13.22
15	8.99	9.13	9.28	9.42	9.56	9.71	9.85	10.00	10.15	10.30	10.45	10.60	10.75
20	7.76	7.91	8.06	8.21	8.37	8.53	8.68	8.84	9.00	9.16	9.33	9.49	9.66
25	7.07	7.23	7.39	7.56	7.72	7.89	8.06	8.23	8.40	8.57	8.74	8.92	9.09
30	6.66	6.83	7.00	7.17	7.34	7.52	7.69	7.87	8.05	8.23	8.41	8.60	8.78
35	6.39	6.57	6.75	6.93	7.11	7.29	7.47	7.66	7.84	8.03	8.22	8.41	8.60
40	6.22	6.40	6.59	6.77	6.96	7.15	7.34	7.53	7.72	7.91	8.11	8.30	8.50

To compute the monthly mortgage payment:

1. Find the column for r the interest rate in Table 8–3.
2. Read down the column to the row for the number of years n of the loan to find $L_{n,\,r}$.
3. Multiply this number by the number of $1000's borrowed.

Notice that we call the numbers in Table 8–3 $L_{n,\,r}$. L stands for loan.

Problem   Find $L_{25,\,8\%}$. What does it mean?

ANSWER   To find $L_{25,\,8\%}$, we go to Table 8–3. We first find the 8% column; then we read down to the 25-year row. We see that they meet at $L_{25,\,8\%} = 7.72$.

$$8\%$$
$$\downarrow$$
$$25 \longrightarrow 7.72$$

This number, $7.72, means it costs $7.72 a month for 25 years to borrow $1000. To borrow $2000, it costs twice as much, $15.44 a month. To borrow $3000, it costs three times as much, $23.16 a month. And so on.

We have a formula to find the monthly payment, M:

Formula 8–3–2

$$M = L_{n,\,r}\left(\frac{P}{1000}\right)$$

where M = monthly payment, P = amount borrowed, n = years of loan, r = interest rate, and $L_{n,r}$ is given in Table 8–3.

The term $\dfrac{P}{1000}$ gives the number of $1000's borrowed. For example, if we borrow $P = 25,000$, then $\dfrac{P}{1000} = \dfrac{25,000}{1000} = 25$. This means we borrowed twenty-five $1000's.

Problem   Ralph and Alice borrow $22,000 for a house. They are to pay 9% interest for a 30-year period. What is their monthly payment on this mortgage?

ANSWER   We can use Table 8–3 and Formula 8–3–2. The unknown is the monthly payment, M.

Identify terms

$M = ?$
$r = 9\%$
$n = 30\,yr$
$P = 22,000$

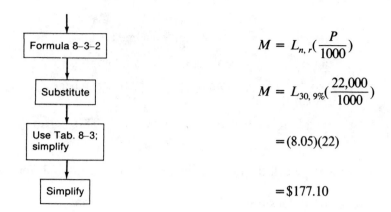

Formula 8-3-2	$M = L_{n,\,r}\left(\dfrac{P}{1000}\right)$
Substitute	$M = L_{30,\,9\%}\left(\dfrac{22{,}000}{1000}\right)$
Use Tab. 8-3; simplify	$= (8.05)(22)$
Simplify	$= \$177.10$

This tells us that their monthly mortgage payments are $177.10. The number 8.05 tells us that it costs $8.05 a month for each $1000 that they borrow. Since they borrowed twenty-two $1000's, we multiplied by 22.

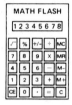

Hand Calculator Instant Replay

PUNCH	DISPLAY	MEANING
C	0.	Clear
8 . 0 5	8.05	$L_{30,\,9\%}$ from Table 8–3
× 2 2 0 0 0	22000.	Times P
÷ 1 0 0 0	1000.	Divided by 1000
=	177.1	Answer

Question What does the $177.10 go for?

ANSWER When we borrow people's money we have to pay them interest. We also have to pay them back the original amount. The $177.10 includes some interest and some money to repay the loan.

Problem How much of the $177.10 is interest; how much is repaying the loan?

ANSWER To solve this, we must first convert 9% to a monthly rate. Remember that 9% is for 1 year, so each month is $\dfrac{9}{12}\% = 0.75\% = 0.0075$. (We have to be very careful in computing this as a decimal.)

To find the first month's interest, we simply use the interest formula, 8–1–1.

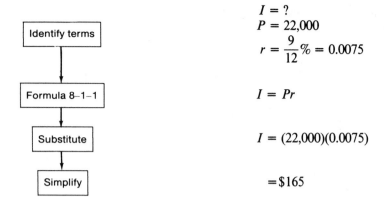

$$I = ?$$
$$P = 22{,}000$$
$$r = \frac{9}{12}\% = 0.0075$$

Identify terms

Formula 8–1–1

$$I = Pr$$

Substitute

$$I = (22{,}000)(0.0075)$$

Simplify

$$= \$165$$

Of the $177.10 payment, $165.00 is interest! Thus only $177.10 − $165.00 = $12.10 is going to repay the debt. That's why it takes 30 years to repay the loan.

Therefore, after 1 month, Ralph and Alice now owe $22,000 − $12.10 = $21,987.90. For the first few years, it seems as if the mortgage goes down by pennies.

Problem Sol and Deena are looking at a $60,000 house. They must put 10% down, and borrow the rest at $9\frac{1}{2}\%$ for 25 years. What will their monthly mortgage payments be?

ANSWER We must first find out how much they have to borrow. The 10% down payment comes to $(0.10)(60{,}000) = \$6000$. So they must borrow $P = 60{,}000 - 6000 = \$54{,}000$.

To find the mortgage payments on $54,000, we use Formula 8–3–2.

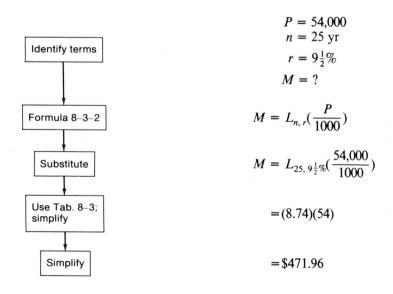

$$P = 54{,}000$$
$$n = 25 \text{ yr}$$
$$r = 9\tfrac{1}{2}\%$$
$$M = ?$$

Identify terms

Formula 8–3–2

$$M = L_{n,\,r}\left(\frac{P}{1000}\right)$$

Substitute

$$M = L_{25,\,9\frac{1}{2}\%}\left(\frac{54{,}000}{1000}\right)$$

Use Tab. 8–3; simplify

$$= (8.74)(54)$$

Simplify

$$= \$471.96$$

Their monthly mortgage payments are $471.96. This is a lot of money (and it does not include taxes, utilities, upkeep, and so on).

Problem Suppose that Sol and Deena take their $6000 down payment and buy an old $30,000 house. What are their monthly mortgage payments on this house (assuming the same $9\frac{1}{2}\%$ interest for 25 years)?

ANSWER If they put $6000 down on a $30,000 house, they must borrow

$$P = \$30{,}000 - \$6000 = \$24{,}000.$$

To find the monthly mortgage payments, we again use Formula 8–3–2.

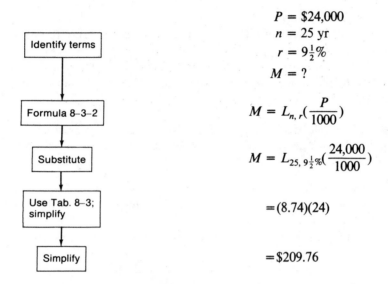

$$P = \$24{,}000$$
$$n = 25 \text{ yr}$$
$$r = 9\tfrac{1}{2}\%$$
$$M = ?$$

Identify terms

Formula 8–3–2
$$M = L_{n,\,r}\left(\frac{P}{1000}\right)$$

Substitute
$$M = L_{25,\,9\frac{1}{2}\%}\left(\frac{24{,}000}{1000}\right)$$

Use Tab. 8–3; simplify
$$= (8.74)(24)$$

Simplify
$$= \$209.76$$

Now their mortgage payments are only $209.76, which is less than half the previous figure of $471.96 for the $60,000 house.

This is the main reason people often buy an old $30,000 house when they could get a new $60,000 house. They cannot afford the big mortgage and upkeep on the expensive house.

Problem Marcy and Leigh feel that they can afford a monthly mortgage payment of $230. Mortgages are currently $8\frac{1}{2}\%$ for 20 years.
   (a) How much money can they afford to borrow?
   (b) If they will have to put 20% down on a house, how much house can they buy?
   (c) How much down payment will they need?

ANSWER   (a) We first solve for the amount that they can borrow, P. $M = 230$ is given.

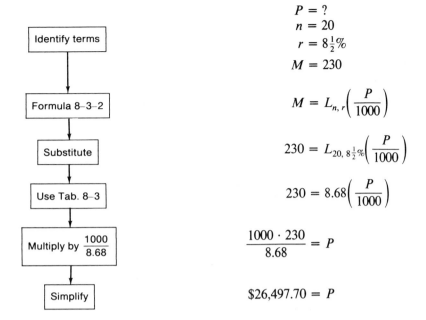

Identify terms

$$P = ?$$
$$n = 20$$
$$r = 8\tfrac{1}{2}\%$$
$$M = 230$$

Formula 8–3–2

$$M = L_{n,\,r}\left(\frac{P}{1000}\right)$$

Substitute

$$230 = L_{20,\,8\frac{1}{2}\%}\left(\frac{P}{1000}\right)$$

Use Tab. 8–3

$$230 = 8.68\left(\frac{P}{1000}\right)$$

Multiply by $\dfrac{1000}{8.68}$

$$\frac{1000 \cdot 230}{8.68} = P$$

Simplify

$$\$26{,}497.70 = P$$

Thus they can afford to borrow $26,497.70. Let us round this to $26,500 since banks usually give loans in multiples of $100.

(b) How expensive a house can this $26,500 mortgage buy? We know that there is a 20% down payment. But the $26,500 is not the 20% down, but the 80% borrowed. Remember, in a percentage problem, that the A and P terms must be the same.

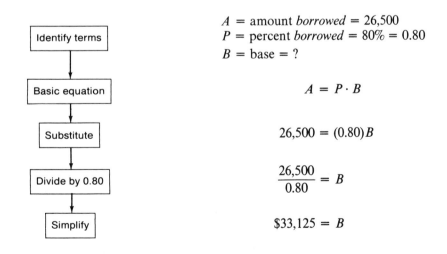

Identify terms

$$A = \text{amount } borrowed = 26{,}500$$
$$P = \text{percent } borrowed = 80\% = 0.80$$
$$B = \text{base} = ?$$

Basic equation

$$A = P \cdot B$$

Substitute

$$26{,}500 = (0.80)B$$

Divide by 0.80

$$\frac{26{,}500}{0.80} = B$$

Simplify

$$\$33{,}125 = B$$

They can afford a $33,125 house.

(c) The down payment is 20% of 33,125, or (0.20)(33125) = $6625.

Hand Calculator Instant Replay

	PUNCH	DISPLAY	MEANING
(a)	C	0.	Clear
	1 0 0 0	1000.	1000
	× 2 3 0	230.	Times M
	÷ 8 . 6 8	8.68	Divided by $L_{20,\,8\frac{1}{2}\%}$
	=	26497.695	Total loan (answer)
(b)	÷ . 8 0	0.80	Divide by 0.80
	=	33122.118	Total price (answer)
(c)	× . 2 0	0.20	Times 20%
	=	6624.4236	Down payment (answer)

Notice that the answer is slightly different (by a few cents) since we do not round off with the calculator; we just keep going.

PROBLEM SET 8–3–2

Use Formula 8–2–3 and Table 8–3 to complete the following table.

	Amount borrowed P	Time n	Interest $r\,(\%)$	Monthly Payments M
1.	$24,000	25	8	?
2.	17,000	20	$9\frac{1}{2}$	?
3.	32,000	30	9	?
4.	38,000	20	8	?
5.	13,000	15	$8\frac{3}{4}$	?
6.	?	25	$9\frac{1}{4}$	220
7.	?	20	$8\frac{1}{2}$	309
8.	21,000	20	?	179
9.	15,000	25	?	126

Ken and Lisa buy a $29,000 house. They put 20% down and borrow the rest at $8\frac{1}{2}\%$ for 25 years.

10. What is the down payment?

11. What is the amount borrowed?

12. What are the monthly mortgage payments?

Pam and Hank buy a new $42,000 house. They put 10% down and borrow the rest at 9% for 30 years.

13. What is the down payment?

14. What is the amount borrowed?

15. What are the monthly mortgage payments?

Roger and Denice can afford a monthly mortgage payment of $255. The current mortgage rate in their area is $9\frac{1}{4}\%$ for 20 years.

16. How much of a mortgage loan can they get?

17. They have to put 20% down on a house. The mortgage is 80% of what total price?

18. What is the down payment?

Cindy and Frank are interested in a $36,000 house. The mortgage terms are $8\frac{3}{4}\%$ for 25 years.

19. What are their monthly payments if they put 10% down?

20. If they put 20% down?

21. If they put 30% down?

22. If they put 50% down?

Sharon and Willie have saved $4700 to put down on a house. Loans are $8\frac{1}{2}\%$ for 20 years.

23. If the bank requires 10% down, how expensive a house can they buy?

24. How big will their loan be?

25. How much will their monthly mortgage payments be?

26. Suppose that the bank requires 25% down; how expensive a house can they buy?

27. Now how big will their loan be?

28. What will their monthly mortgage payment be?

Taxes

Question What are real estate taxes?

ANSWER As if the federal tax, state tax, and sales tax weren't enough, we have to pay a tax (a big tax) on our house and land. This is called a **real estate tax**. This tax varies from town to town. It depends on the value of the house and land. It depends on whether the town is growing (needing schools, road, sewers, and the like) or not. Usually, most real estate taxes run about 2 to 3% of the house's value per year.

Problem The tax rate where the Bakers live is 3% of the value of the house. The house is worth $35,000.
    (a) What is the yearly real estate tax?
    (b) What is the monthly real estate tax?

ANSWER (a) The annual tax is 3% of 35,000, which is (0.03)(35,000) = $1050 per year.

    (b) This means that each month they will have to pay $\frac{\$1050}{12}$ = $87.50 per month toward the real estate tax. Usually, the bank collects the tax every month with the mortgage.

Utilities

Question What are utilities?

ANSWER **Utilities** are the services such as electricity, gas, telephone, and water that we need to run a normal household.

Problem Shirley and Ned have a solid-state color TV marked 200 watts. They watch about 6 hours of TV a day. How much will this cost them per month?

ANSWER Electrical use is measured in kilowatt-hours (kwh). But we have to be careful since most electrical appliances are marked in watts (w). We use the conversion fact 1 kilowatt = 1000 watt.
    In the TV problem, we have

$$200 \, w \times \frac{1 \text{ kw}}{1000 \, w} = 0.2 \text{ kw}$$

Thus the TV uses 0.2 kilowatts of power. To find the kilowatt-hours, we multiply this by the number of hours that the TV is in use.

In one day, $(0.2 \text{ kw}) \cdot (6 \text{ hr}) = 1.2 \text{ kwh}$

In 30 days, $(1.2 \text{ kwh}) \cdot (30) = 36 \text{ kwh}$

How much will this cost? Like taxes and interest rates, this varies from town to town. Because of America's energy crisis, it may cost as much as 4.5 cents per kilowatt-hour. So

$$\frac{4.5¢}{kwh} \times 36\ kwh = 162¢ = \$1.62$$

Thus it costs them about $1.62 per month to operate their TV.

> To find the cost of operating an electrical appliance each month:
>
> 1. Find the wattage (usually marked on the appliance itself).
> 2. Divide by 1000 to get kilowatts (kw).
> 3. Multiply kilowatts by the number of hours the appliance is used in a day.
> 4. Multiply by 30 to get the kilowatt-hours use per month.
> 5. Multiply by the cost rate of a kilowatt-hour (usually between 3 and 5 cents).

We can put this into a formula.

Formula 8–3–3

$$C = \left(\frac{W}{1000}\right) \cdot T \cdot 30 \cdot \left(\frac{R}{100}\right) = \frac{3 \cdot W \cdot T \cdot R}{10,000}$$

where C = total monthly cost, W = wattage of appliance, T = time used each day (in hours), and R = kilowatt-hour rate in cents.

Problem   Mary and Glen have 22 light bulbs in their house. The average wattage is 75 watts. (Some are more, some are less, but the average is 75 watts.) Each bulb is used an average of 4 hours. If electricity costs them 4.2 cents per kilowatt-hour, how much will it cost them to burn the light bulbs each month?

ANSWER   First, let's find the total time T that all the lights are on in 1 month. There are 22 bulbs × 4 hours per bulb = 88 total hours that all the bulbs burn each day. Now we use Formula 8–3–3 for the cost C:

$$C = ?$$
$$W = 75w$$
$$T = 88\ hr$$
$$R = 4.2¢/kwh$$

Identify terms

$$C = \frac{3 \cdot W \cdot R \cdot T}{10,000}$$

Formula 8–3–3

Substitute

$$C = \frac{(3)(75)(4.2)(88)}{10,000}$$

Simplify

$$= \$8.32$$

Thus it costs them about $8.32 per month for their lights.

Hand Calculator Instant Replay

PUNCH	DISPLAY	MEANING
C	0.	Clear
3	3.	
× 7 5	75.	
× 4 . 2	4.2	Formula 8–3–3
× 8 8	88.	
÷ 1 0 0 0 0	10000.	
=	8.316	Answer

Question What are the other utilities?

ANSWER We also have to pay for water use and telephone service. To heat our house, we usually use gas, oil, or electricity. Experts estimate that people can save about 3% on their heating bill for every degree that they lower their thermostat.

Problem Jim lowers his thermostat from 72° to 66°. His heating bill was about $50 per month. Now what might it be?

ANSWER Since Jim lowers his thermostat by 6°, he can save about 6 × 3% = 18% on his bill. This is a total savings of 18% of $50, which is (0.18)(50) = $9. So his new bill might be $41.

Problem Lisa used to keep her thermostat at 72°. Now she puts it at 68° during the day and

$63°$ at night when the family sleeps. What percent of her heating bill can she save?

ANSWER For 16 hours $\left(\dfrac{2}{3}\text{ day}\right)$, she has lowered the thermostat by $4°$, for a 12% savings.

For the other 8 hours $\left(\dfrac{1}{3}\text{ day}\right)$ she has lowered the heat $9°$, for a 27% savings. Together we get

$$\text{Savings} = \frac{2}{3}(12\%) + \frac{1}{3}(27\%)$$

$$= 8\% + 9\% = 17\%$$

Notice that we multiply each saving by the fraction of a day that she saves that heat.

Question What are the other expenses of home owning?

ANSWER We have to buy insurance to protect the building and our possessions against fire and theft. We will study this in Section 8.4.

Another big expense is maintenance. This includes painting inside and outside, tuckpointing the mortar, repairing the roof, fixing the furnace, mowing the lawn, cleaning the gutters, rodding the sewers and pipes, replacing bad electrical wiring, and so on.

The maintenance list for a house can become very long and expensive. One estimate is that yearly maintenance is about 1% of the cost of the house. For example, a $30,000 house might cost $300 per year to maintain.

PROBLEM SET 8-3-3

1. Define or discuss:
   (a) Real estate tax.
   (b) Utilities.
   (c) Maintenance.

Complete the following table for real estate tax.

	Value of property	Tax rate (%)	Tax in dollars
2.	$23,000	?	$ 550
3.	32,000	2.75	?
4.	51,000	?	1400
5.	?	2.8	1220
6.	44,000	3.1	?
7.	?	2.7	1510

8. Mike lowers his thermostat from 70° to 65°. His heating bill was $60 per month. What might it be now?

Ellen lowers the thermostat from 72° to 65° in the day, and from 68° to 60° the 8 hours that the family sleeps.

9. What percent savings will her family have?

10. If their heating bill was $90, what will it be now?

Stan and Barb used to leave their thermostat at 72°. Since they both work, they leave it at 60° from 11 P.M. to next 5 P.M. (while they sleep and work). They put it at 66° from 5 P.M. to 11 P.M. (while they are at home after work).

11. What percent can they save on their heating bill?

12. If the bill was $55 a month, what will it be now?

13. In the following table, you are to compute your (or your family's) electrical use and cost. For each appliance, use the wattage given, or get a more correct number off the appliance itself. Under Column T, put the average usage. (For example, a vacuum cleaner used 1 hour per week is the same as $\frac{1}{7}$ hour per day.) Call your electric company to find the kilowatt-hour rate, R.

Appliance	Number in house (N)	Average wattage (W)	Hours used per day (T)	Kilowatt-hour $=\dfrac{N \cdot W \cdot T}{1000}$	Cost per month $30 \cdot (\text{kwh}) \cdot R$
Air cleaner		50			
Air conditioner (room)		860			
Blanket, electric		177			
Blender		386			
Broiler		1,436			
Clock		2			
Clothes dryer (electric)		4,856			
Coffee maker		894			
Deep frier		1,448			
Dishwasher		1,200			
Fan (window)		200			
Freezer		340			
Frying pan		1,196			
Hair dryer		1,000			
Heater (space)		1,322			
Hot plate		1,257			
Iron		1,000			
Microwave oven		1,450			

Appliance	Number in house (N)	Average wattage (W)	Hours used per day (T)	Kilowatt-hour $= \dfrac{N \cdot W \cdot T}{1000}$	Cost per month $30 \cdot (\text{kwh}) \cdot R$
Range with oven		12,200			
Radio		70			
Record player		110			
Refrigerator/freezer		325			
Refrigerator/freezer (frostless)		615			
Sandwich grill		1,161			
Sewing machine		75			
TV (B&W tube)		160			
TV (B&W solid state)		55			
TV (color tube)		300			
TV (color, solid state)		200			
Toaster		1,146			
Vacuum cleaner		630			
Vibrator		40			
Washing machine		512			
Waste disposer		445			
Water heater		2,475			
Others _____					
_____					
_____					
_____					
_____					

Total

Benefits of Homeowning

Question   What are the benefits of owning a home?

ANSWER   This is a mathematics book, so we will not dwell on personal benefits like pride in ownership, freedom from landlords, and the like. We will study only the money benefits of owning a home.

The main financial benefits of home owning are the following:

1. Equity
2. Appreciation
3. Tax savings

Question   What is equity?

ANSWER   Each month, we make a mortgage payment. For many years, most of it is interest, but a little of it goes to repay the loan. This little bit that we repay is *less* that we owe, so it is really money that we are saving. [Two negatives = a positive: less ($-$) debt ($-$) = savings ($+$).]

This money is called **equity**. For example, suppose that the Brown's mortgage payment is $170 a month, and $150 of this is interest. Then the difference, $20, goes toward the equity. After several years this builds up. Ultimately, they own the house.

Formula 8–3–4

$$E = T - I$$

where E = equity, T = total payment, and I = interest.

Problem Fred and Mary borrow $20,000 at $9\frac{1}{2}$% interest for 20 years. How much equity do they get in the first year?

ANSWER We must first compute the monthly payments.

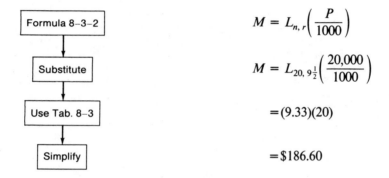

Formula 8–3–2	$M = L_{n,\,r}\left(\dfrac{P}{1000}\right)$
Substitute	$M = L_{20,\,9\frac{1}{2}}\left(\dfrac{20{,}000}{1000}\right)$
Use Tab. 8–3	$= (9.33)(20)$
Simplify	$= \$186.60$

Their monthly payments are $186.60, and their yearly payments are $(12)(\$186.60) = \$2239.20$. Their yearly interest is

Formula 8–1–1	$I = Pr$
Substitute	$I = (20{,}000)(0.095)$
Simplify	$= \$1900$

The equity is the difference between the total payment and the interest. This is $\$2239.20 - \$1900 = \$339.20$, for the first year. Unfortunately, Fred and Mary won't get this money until they sell their house. But it is like a savings account. In the second year they earn even more equity.

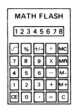

Hand Calculator Instant Replay

PUNCH	DISPLAY	MEANING
C MC	0.	Clear
9 . 3 3	9.33	$L_{n, r}$
× 2 0	20.	$P/1000$
=	186.6	M
× 1 2	12.	12 months
=	2239.2	T
M+	2339.2	Store T in memory
2 0 0 0 0	20000.	P
× . 0 9 5	0.095	Times r
=	1900.	I
M −	1900.	Subtract I from T (from memory)
MR	339.2	Answer

Question What is appreciation?

ANSWER When we buy a car, the value usually goes down. This is called depreciation. When we buy a house, the value usually goes up. This is called **appreciation**.

When we buy a house, the value usually goes up. This is called **appreciation**.

If a house is well built and in a good location, it will probably appreciate (go up in value) as fast or faster than inflation. In fact, appreciation works like compound interest.

Problem Marion and Dick buy a $32,000 house. It appreciates in value 5% per year. What is the house worth in
  (a) One year?
  (b) Five years?
  (c) Twenty years?

ANSWER In Section 8.1, we saw how the compound interest formulas and tables could be used to handle problems in which something (such as value) is growing. We will use Formulas 8–1–2 and 8–1–3 and Table 8–1 on page 444.

(a) For 1 year, we have

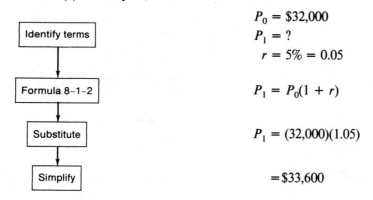

Identify terms	$P_0 = \$32{,}000$
	$P_1 = ?$
	$r = 5\% = 0.05$
Formula 8–1–2	$P_1 = P_0(1 + r)$
Substitute	$P_1 = (32{,}000)(1.05)$
Simplify	$= \$33{,}600$

(b) For 5 years, we use the compound interest formula, 8–1–3.

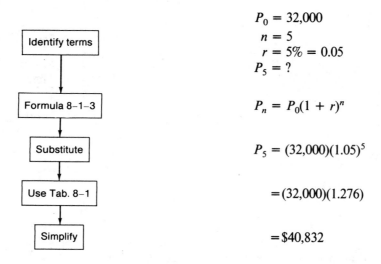

Identify terms	$P_0 = 32{,}000$
	$n = 5$
	$r = 5\% = 0.05$
	$P_5 = ?$
Formula 8–1–3	$P_n = P_0(1 + r)^n$
Substitute	$P_5 = (32{,}000)(1.05)^5$
Use Tab. 8–1	$= (32{,}000)(1.276)$
Simplify	$= \$40{,}832$

After 5 years, their house has gone up over $8000 in value. They have made $8000 by doing nothing but living in the house and meeting their mortgage payments.

(c) For 20 years, the growth is unbelievable.

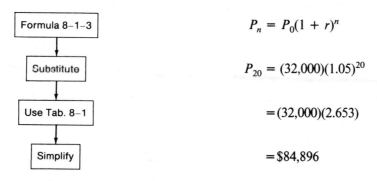

Formula 8–1–3	$P_n = P_0(1 + r)^n$
Substitute	$P_{20} = (32{,}000)(1.05)^{20}$
Use Tab. 8–1	$= (32{,}000)(2.653)$
Simplify	$= \$84{,}896$

That's right, the house will be worth about \$84,896! If they have their 20-year mortgage paid off, they will own the whole \$84,896 house.

Like equity, they do not get the appreciation until they sell the house. Also, they will have to pay some tax on the profit that they made on the house. But this is taxed at about *half* the usual rate.

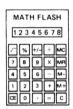

Hand Calculator Instant Replay

	PUNCH	DISPLAY	MEANING
(a)	C	0.	Clear
	3 2 0 0 0	32000.	P_0
	× 1 . 0 5	1.05	Times $(1 + r)$
	=	33600	P_1 (answer)
(b)	C	0.	Clear
	1 . 0 5	1.05	$(1 + r)$
	×	1.05	Start multiplying
	=	1.1025	$(1 + r)^2$
	=	1.157625	$(1 + r)^3$
	=	1.2155062	$(1 + r)^4$
	=	1.2762815	$(1 + r)^5$
	× 3 2 0 0 0	32000.	Times P_0
	=	40841.008	P_5 (answer)
(c)	C	0.	Clear
	1 . 0 5	1.05	$(1 + r)$
	×	1.05	Start multiplying
	=	1.1025	$(1 + r)^2$
	⋮	⋮	⋮
	=	2.526949	$(1 + r)^{19}$
	=	2.6532964	$(1 + r)^{20}$
	× 3 2 0 0 0	32000	Times P_0
	=	84905.484	P_{20} (answer)

Notice that the calculator answers are slightly different because the calculator doesn't round off the way we do.

Question What are the tax savings of home ownership?

ANSWER Remember, in Chapter 5 we saw that real estate taxes and mortgage interest were tax deductible. So homeowners tend to pay less taxes. This is called **tax saving**.

Problem Bob and Judy buy a $32,000 house. They pay a 3% real estate tax and $2300 in interest. If they are in the 22% tax bracket, how much tax savings will they get?

ANSWER The real estate tax is 3% of $32,000, which is (0.03)(32,000) = $960. Together, we have

$$
\begin{array}{rl}
\text{Real estate tax} =& \$\ 960 \\
\text{Mortgage interest} =& \underline{\ \ 2300} \\
\text{Total deduction} =& \$3260
\end{array}
$$

Recall from Chapter 5 that, since their tax bracket is 22%, they get 22% of the deduction back. Thus their tax saving is 22% of $3260, which is (0.22)(3260) = $717.20.

   The mortgage payment is in two parts: interest and repayment of the loan, as illustrated in the figure. We get all the repayment back as equity. And if we are in the $T\%$ tax bracket, we get $T\%$ of the interest back. This certainly takes a lot of the sting out of the mortgage payment.

Problem Ron and Paula buy a $38,000 house with 20% down. Their mortgage is for 25 years at $8\frac{3}{4}\%$. Their tax bracket is 25%.
   (a) What is their down payment and mortgage?
   (b) What are their monthly mortgage payments?
   (c) How much interest do they pay the first year?
   (d) How much equity do they build up the first year?
   (e) How much is their real estate tax if they pay $2\frac{1}{2}\%$ of their house's value?
   (f) If the house appreciates 5% a year, what is the house worth in 1 year?
   (g) What are their tax savings?

ANSWER These are a lot of questions, but they are the important questions every homeowner should be able to answer.
   (a) The down payment is 20% of $38,000, which is (0.20)(38,000) = $7600. They must come up with $7600 before they can buy the house. The

mortgage is the difference that they must borrow from the bank. This is $38,000 − 7600 = $30,400.

(b) We use Formula 8–3–1 to compute the monthly mortgage payments.

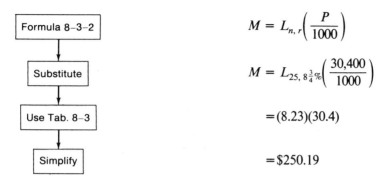

$$M = L_{n,\,r}\left(\frac{P}{1000}\right)$$

$$M = L_{25,\,8\frac{3}{4}\%}\left(\frac{30{,}400}{1000}\right)$$

$$= (8.23)(30.4)$$

$$= \$250.19$$

(c) To get an idea of the interest for the first year, we take $8\frac{3}{4}\%$ of the $30,400 debt. This is $(0.0875)(30{,}400) = \$2660$.

(d) To compute the equity we need to know their total payments for the year. This is $12 \times \$250.19 = \$3002.28$. The equity is the difference between the total payments and the interest. This is $\$3002.28 − 2660.00 = \$342.28$.

(e) The real estate tax is $2\frac{1}{2}\%$ of the $38,000 value. This is $(0.025)(38{,}000) = \$950$.

(f) The value of the house after 1 year is

$$P_1 = P_0(1 + r)$$

$$P_1 = (38{,}000)(1.05)$$

$$= \$39{,}900$$

After 1 year, their house is worth $39,900, and they have made $39,900 − 38,000 = $1990 in appreciation.

(g) Finally, we want to know what their tax savings are. We first find their deductions: real estate tax and interest.

$$\begin{aligned}
\text{Real estate tax} &= \$\ 950 \\
\text{Interest} &= \ \ 2660 \\
\hline
\text{Total deduction} &= \$3610
\end{aligned}$$

Since they are in the 25% tax bracket, they save on their income tax 25% of $3610, which is $(0.25)(3610) = \$902.50$.

1. Define or discuss:
   (a) Equity.
   (b) Appreciation.
   (c) Tax savings.

Complete the following table to determine the equity one gets in the first year.

	Mortgage P	Rate $r\,(\%)$	Years n	Monthly payments M	Yearly payments $Y=12M$	First year's interest $I=Pr$	First year's equity $E=Y-I$
2.	$20,000	$8\frac{3}{4}$	25	?	?	?	?
3.	32,000	9	20	?	?	?	?
4.	16,500	$8\frac{1}{2}$	20	?	?	?	?
5.	22,600	$9\frac{1}{2}$	30	?	?	?	?
6.	27,000	$8\frac{1}{4}$	25	?	?	?	?
7.	19,100	$9\frac{1}{4}$	25	?	?	?	?

Use Formula 8–1–3 and Table 8–1 to complete the following table on appreciation.

	Original price P_0	Growth rate $r\,(\%)$	After 1 year P_1	After 5 years P_5	After 20 years P_{20}
8.	$25,000	5	?	?	?
9.	32,000	$4\frac{1}{2}$	?	?	?
10.	44,000	5	?	?	?
11.	?	$5\frac{1}{2}$	21,840	?	?
12.	29,000	?	?	37,013	?
13.	25,000	?	?	?	60,000
14.	37,000	6	?	?	?
15.	?	5	?	49,776	?

Complete the following table for tax saving.

	Real estate tax R	Interest I	Tax bracket $T(\%)$	Tax saving TS
16.	$1100	$2100	25	?
17.	800	2080	22	?
18.	1080	2810	25	?
19.	1300	3190	28	?

	Real estate tax R	Interest I	Tax bracket T(%)	Tax saving TS
20.	1310	3570	32	?
21.	1560	4000	34	?
22.	680	1540	19	?

For Problems 23–27, answer each of the following questions.

(a) Can they afford the house? If the answer is *no*, go to the next problem. If the answer is *yes*, continue.

(b) What is their down payment?

(c) What is their mortgage?

(d) How much are their monthly mortgage payments?

(e) How much interest do they pay in the first year?

(f) How much equity do they build up in the first year?

(g) What is the house worth in 1 year? 5 years? 10 years?

(h) How much is their real estate tax?

(i) What is their tax saving?

(j) If they sell it in 5 years, how much will they receive after paying 9% commission and fees to a realtor?

23. Albert and Bonnie earn $21,000 together. They want to buy a $35,000 house with 20% down. The mortgage terms are 9% interest for 25 years. The property values are increasing $4\frac{1}{2}\%$ per year. The real estate taxes are 2.7% per year. They are in the 25% tax bracket.

24. Carol and Dave earn $18,000 together. They want to buy a $29,000 house with 15% down. The mortgage terms are $8\frac{3}{4}\%$ for 20 years. The property values are increasing 5% per year. The real estate taxes are 2.9% per year. They are in the 22% tax bracket.

25. Eileen and Fred earn $24,000 together. They are interested in a $44,000 house with 10% down. The mortgage terms are $9\frac{1}{2}\%$ for 30 years. The property values are increasing 5% per year. The real estate tax is 3.1% per year. They are in the 28% tax bracket.

26. Ginny and Hank earn $16,000 together. They want to buy a $35,000 house with 20% down. The mortgage terms are 9% interest for 20 years. The property value is increasing 6% per year. The real estate tax is 2.8% per year. They are in the 18% tax bracket.

27. Ike and Judy earn $29,000 together. They want to buy a $49,000 house with 25% down. The mortgage terms are $9\frac{1}{4}\%$ for 30 years. The property value is increasing by $4\frac{1}{2}\%$ per year. The real estate taxes are 2.6% per year. They are in the 32% tax bracket.

8–4 INSURANCE

Question What is insurance?

ANSWER Nothing frightens people as much as the thought of a big loss: their house burning, their car being stolen or smashed, a huge hospital bill, their dying and leaving their children no money, and so on. **Insurance** is a form of protection against these and other big losses. Insurance cannot prevent such mishaps, but it can help compensate for the losses.

In a way, insurance can be thought of as a bet against yourself. When we buy life insurance, we are betting that we will die within the year. Of course, we hope to live and lose the bet. (In fact, R. Shulman has written an interesting book about life insurance companies called *Billion Dollar Bookies*.)

The accompanying illustration shows how insurance pays off in the small likelihood that we die between 30 and 31.

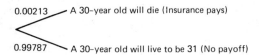

Another example of insurance is fire insurance on the house. Let us say, for example, that our house is worth $24,000. (Not counting the land under it, just the house.) Then we might have the following very simple cases.

Outcome	Probability	Payoff	Product
No damage	$\frac{995}{1000}$	$0	$0
House half-burned	$\frac{4}{1000}$	12,000	48
House all burned	$\frac{1}{1000}$	24,000	24
			$E = \overline{\$72}$

This is illustrated in the figure.

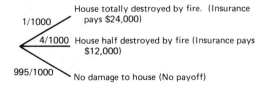

Recall from Chapter 6 that E is the expected value or payoff of a game or situation. This is like an average payoff for many players. This tells us that our expected payoff from the "fire insurance bet" is $72. Then, to be a fair game, we should pay a $72 fee (or premium) to play the game. Since insurance companies also have expenses and sales commissions to pay, we will probably have to pay more than $72, possibly even $100 for this protection.

Of course, insurance is not a game; it is a serious matter. But it is instructive to look at insurance as a bet against yourself. The fee we pay to play the game is called the **premium**. The payoff is called the **face value**.

For example, suppose that Tom pays $170 a year for a $10,000 life insurance policy; $170 is the premium, and $10,000 is the face value (the amount his family gets if he dies).

Automobile Insurance

Question   What does automobile insurance cover?

ANSWER   As we saw in Chapter 5, owning an automobile is very expensive. Automobile insurance protects us against the big expenses. The following is a list of the standard items in an automobile insurance policy.

1. 10/20 B.I. This is *bodily injury* coverage. It will pay up to $10,000 for injured person and up to $20,000 for all injured people in the accident. This protects you against a lawsuit for an accident that you cause. You can also get higher payoff limits, such as 50/100 (meaning $50,000 for one person, $100,000 for everyone involved), but this will cost more.

2. 5 P.D. This is *property damage* coverage. This will pay up to $5000 for any damage we do by smashing into another car, a house, a street light, and so on.

3. 10/20 U.M. This is *uninsured motorist* coverage. This protects us up to $10,000 per person and $20,000 per accident if we are hit by a driver who does not carry his or her own insurance.

4. 500 MED. This is *medical payments* coverage. This pays up to $500 per person in medical fees resulting from an accident in which we are to blame.

5. 50 DED. COMP. This is a *comprehensive* coverage. This protects our car from fire, theft, vandalism, and the like. It has $50 deductible. Suppose that vandals do $200 damage to our car. We have to pay the first $50. The insurance company then pays the other $150.

6. 100 DED. COLL. This is a *collision* coverage. This pays to repair our car for a collision that is our fault. It has $100 deductible. Suppose that we smash up our car, doing $550 damage. We must pay the first $100. The insurance company pays the other $450.

Table 8-4   Typical Premium Rates (per year)

Basic rates

10/20 B.I.	5 P.D.	10/20 U.M.	500 MED.	50 DED. COMP.	100 DED. COLL.
$245	$78	$6*	$11	$39	$206

Rate Factors

	Situation			
Residence	Man (married, over 25)	Woman (married, over 25)	Man (unmarried, around 20)	Woman (unmarried, around 20)
City	1.27	0.71	3.28	1.84
Suburb	1.00	0.56	2.59	1.46

*Always $6 (no change because of situation).

Question How do we compute auto insurance premiums?

ANSWER Table 8–4 gives some typical rates for typical situations. Obviously, not all situations can be listed here; also, rates vary from company to company.

> To compute an insurance premium:
>
> 1. Locate the basic rate for this coverage.
> 2. Locate the rate factor for your residence/situation.
> 3. Multiply the basic rate by the rate factor (except for 10/20 U.M., which is always $6).

Problem Jack is a 32-year-old, married, and lives in the city. What will his insurance be for all six coverages?

ANSWER Since Jack is a married man over 25, living in the city, his rate factor is 1.27. Now we multiply this by the basic rates.

Coverage	Basic rate	Rate factor	Total rate
10/20 B.I.	245	1.27	$311.15
5 P.D.	78	1.27	99.06
10/20 U.M.	6	—	6.00
500 MED.	11	1.27	13.97
50 DED. COMP.	39	1.27	49.53
100 DED. COLL.	206	1.27	261.62
			Total rate = $741.33

Thus Jack's total rate is $741.33 per year.

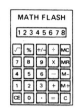

MATH FLASH

Hand Calculator Instant Replay

PUNCH	DISPLAY	MEANING
C MC	0.	Clear everything
2 4 5 × 1 · 2 7 =	311.15	10/20 B.I.
M+	311.15	Store in memory
7 8 × 1 · 2 7 =	99.06	5 P.D.
M+	99.06	Add to memory
6	6.	10/20 U.M.
M+	6.	Add to memory
1 1 × 1 · 2 7 =	13.97	500 MED.
M+	13.97	Add to memory
3 9 × 1 · 2 7 =	49.53	50 DED. COMP.
M+	49.53	Add to memory
2 0 6 × 1 · 2 7 =	261.62	100 DED. COLL.
M+	261.62	Add to memory
MR	741.33	Recall memory (answer)

Problem Sally is 20, unmarried, and lives in the suburbs. She drives a 1969 car. Is it worth it to her to get the comprehensive and collision coverages?

ANSWER Since Sally's car is a 1969 model, it is probably worth no more than $300. So this is all that insurance will pay on a loss.

Let us see what her premium is. For collision, the basic rate is $206 (with a $100 deductible). Since Sally is an unmarried 20-year-old woman living in the suburbs, her rate factor is 1.46. Thus her premium for 100 DED. COLL. is (1.46)(206) = $300.76.

Suppose that Sally's car is smashed, doing $800 damage. The insurance company will say, "Sally, your car is only worth $300. We're not going to pay $800 to fix a $300 car. You pay the first $100 deductible, and here's a check for the other $200." That's right, Sally only gets $200 from the insurance company.

Obviously, it is foolish for her to pay a $301 premium for the $200 protection. Therefore, she should not get the collision. By the same reasoning, buying theft protection (50 DED. COMP.) for a 10-year-old car is foolish. Some people stop buying collision insurance even when the car is only 5 years old.

1. Define or discuss:
   (a) Premium.
   (b) Face value.

Find the total premium for all six coverages in the following cases.

2. Steve, age 20, living in the city, not married.

3. Annie, married, age 30, living in the suburbs.

4. Chuck, age 35, married, living in the suburbs.

5. Janice, age 32, married, living in the city.

6. Mike, age 20, unmarried, living in the suburbs.

7. Claudia, age 20, unmarried, living in the city.

Suppose that you have a $100 deductible collision. Complete the following table for how much the insurance company pays and how much you pay on an accident. (Remember that a claim cannot be more than the value of the car.)

	Value of car	Damage	Insurance pays
	$2000	$700	$600
	400	850	300
8.	4600	1200	?
9.	1500	75	?
10.	250	300	?
11.	2200	1500	?
12.	320	175	?
13.	3500	Total	?
14.	1750	Total	?

Homeowner's Insurance

Question   What is homeowner's insurance?

ANSWER   **Homeowner's insurance** is a package of protections for a homeowner. It includes:

1. Protection against fire, windstorm, hail, and other damage to the building.
2. Protection against fire, damage to, and theft of the contents (furniture, clothes, and so on).
3. Protection against law suit if someone gets hurt on your property.

The cost of this insurance depends basically on three things: (1) the value of your house (this is the total cost to have it rebuilt), (2) whether the house is brick or frame (wood), and (3) how good your town's fire department is.

In Table 8–5 we have given six formulas that might be used to find the cost of a 3-year premium for homeowner's.

To find the premium for homeowner's insurance:

1. Determine where the community type (A, B or C) and the house type (brick or frame) meet in Table 8-5.
2. Write down the formula given in the box.
3. For V, substitute the value (cost to rebuild) of the house, and compute the 3-year premium, P
4. The 1-year premium is 35% of P.

Table 8-5 Three-year Premiums for Homeowner's Insurance

House	Community class		
	A	B	C
Brick	$8.5\left(\dfrac{V}{1000}\right) + 47$	$10.2\left(\dfrac{V}{1000}\right) + 52$	$14.3\left(\dfrac{V}{1000}\right) + 57$
Frame	$9.4\left(\dfrac{V}{1000}\right) + 47$	$12.1\left(\dfrac{V}{1000}\right) + 52$	$15.3\left(\dfrac{V}{1000}\right) + 57$

Problem Tom and Jeni live in a class C community. They have a $20,000 frame house. What is their 3-year homeowner's insurance premium? What is the annual premium?

ANSWER They live in a frame house in a class C community, so we find the proper formula in Table 8-5. Then we substitute $V = 20{,}000$, the value of the house.

Find formula in Tab. 8-5

$$P = 15.3\left(\frac{V}{1000}\right) + 57$$

Substitute

$$P = 15.3\left(\frac{20{,}000}{1000}\right) + 57$$

Simplify

$$= \$363$$

Their 3-year premium is $363. The yearly premium is 35% of this, or $(0.35)(363) =$ $127.05. (You can save a little money by paying all three years at once.)

Problem Ron and Paula live in a $55,000 brick home (just the home, not the land) in a class A community. What is their 3-year premium?

ANSWER They live in a brick home in a class A community, so we find the formula for this in the Table 8-5. We then substitute $V = 55{,}000$.

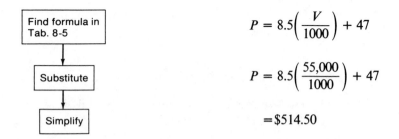

Find formula in Tab. 8-5	$P = 8.5\left(\dfrac{V}{1000}\right) + 47$
Substitute	$P = 8.5\left(\dfrac{55{,}000}{1000}\right) + 47$
Simplify	$= \$514.50$

The 3-year premium is $514.50. The yearly premium is 35% of this, or $(0.35)(514.50) = \$180.08$.

Question What is coinsurance?

ANSWER This is a tricky but important concept, since it is in almost every homeowner's policy. This is sometimes called the 80% *clause*. If your house is insured for less than 80% of its value, the insurance will not pay off *any* damage in full. This is called **coinsurance.**

For example, suppose that Ted and Linda live in a $30,000 house, but only insure it for $20,000. This is less than 80%. Now suppose that they have a $3000 fire damage. Do they get a check for $3000 from the insurance company? No! They get a check for only $2000.

Why did this happen? Since they were insured for less than 80%, they will only be paid in the proportion that they are covered. Here they are $\dfrac{20{,}000}{30{,}000} = \dfrac{2}{3}$ covered. So they can only get $\dfrac{2}{3}$ of $3000, or $2000.

We can do this as a proportion problem:

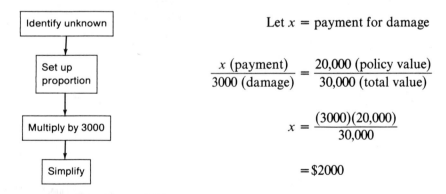

Identify unknown	Let $x =$ payment for damage
Set up proportion	$\dfrac{x \text{ (payment)}}{3000 \text{ (damage)}} = \dfrac{20{,}000 \text{ (policy value)}}{30{,}000 \text{ (total value)}}$
Multiply by 3000	$x = \dfrac{(3000)(20{,}000)}{30{,}000}$
Simplify	$= \$2000$

Question How do we determine how much an insurance company will pay on a damage?

ANSWER The rule for this is as follows.

To determine what the insurance company will pay on a loss or damage:

1. Determine the total value of the house (just the building, not the land).

2. Compute 80% of this value.

3a. If you are insured for *more* than this 80%, the company will pay the full loss up to the face value of the policy.

3b. If you are insured for less than this 80%, the company will only pay a fraction of the loss according to Formula 8–4–1.

Formula 8–4–1

$$P = L\left(\frac{I}{V}\right)$$

where L = amount of loss, I = amount of insurance, V = value of house structure, and P = payment.

Problem   Al and Ruth live in a house worth $32,000. They have $30,000 insurance on it. They have a $6000 fire. How much will the insurance company pay?

(Courtesy of Star Publications. Williams Press Newspaper Group)

ANSWER   The value of the building is $32,000. We compute 80% of this, which is (0.80)(32,000) = $25,600. Since their policy of $30,000 is *over* this, they are fully insured (up to $30,000). Therefore, the insurance company will pay the full $6000 loss. If Al and Ruth had had a $31,000 loss, they would only collect $30,000 for the insurance since this is all they were insured for.

Problem   Neal and Kay own a $25,000 house. But they only have $15,000 insurance on it. They have a $7000 fire. How much will the insurance company pay?

ANSWER   The value of the building is $25,000. We compute 80% of this, which is (0.80)(25,000) = $20,000. The policy of $15,000 is *less* than this 80%, so they fall under the coinsurance terms of the policy. They cannot collect the full loss, but instead they must use Formula 8–4–1 to find the partial payment.

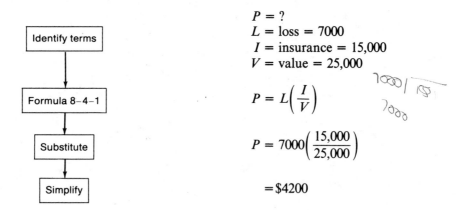

Identify terms

Formula 8–4–1

Substitute

Simplify

$P = ?$
$L = \text{loss} = 7000$
$I = \text{insurance} = 15,000$
$V = \text{value} = 25,000$

$$P = L\left(\frac{I}{V}\right)$$

$$P = 7000\left(\frac{15,000}{25,000}\right)$$

$$= \$4200$$

Neal and Kay are going to have quite a surprise when they only get a check for $4200, and they have to pay the other $2800 themselves. This is why it is very important to be at least 80% insured. Many insurance salespeople will not even sell coinsurance (under 80%). How does it happen that many people fall below the 80% figure? Read the next problem.

Problem   Roy and Cecelia buy a house in 1968 that is worth $22,000. They wisely buy $22,000 worth of insurance. Ten years later, inflation in construction costs makes the house cost $35,000 to rebuild. But Roy and Cecelia neglect to increase their insurance. They have an $8000 fire in 1978. What will they collect?

ANSWER   The value (or cost to rebuild) is $35,000; 80% of this is (0.80)(35,000) = $28,000. Since they still have $22,000 insurance they fall into coinsurance and must pay a fraction of the loss.

Identify terms

$P = ?$
$L = \text{loss} = 8000$
$I = \text{insurance} = 22,000$
$V = \text{value} = 35,000$

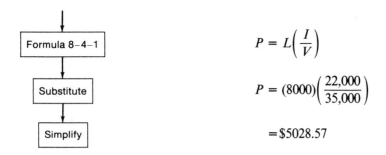

			$P = L\left(\dfrac{I}{V}\right)$
Formula 8–4–1			
Substitute			$P = (8000)\left(\dfrac{22{,}000}{35{,}000}\right)$
Simplify			$= \$5028.57$

Roy and Cecelia are going to have a surprise also. They will have to pay almost $3000 out of their own pockets just because they didn't keep their insurance policy current at the full reconstruction costs of the house.

PROBLEM
SET 8–4–2

1. Define or discuss:
   (a) Homeowner's insurance.
   (b) Coinsurance.

Use Table 8-5 to find the 3-year premium (P) and yearly premium ($0.35P$) for the following situations.

	Community class	House type	Value	P (3-year)	$0.35P$ (1-year)
2.	C	Brick	$25,000	?	?
3.	B	Frame	19,000	?	?
4.	A	Frame	31,000	?	?
5.	A	Brick	44,000	?	?
6.	C	Frame	22,500	?	?
7.	B	Brick	37,800	?	?
8.	B	Frame	27,100	?	?
9.	A	Brick	60,000	?	?
10.	C	Frame	14,200	?	?

11. Using Table 8-5, make a graph of the 3-year premiums for the following. Put all of them on one graph with V on the x-axis and P on the y-axis.
    (a) Brick house, community class A.
    (b) Brick house, community class B.
    (c) Brick house, community class C.

12. Using Table 8-5, make a graph of the 3-year premiums of the following. Put all of them on one graph with V on the x-axis and P on the y-axis.
    (a) Frame house, community class A.
    (b) Frame house, community class B.
    (c) Frame house, community class C.

Complete the following table to determine the amount that the insurance company will pay for each loss.

	Value of house	80% of value	Amount of insurance	Amount of loss	Insurance payment
13.	$30,000	?	$27,000	$12,000	?
14.	40,000	?	35,000	38,000	?
15.	25,500	?	20,000	5,500	?
16.	22,100	?	17,000	22,100	?
17.	36,000	?	31,000	24,000	?
18.	29,000	?	21,000	11,000	?
19.	16,500	?	14,000	15,000	?
20.	75,000	?	55,000	65,000	?
21.	49,000	?	45,000	44,000	?

Mike and Kathy own a home worth $25,000 (just the house). They insured it for $25,000. In the next 5 years, construction costs go up, and their house would cost 40% more to rebuild. But they do not increase their insurance.

22. What will their house cost to rebuild?

23. If they have a $6000 fire, how much will the insurance company pay?

Life Insurance

Question What are the different types of life insurance?

ANSWER There are many, many different life insurance packages offered by all the different insurance companies. But two basic types of policies are most commonly sold by most companies: (1) term, and (2) whole life. We will now discuss each policy type.

Question What is term insurance?

ANSWER **Term insurance** is just like automobile or homeowner's insurance. It only provides protection. You pay your premium: if you die, your family collects; if you live, you lose the premium. It is the cheapest of the insurance policies.

Since people are more and more likely to die as they age, the cost of protection has to go up as people age. There are two ways to buy this:

1. **Five-year renewable level term.** With this policy we keep the same face value, for example $25,000. But every 5 years we have to renew the policy and pay a higher premium. Often, we also have to take a physical exam to show that we are still healthy. Part (a) of the figure indicates roughly how the coverage stays the same, but the premiums go up.
2. **Decreasing term.** The other type of term insurance keeps the same premium for a

period of years, for example 20 years. But every year the protection goes down. Part (b) of the figure shows how the premium stays the same, while the coverage goes down.

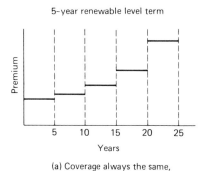

5-year renewable level term

(a) Coverage always the same,
premium goes up

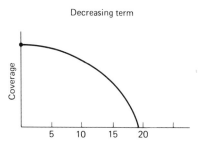

Decreasing term

(b) Premium always the same,
coverage goes down

Question What are the premiums for these policies?

ANSWER Table 8–6 gives the premiums for the two policies.

1. Twenty-year decreasing term has the same premium for 20 years, but protection keeps going down.
2. Five-year renewable term has the same protection and coverage for 5 years. This one can usually be converted to a whole-life policy.

Notice that the premiums depend on the age at which you buy the policy.

Table 8-6 Term Insurance Premiums

| | | Premium per $1000 coverage | |
| | | Policy | |
Age Men	Women	20-year decreasing term	5-year level term
25	30	$3.17	$4.95
30	35	3.70	5.51
35	40	4.72	6.55
40	45	6.41	8.36
45	50	9.13	11.39
50	55	13.42	16.30
55	60	—	24.34

Problem Tom is 30 years old and wants to buy a $35,000, 20-year decreasing term policy. What will the premium be?

ANSWER The decreasing term policy is also called *mortgage insurance*, since it declines the way the balance of a mortgage declines. That way, if Tom dies the mortgage on the house automatically can be paid off.

To compute the premium, we find where the 20-year decreasing term column meets the age 30 (men) row. This number is $3.70 per $1000. Since Tom wants to buy $35,000, we multiply (3.70)(35) = $129.50. This is Tom's premium for all 20 years.

Problem Suppose that Tom dies when he is 39. How much will his family receive from the insurance company?

ANSWER To do this, we must use Table 8–7. This tells us how $1000 declines every year. Tom will be 39 in 9 years. In Table 8–7, we see that after 9 years $1000 coverage is down to $711. We multiply this by 35; so his coverage is (35)(711) = $24,885.

Thus his family gets only $24,885. But this is because he is still paying the same premium that he did when he was 30.

Table 8–7 Declining Value of $1000 of 20-year Decreasing Term

After year	Face value	After year	Face value
1	$946	11	$619
2	950	12	568
3	923	13	513
4	894	14	454
5	864	15	391
6	828	16	323
7	792	17	250
8	753	18	172
9	711	19	89
10	667	20	0

Problem Joanne is 30. She wants to buy a $20,000, 20-year decreasing term policy.
(a) What will the annual premium be?
(b) What will the face value be when she is 42?

ANSWER (a) To find the premium, we use Table 8–6. A 20-year decreasing term policy for a 30-year-old woman is $3.17 per $1000. Since she wants $20,000 coverage, it will cost (20)(3.17) = $63.40.
(b) She will be 42 in 12 years. From Table 8–7, we see that in 12 years a $1000 policy has declined to $568. So her $20,000 is down to (20)(568) = $11,360. When she is 42, her policy is only worth $11,360.

Problem Mike is 35 years old. He wants to buy a $40,000 term policy, but he does not want it to decline. What premiums will he have to pay?

ANSWER If Mike wants his policy to stay at $40,000, he will have to buy 5-year level term policies and increase the premiums every 5 years. We look at Table 8–6. For a 35-year-old man, $1000 of 5-year level term is $6.55. Since Mike wants $40,000, he will have to pay (40)(6.55) = $262.

When Mike is 40 years old, he has to renew the policy (since the policies are only good for 5 years). Table 8–6 tells us that $1000 of coverage costs $8.36. Now $40,000 of coverage will be (40)(8.36) = $334.40.

When he is 45, the same policy will cost (40)(11.39) = $455.60. Notice that every 5 years it costs more and more to buy the same policy. Why? Because every year, Mike is more and more likely to die.

Problem   George is 35 years old. He can afford $250 per year for insurance.
         (a) How much 20-year decreasing term can he buy?
         (b) How much 5-year level term can he buy?

ANSWER   Here we are given the premium, $250, and we wish to know the face value.
         (a) Since George is 35, $1000 of 20-year decreasing term costs $4.72. Call the unknown amount of insurance x thousands. His premium of $250 is $4.72 times x.

$$250 = 4.72x$$

$$\frac{250}{4.72} = x$$

$$52.966 = x$$

This is the number of $1000's. So George can buy $52,966, or about $53,000 of 20-year decreasing term.
         (b) Let y be the face value (in $1000's) of his 5-year level term policy. Since this costs $6.55 per $1000, the $250 premium is $6.55 times y.

$$250 = 6.55y$$

$$\frac{250}{6.55} = y$$

$$38.168 = y$$

So George can only buy about $38,000 of 5-year level term.

PROBLEM
SET 8–4–3
   1. Define or discuss:
       (a) Twenty-year decreasing term.
       (b) Five-year level term.

Complete the following table using Table 8–6 (DT = 20-year decreasing term; LT = 5-year level term).

	Name	Age	Type of insurance	Face value	Annual premium
2.	Ralph	40	DT	25,000	?
3.	Jim	30	LT	55,000	?
4.	Sally	35	DT	32,000	?
5.	Roger	25	LT	30,000	?
6.	Ruth	45	LT	20,000	?
7.	Ed	45	DT	37,000	?
8.	Harry	35	LT	?	400
9.	Diane	30	LT	?	150
10.	Dan	35	DT	?	320
11.	John	25	DT	?	200
12.	Karen	40	LT	?	550

Use Table 8–7 to complete the following table for the declining value of decreasing term insurance.

	Age when policy was bought	Present age	Original face value	Present face value
13.	28	36	20,000	?
14.	32	43	45,000	?
15.	29	41	50,000	?
16.	39	47	32,000	?
17.	41	43	43,000	?
18.	48	55	10,000	?
19.	37	41	?	21,456
20.	31	43	?	19,880
21.	26	?	28,000	19,908
22.	?	41	33,000	12,903

Douglas wants to buy a $45,000, 20-year decreasing term policy. He is 35 years old.

23. How much will the policy cost him annually?

24. What will the face value be in 5 years? 10 years? 15 years?

Victor can afford $150 per year for life insurance. He is 25 now.

25. How much 20-year decreasing term can he buy?

26. How much 5-year level term can he buy?

27. What will his decreasing term be worth in 5 years?

28. How much 5-year level term will he be able to buy in 5 years with $150?

Linda is 35 years old. She can afford $175 per year for insurance.

29. How much 20-year decreasing term can she buy?

30. How much 5-year level term can she buy?

31. How much will her decreasing term be worth in 5 years?

32. How much 5-year level term will she be able to buy in 5 years with $175?

Graph the values in Table 8–6.

33. With age on the *x*-axis and the 20-year decreasing term premium on the *y*-axis.

34. With age on the *x*-axis and the 5-year level term premium on the *y*-axis.

35. Graph the values in Table 8–7 with the number of years on the *x*-axis and the face value on the *y*-axis.

Whole Life

Question   What is whole-life insurance?

ANSWER   **Whole life** (also called *ordinary life* or *straight life*) is different from just about all other insurance policies. In addition to protection (death benefits), whole life also builds up a savings account or **cash value** on the side. Thus, not only are you insured against death, but you are also saving money that you can borrow or eventually have.

    This is a very popular form of insurance. In fact, it is the one most insurance salesmen push the hardest. Not surprisingly, the commission is also very high.

Question   What does whole life insurance cost?

ANSWER   Table 8–8 tells what the typical rates are per $1000 coverage.

Table 8–8   Whole-Life Premiums

Age		Premium per $1000
Men	Women	
25	30	$17.07
30	35	19.77
35	40	23.04
40	45	26.96
45	50	31.83
50	55	37.95
55	60	45.78

Problem   Chuck is 30 years old. He wants to buy a $20,000 whole-life policy. What will the annual premium be?

ANSWER   We see in Table 8–8 that the premium for a 30-year-old man is $19.77 per $1000. Since Chuck wants $20,000, we multiply (20)(19.77) = $395.40. This is the premium as long as Chuck keeps the policy.

Problem   Barbara is 30 years old. She wants to buy a $15,000 straight life policy. What will the annual premium be?

ANSWER   We see in Table 8–8 that the premium for a 30-year-old woman is $17.07. Since she wants $15,000, we multiply (17.07)(15) = $256.05. Notice that straight life and whole life are the same thing and use the same table.

Question   How does the cash value work?

ANSWER   The money in the cash value depends on how long you have had the policy, when you bought it, and how big the face value is. Table 8–9 shows how the typical cash value grows on a $1000 policy. Some of these figures include dividends as well as cash values.

Table 8–9 Cash Values and Dividends

| Age when policy was bought | | Typical cash values plus dividends per $1000 face value (whole-life policy) | | | |
| Men | Women | Years later | | | |
		5	10	15	20
25	30	$55	$154	$265	$395
30	35	66	179	307	450
35	40	80	208	348	510
40	45	95	243	402	576
45	50	112	278	455	648
50	55	129	320	519	730

Use Table 8–9 to find cash values on a policy:

1. Find the row for the age at which the policy was bought.
2. Find the column for the number of years later.
3. Where these meet is the cash value for $1000.
4. Multiply by the number of $1000's in your policy to get cash value.

Problem   Frank buys a $25,000 whole-life policy at age 30. What is the cash value at age 40?

ANSWER   Frank bought the policy at age 30. Age 40 means 10 years later. In Table 8–9, these meet at $179. This means a $1000 policy has a $179 cash value. Since Frank has a $25,000 policy, we multiply $(25)(179) = \$4475$. Thus, when Frank is 40, he has a $4475 cash value saved up.

Question   Does this mean that if Frank dies his family gets $25,000 *plus* $4475, or a total of $29,475?

ANSWER   No. If Frank dies, his family only gets $25,000. The $4475 is his *only* if he lives. (Many insurance salesmen forget to mention this.) Frank can get the $4475, but he has to surrender or give up the insurance policy. Or, Frank can borrow his own money, the $4475, and pay $5\frac{1}{2}\%$ interest. This is one "catch" to whole-life insurance.

Problem   Bryan buys $30,000 of whole-life insurance when he is 25 years old.
   (a)   What is his yearly premium?
   (b)   What is the total amount of premiums he has paid over 20 years?
   (c)   What is the cash value in 20 years?

ANSWER   (a)   To find the premium, we look at Table 8–8. A 25-year-old man must pay $17.07 per 1000. Since Bryan wants $30,000 coverage, we multiply $(30)(17.07) = \$512.10$. Thus his yearly premium is $512.10.
   (b)   Over 20 years, he will have paid a total of $(20)(512.10) = \$10,242$.
   (c)   To find his cash value after 20 years, we use Table 8–9. We see that a policy bought at age 25 has a cash value of $395 per 1000 twenty years later. Since Bryan has $30,000 coverage, we multiply $(30)(395) = \$11,850$.
         Notice that this cash value is more than all his total premiums. In fact, he has $11,850 - 10,242 = \$1608$ *more* in cash value than he paid in total premiums.

Question   Does this mean that Bryan has had free insurance plus getting a $1608 bonus?

ANSWER   This is exactly the pitch every insurance salesperson will give you: "You pay in $10,242. You have $11,850 built up in cash value. That's a $1608 profit. Plus, you've been insured for 20 years. In other words, we've paid you $1608 to be insured for 20 years! That's $80.40 a year *we* paid you to be insured. You can't beat that!"

Question   Can we beat that?

ANSWER   Perhaps. According to many experts, people should buy the cheaper term insurance and put the rest of the money in the bank, mutual funds, and the like.

Question   Which is better, whole life or term?

ANSWER   There is a raging debate between consumer advocates (for term) and the life insurance people (for whole life). Let us look at the relative advantages of both.
         Term insurance provides very inexpensive protection when a person is young and needs a lot of inexpensive protection. If you have the discipline, you can put

the saved premiums in the bank. This money will probably grow faster than the cash value of whole life. Plus, it is your money whether you live or die.

Whole life is forced saving. Since many people are weak willed about saving money, whole life is good because it forces them to save while paying their regular insurance bill. Also, whole-life premiums do not increase as you grow older the way that term premiums do.

PROBLEM SET 8–4–4

1. Define or discuss:
   (a) Whole life insurance.
   (b) Cash value.

Complete the following table using Table 8–8 for whole-life insurance.

	Name	Age	Face value	Annual premium
2.	Chuck	25	40,000	?
3.	Anna	30	20,000	?
4.	Margaret	45	25,000	?
5.	Henry	30	15,000	?
6.	Mitchell	45	50,000	?
7.	Bob	35	?	350
8.	Ilene	40	?	250
9.	Albert	30	?	600
10.	Monica	55	?	500

	Name	Age when policy was bought	Present age	Face value of policy	Cash value
11.	Keith	25	35	30,000	?
12.	Karyn	35	40	25,000	?
13.	John	25	45	50,000	?
14.	Ralph	35	50	15,000	?
15.	Judy	40	50	21,000	?
16.	Sid	45	60	40,000	?
17.	Louise	45	65	35,000	?
18.	Jennifer	30	?	20,000	1,100
19.	Bob	30	?	30,000	13,500
20.	Susan	40	60	?	20,400

	Name	Age when policy was bought	Present age	Face value of policy	Cash value
21.	Karl	45	65	?	100,440
22.	Linda	?	55	25,000	12,750
23.	Wally	?	45	40,000	8,320

Roger is 30 years old and wishes to buy a $30,000 whole-life policy.

24. What is the annual premium?

25. What is the semiannual premium (52% of annual)?

26. What is the quarterly premium (26.5% of annual)?

27. What is the cash value when he is 40?

28. What is the cash value when he is 50?

Tina is 40 and wants to buy a $25,000 whole-life policy.

29. What is the annual premium?

30. What is the semiannual premium (52% of annual)?

31. What is the quarterly premium (26.5% of annual)?

32. What is the cash value when she is 45?

33. What is the cash value when she is 55?

Denny is 25 years old. He can afford to spend $300 per year for life insurance.

34. How much whole life can he buy for the $300?

35. How much 5-year level term can he buy for the $300?

Melissa is 35 years old. She can spend $350 per year for life insurance.

36. How much whole life can she buy for the $350?

37. How much will the same face value of 5-year level term cost per year?

38. How much premium can she save by buying the term?

39. If she puts this premium savings in the bank at $6\frac{1}{2}\%$ every year, what will she accumulate after 5 years? (Use Formula 8–2–1.)

40. What is the cash value after 5 years? (Use face value from Problem 36.)

Bill is 25. He buys a $50,000 whole-life policy.

41. What is the annual premium?

42. What is the cash value when he is 25? 30? 35? 40? 45?

43. Make a graph with cash value on the y-axis and age on the x-axis.

Ray wants to buy life insurance. He is trying to choose between whole life and decreasing term. He is 30 years old.

44. How much is the premium on a $40,000 whole-life policy?

45. How much is the premium on a $60,000, 20-year decreasing term policy?

46. How much does he save every year by buying the term insurance? (Call this savings S.)

Complete the following tables. (Use S from Problem 46.)

	Age	Face value of term (use Tab. 8–7)	Face value of whole life	Accumulation of S at 7% (use Formula 8–2–1)	Cash value
	30	$60,000	$40,000	0	0
47.	35	?	40,000	?	?
48.	40	?	40,000	?	?
49.	45	?	40,000	?	?
50.	50	?	40,000	?	?

IMPORTANT WORDS

annuity (8–2)
appreciation (8–3)
cash value (8–4)
coinsurance (8–4)
compound interest (8–1)
doubling time (8–1)
down payment (8–3)
equity (8–3)
face value (8–4)
five-year level term (8–4)
homeowner's insurance (8–4)

insurance (8–4)
interest (8–1)
maintenance (8–3)
mortgage (8–3)
premium (8–4)
real estate tax (8–3)
tax savings (8–3)
twenty-year decreasing term (8–4)
utilities (8–3)
whole-life insurance (8–4)

REVIEW EXERCISES

Complete the following tables.

	I	P	r (%)
1.	?	650	$5\frac{1}{2}$
2.	71.25	1250	?
3.	69.35	?	7.3

$I = P \kappa B$

	P_0	r (%)	n	P_n
4.	2400	$5\frac{1}{2}$	7	?

	P_0	r (%)	n	P_n
5.	?	7	20	10,000
6.	400	$6\frac{1}{2}$	?	1,000
7.	3000	?	10	6,000

$$P_0 = P_N (r\% + N)$$
$$P_N = P_0 (1+r)^N$$

	Rate of interest (growth; %)	Doubling time (yr)
8.	5	?
9.	8	?
10.	?	10
11.	?	5

$$DT = 72/r$$

12. Lynn and Steve buy a house for $32,000. Eight years later they sell it for $51,000. This is equivalent to what rate of bank interest?

13. The Rockes buy a refrigerator for $350, which depreciates 10% per year. What is the refrigerator worth in 3 years?

Complete the following table using Formula 8–2–1.

	D	n	r (%)	A
14.	1200	30	$6\frac{1}{2}$	?
15.	?	20	7	100,000
16.	500	?	$5\frac{1}{2}$	7,000
17.	300	10	?	4,000

$$A = D \cdot S_{\overline{N}|r}$$

18. The Millers want to accumulate $200,000 in 25 years. If they can get 7% interest, how much do they have to save each year?

19. If the Gerbers have a family income of $23,000, what price house can they afford?

Complete the following tables.

$$2 \times 23,000$$

	Price of house	Down payment (%)	Down payment ($)	Mortgage
20.	37,000	25	$A = P \times B$?	?
21.	42,000	?	9800	?
22.	?	20	7000	?

	Amount borrowed P	Time n (yr)	Interest r (%)	Monthly payments M
23.	29,000	25	$8\frac{1}{2}$	?
24.	?	30	9	280
25.	21,000	20	$9\frac{1}{4}$	?

Robin and Alex buy a $49,000 house. They put 20% down and borrow the rest at $9\frac{1}{4}$% for 25 years.

26. What is the down payment?

27. What is the amount borrowed?

28. What are the monthly mortgage payments?

29. The Hensleys pay $840 per year real estate tax on their $39,000 house. What is this rate?

30. The Edelsons lower their thermostat by 6° for half the day and 2° for the other half. By what percent might they lower their heating bill?

31. Jerry runs his 900-watt room air conditioner 16 hours per day in the summer. Electricity costs 4.3 cents per kilowatt-hour. What does it cost him per month to run the air conditioner?

Sherri and Shelly buy a $42,000 house with 20% down. Their mortgage terms are $8\frac{3}{4}$% interest for 25 years. Property values are increasing at a rate of $4\frac{1}{2}$% per year. Their real estate tax is $900 and they are in a 24% tax bracket.

32. What is their down payment?

33. What is their mortgage?

34. What are their monthly mortgage payments?

35. How much interest do they pay in the first year?

36. How much equity do they build up in the first year?

37. What is the house worth in 5 years? 10 years?

38. What are their tax savings?

39. If they sell their house in 5 years, how much will they clear after paying 9% sales commission and fees?

Find the automobile insurance premium for

40. Barry, age 40, married, living in the suburbs.

41. Kathy, age 20, unmarried, living in the city.

Complete the following table for an accident with $100 deductible.

	Value of car	Damage	Insurance pays
42.	3500	$850	?
43.	500	650	?

Complete the following tables.

	Community class	House type	Value	3-year premium
44.	B	Brick	$42,000	?
45.	A	Frame	35,000	?

	Value of house	80% of value	Amount of insurance	Amount of damage	Insurance payment
46.	$40,000	?	$35,000	$27,500	?
47.	32,000	?	25,000	4,200	?
48.	27,000	?	24,000	25,500	?

	Name	Age	Type of insurance	Face value	Annual premium
49.	Wally	35	5-yr level term	45,000	?
50.	Lisa	30	5-yr level term	35,000	?
51.	Tim	40	20-yr decr. term	?	250
52.	Cynthia	35	Whole life	25,000	?
53.	Elmer	45	Whole life	?	600

	Age when policy was bought	Present age	Original decreasing term face value	Present face value
54.	24	38	30,000	?
55.	32	41	?	35,500

	Age when whole-life policy was bought	Present age	Face value	Cash value
56.	45 (woman)	55	$30,000	?
57.	25 (man)	40	45,000	?

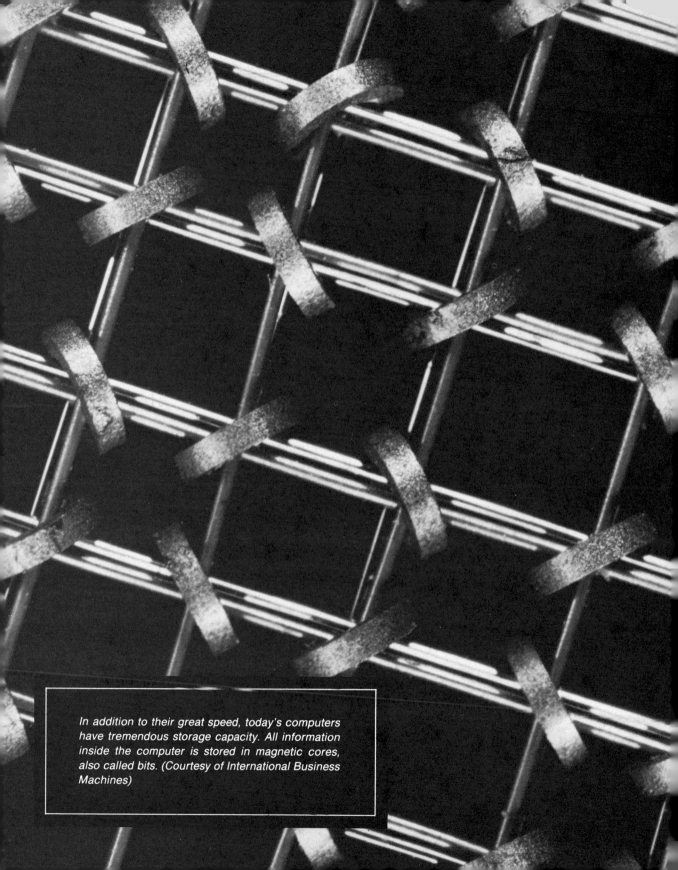

In addition to their great speed, today's computers have tremendous storage capacity. All information inside the computer is stored in magnetic cores, also called bits. (Courtesy of International Business Machines)

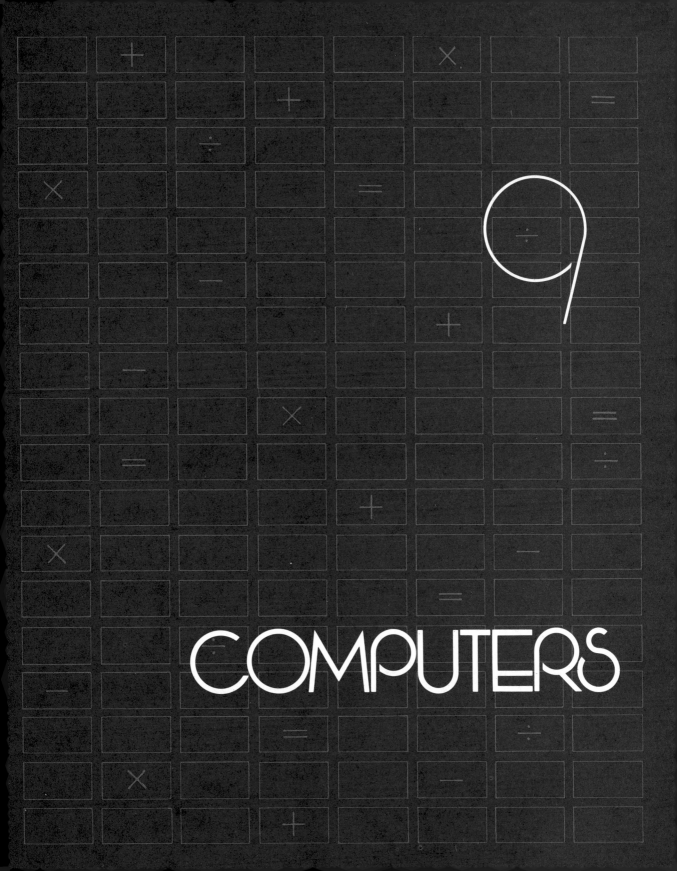

9

COMPUTERS

9-1   INTRODUCTION TO COMPUTERS

Question   What is a computer?

ANSWER   A **computer** is any device that helps us to compute. This definition is probably not too exciting, since most people think of a computer as being an electronic beast with flashing lights, tapes, wheels, and IBM cards. But these scary devices are very new, and people have been computing for thousands of years.

The earliest computers were probably man's fingers. Then he used pebbles and sticks to help him add and subtract. Then came other hand devices, such as the abacus, Napier's rods, the slide rule, and today's $10 electronic calculator.

In 1640, Blaise Pascal invented the first mechanical adding machine with little wheels and dials. In 1812, Charles Babbage began to construct various machines, most of which he could not finish. In the 1890s, Herman Hollerith devised a punch-card system to speed the counting of the census.

In 1944, Howard Aiken and International Business Machines completed the Mark I computer. About the same time, the Army developed the ENIAC for use in World War II. At this time, John von Neumann made the simple, but brilliant, suggestion that the computer store programs as well as numbers. Present-day computers are basically just bigger, faster, transistorized versions of these computers of the 1940s.

Question   Can a computer think?

ANSWER   No, not in the usual sense of thinking. A computer can only follow directions. Sometimes the computer does such impressive things that it appears that it can think. But this is only the programmer's thinking.

Question   What can the computer do?

ANSWER   The computer cannot do any more than most 10-year old children can do. Here is what a computer can do.

1. *Read numbers.* The computer can read and record a set of numbers that we give it.
2. *Print.* The computer can print out numbers or letters that it has in its memory.
3. *Store numbers.* The computer can take a number and store it away for later use. If we tell the computer $A = 10$, it will put 10 in a spot called A. This is just like the memory (M+ or STO buttons) on your $10 calculator.
4. *Retrieve numbers.* The computer can fetch any number that it was storing. This is just like the recall (MR or RCL buttons) on your $10 calculator.
5. *Do arithmetic.* The computer can do basic arithmetic: add, subtract, multiply, divide, and do exponents.
6. *Decide number sizes.* The computer can tell which of two numbers is bigger, or if they're equal.

As you read through this list, you can see that indeed there is nothing a normal 10-year-old child could not do.

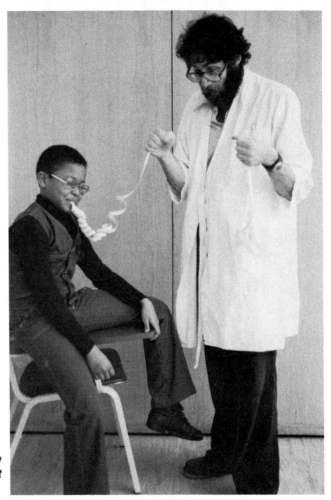

The computer is capable of no more than a bright ten-year-old child.

Question Why then do businesses, industries, governments, and scientists use big computers instead of 10-year-old children?

ANSWER A 10-year-old child can do *what* a computer can do. But the difference is in *how* the computer does what it does:

1. The computer works millions of times *faster* than a 10-year-old.
2. The computer works *much more accurately* than a 10-year-old.
3. The computer *never gets bored* by repeating a process over and over as a 10-year-old would.
4. The computer *never complains*, "Can I watch TV now?" as a 10-year-old would.

Question How does the computer do all this?

ANSWER The sketch gives a rough idea of the organization inside a computer.

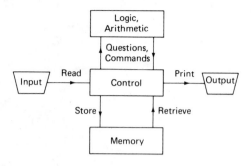

The main components are the following:

1. **Control**. This is the brain of the computer, which *processes* all of the information to or from the other parts of the computer.
2. **Input**. This is where the data are entered. Usually, we *read* the data in, and the control sends it where it belongs.
3. **Memory**. This is like the human memory, except it is bigger and less likely to forget. The control *stores* information in the memory and *retrieves* it from the memory.
4. **Arithmetic, logic**. Here is where the computer does all its *arithmetic*: adding, subtracting, multiplying, dividing, and so on. Here is also where the computer *makes decisions* telling which of two numbers is larger.
5. **Output**. This is where the computer *prints* out answers and/or information that it has stored.

PROBLEM SET 9-1-1

1. Define or discuss:

 (a) Input.   (b) Output.   (c) Memory.
 (d) Control.   (e) Store.   (f) Retrieve.
 (g) Computer.

2. Discuss what the computer can and cannot do.

3. Get a more detailed history of the computer from your library.

4. Compare and contrast a computer with an average 10-year-old child.

9–2 BINARY SYSTEM

Question How does the computer store its information?

ANSWER All information inside the computer is stored in **magnetic cores**, which are also called **bits**. Each bit looks like a tiny metal doughnut; it is very easily magnetized when enough electrical current is sent through the hole. Each core has two possible magnetic spins.

Each core has two wires running through it. It takes current in *both* wires to change the magnetic spin direction. The figure shows a core spinning clockwise. If there is current in just one wire, there is no change in the spin. If both wires have current, the spin direction changes.

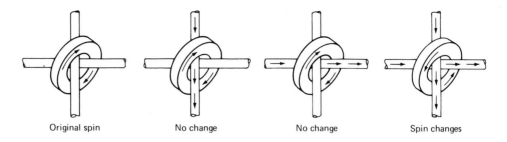

Original spin   No change   No change   Spin changes

The storage is made up of lots of these magnetic cores wired together, as illustrated.

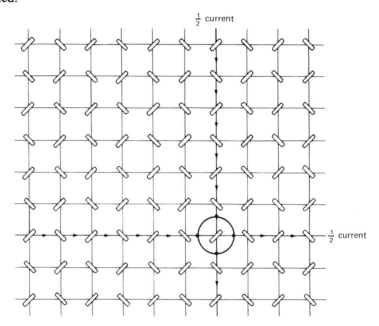

The circled bit is the only one that will change since it is the only one getting enough current ($\frac{1}{2} + \frac{1}{2} = 1$). All the other bits either get zero or half of the current, which isn't enough.

Question How does the computer use these cores or bits to store numbers?

ANSWER For the past 800 years or so, man has used the base ten (or decimal) system to write numbers. Each column has its own value, which is a power of 10:

1,000,000	100,000	10,000	1,000	100	10	1

The base-ten system uses the digits, 0 through 9, in these columns. But the base-ten system is not the best system for the computer. Remember that each bit has only two possible spins. For this reason, the computer uses a system with just two digits, 0 and 1. This is called the **base-two system,** or the **binary system.**

Like the base-ten system, each column has its own value. But now they are powers of 2:

1024	512	256	128	64	32	16	8	4	2	1

Each column can only use a 0 or 1 digit. The figure shows how this might appear in the magnetic core. This means we either count the column value (if it has a 1 in it) or we don't (if it has a 0 in it).

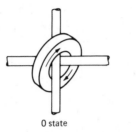

0 state

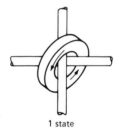

1 state

For example, consider the following numbers in base two (on the left) and their base ten equivalents (on the right).

64	32	16	8	4	2	1	Base-ten equivalents
1	0	1	1	1	0	0	$64 + 16 + 8 + 4 = 92_{ten}$
1	0	0	1	0	0	1	$64 + 8 + 1 = 73_{ten}$
	1	0	1	1	1	0	$32 + 8 + 4 + 2 = 46_{ten}$
		1	0	0	1	1	$16 + 2 + 1 = 19_{ten}$

Notice that all we did was to add up the column values where a 1 appeared. We ignored the columns with a 0 in them.

Abacus, an ancient calculator. (Courtesy of International Business Machines)

The next figure shows the first 15 binary numbers. In the left column we can view each of the digits as a bit that is either on (1) or off (0). In the middle, the numbers are written out in the usual binary form, with 0's and 1's. In the right column, we have changed these into base ten numbers by adding up the columns where a 1 appears.

8	4	2	1	Binary	Base Ten Equivalent
0	0	0	0	0	0
0	0	0	●	1	1
0	0	●	0	10	2
0	0	●	●	11	2 + 1 = 3
0	●	0	0	100	4
0	●	0	●	101	4 + 1 = 5
0	●	●	0	110	4 + 2 = 6
0	●	●	●	111	4 + 2 + 1 = 7
●	0	0	0	1000	8
●	0	0	●	1001	8 + 1 = 9
●	0	●	0	1010	8 + 2 = 10
●	0	●	●	1011	8 + 2 + 1 = 11
●	●	0	0	1100	8 + 4 = 12
●	●	0	●	1101	8 + 4 + 1 = 13
●	●	●	0	1110	8 + 4 + 2 = 14
●	●	●	●	1111	8 + 4 + 2 + 1 = 15

● = on (1)   0 = off(0)

To change any binary number into base ten:

1. Starting from the right, determine the column values: 1, 2, 4, 8, 16, 32, and so on.

2. Add the column values with a 1 in their columns (ignore the 0 columns).

3. This is the base-ten equivalent.

Problem   What is 110101_{two} in base ten?

ANSWER   Notice that we use "two" as a subscript to the number so that we will know that it is in base two. A number without a subscript is always base-ten.

32	16	8	4	2	1
1	1	0	1	0	1

$$32 + 16 + 4 + 1 = 53$$

$$110101_{two} = 53_{ten}$$

Problem What is 1001110_{two} in base ten?

ANSWER

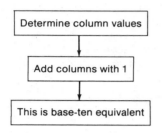

64	32	16	8	4	2	1
1	0	0	1	1	1	0

$$64 + 8 + 4 + 2 = 78$$

$$1001110_{two} = 78_{ten}$$

Question How do we change a base-ten number, such as 57, into a binary number?

ANSWER There are several ways. One way is to use trial and error to determine how many 1's, 2's, 4's, 8's, and so on, are in the number. But there is an easier method. Here is what we do with 57 as an example.

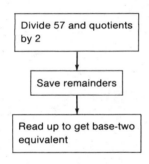

```
2 | 57        Remainders
2 | 28            1 ↑
2 | 14            0
2 | 7             0      Read up
2 | 3             1 ↑
2 | 1             1
    0             1
```

$$57_{ten} = 111001_{two}$$

To change any base-ten number into a base-two (binary) number:

1. Divide 2 into the number and then into all the quotients.
2. Save all the remainders.
3. Continue until quotient is 0.
4. Read *up* the remainders to get binary equivalent.

Problem Write 106_{ten} as a binary number.

ANSWER We divide 106 by 2, and keep dividing until the quotient is 0. We always save the remainders.

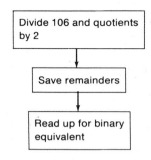

$$\begin{array}{r|l} 2 & 106 \\ \hline 2 & 53 \\ \hline 2 & 26 \\ \hline 2 & 13 \\ \hline 2 & 6 \\ \hline 2 & 3 \\ \hline 2 & 1 \\ \hline & 0 \end{array}$$

Remainders

0
1
0
1
0
1
1

Read up

$$106_{ten} = 1101010_{two}$$

PROBLEM SET 9–2–1

1. Define or discuss:
   (a) Bit.   (b) Magnetic core.   (c) Binary system.

Change the following binary numbers into base ten.

2. 101_{two}

3. 1100_{two}

4. 101011_{two}

5. 1000_{two}

6. 100000_{two}

7. 1111111_{two}

8. 10100101_{two}

9. 110101011_{two}

10. 101101101_{two}

Change the following base-ten numbers into base-two (binary) numbers.

11. 25   12. 19   13. 96

14. 109   15. 54   16. 63

17. 32   18. 129   19. 79

Binary Arithmetic

Question   How does the computer do its arithmetic?

ANSWER   As we just saw, the computer stores all its numbers in base two. Base two has arithmetic rules just as our own base-ten system does. And in some ways the rules are easier.

Remember back in first and second grade when you had to memorize a big addition table. Base-two also has an addition table, but it is much simpler, as the table shows.

+	0	1
0	0	1
1	1	10

The base-ten system has ten digits (0 to 9), so the addition table has 100 facts (10 × 10) to learn. The base-two system has only two digits (0 and 1), so its addition table has only 4 facts (2 × 2) to learn.

The only addition fact in the table that looks strange is $1 + 1 = 10$. But in base two, this is just $1 + 1 = 2$ since $10_{two} = 2_{ten}$. When adding $1 + 1$, we put down a 0 and carry a 1 to the next column, just as we do in base-ten addition.

Problem   Add $1001_{two} + 101_{two}$.

ANSWER   To add these two numbers, we line them up just as we would with base-ten numbers, except we use base-two addition facts.

$$
\begin{array}{rcr}
1 & & \text{Check} \\
1001_{two} = & & 9_{ten} \\
+\ 101_{two} = & + & 5_{ten} \\
\hline
1110_{two} = & & 14_{ten}
\end{array}
$$

In the one's column, we had $1 + 1$. So we put 0 in the one's column and carried a 1 to the ten's column. Notice that we can check the problem by working it in base ten. The problem is really $9 + 5 = 14$ in disguise.

Problem   Add $110111_{two} + 100011_{two}$.

ANSWER

$$
\begin{array}{rcr}
111 & & \text{Check} \\
110111_{two} = & & 55_{ten} \\
+\ 100011_{two} = & + & 35_{ten} \\
\hline
1011010_{two} = & & 90_{ten}
\end{array}
$$

In the one's column, we have $1 + 1$, so we put down 0 and carry the 1. When we carry this 1, we then get $1 + 1 + 1$ in the next column. This is of course 3, which is 11_{two}. So we put down a 1 and carry a 1. And so on. Notice that we can check this as $55_{ten} + 35_{ten} = 90_{ten}$.

Question   How do we subtract in base two?

ANSWER   Subtraction is also similar to that in base ten. Sometimes, we have to "go next door and borrow." The subtraction facts are:

$$
\begin{array}{ll}
0 - 0 = 0 \qquad & 1 - 0 = 1 \\
1 - 1 = 0 \qquad & 10 - 1 = 1 \quad \text{(borrow)}
\end{array}
$$

Problem   Subtract $111010_{two} - 100110_{two}$.

ANSWER   We line this up like a base-ten problem and subtract, except we use base-two facts.

$$
\begin{array}{rr}
\begin{array}{c} 0\ 10 \\ 1\ 1\ \cancel{1}\ 0\ 1\ 0_{two} = \\ -1\ 0\ 0\ 1\ 1\ 0_{two} = \\ \hline 1\ 0\ 1\ 0\ 0_{two} = \end{array} &
\begin{array}{r} \text{Check} \\ 58_{ten} \\ -38_{ten} \\ \hline 20_{ten} \end{array}
\end{array}
$$

In the 4's column (the third from the right) we had to borrow. We went to the next column and borrowed a 1. This made the 4's column a 10 (or 2). Then $10_{two} - 1_{two} = 1_{two}$. Notice that the problem checks as $58_{ten} - 38_{ten} = 20_{ten}$.

Problem   Subtract $1111001_{two} - 1001111_{two}$.

ANSWER

$$
\begin{array}{rr}
\begin{array}{c} 10\ 1 \\ 0\ 0\ \cancel{10}10 \\ 1\ 1\ \cancel{1}\cancel{1}\ 0\ 0\ 1_{two} = \\ -1\ 0\ 0\ 1\ 1\ 1\ 1_{two} = \\ \hline 1\ 0\ 1\ 0\ 1\ 0_{two} = \end{array} &
\begin{array}{r} \\ \\ \text{Check} \\ 121_{ten} \\ -79_{ten} \\ \hline 42_{ten} \end{array}
\end{array}
$$

Here we have to borrow from the 8's column (fourth from the right) to help subtract in the 2's column. Again, the answer checks in base ten.

Question   How do we multiply in the binary system?

ANSWER   Just as in base ten, we have a multiplication table, as shown in the table. Unlike base ten, this table is so simple there is almost nothing to memorize. And there is no carrying.

$\times$	0	1
0	0	0
1	0	1

Problem   Multiply $1101_{two} \times 11_{two}$.

ANSWER   We set this up just like base ten, except we use base-two addition and multiplication facts.

$$
\begin{array}{rr}
\begin{array}{c} 1101_{two} = \\ \times\ \ \ 11_{two} = \\ \hline 1101 \\ 1101 \\ \hline 100111_{two} = \end{array} &
\begin{array}{r} 13_{ten} \\ \times\ 3_{ten} \\ \hline 39_{ten} \\ \\ 39_{ten} \end{array}
\end{array}
$$

Notice that there is no carrying when we multiply, only when we add at the end. Also, we can again check this in base ten to be $3 \times 13 = 39$.

Problem Multiply $11011_{two} \times 1001_{two}$.

ANSWER

$$
\begin{array}{rr}
11011_{two} = & 27_{ten} \\
\times \ 1001_{two} = & \times \ 9_{ten} \\
\hline
11011 & 243_{ten} \\
1101100 & \\
\hline
11110011_{two} = & 243_{ten}
\end{array}
$$

Question How do we divide in base two?

ANSWER Again, it is just like base ten, except easier. We do not have to guess as we do in base ten. Since there are only two digits, the divisor either divides in (1), or it doesn't (0).

Problem Divide 1101100_{two} by 1001_{two}.

ANSWER We set the problem up as we would any other division problem, except we use base-two subtraction facts.

Check

$$
\begin{array}{r}
1100_{two} \\
1001_{two}\overline{)1101100_{two}} \\
1001 \\
\hline
1001 \\
1001 \\
\hline
00
\end{array}
\qquad
\begin{array}{r}
12_{ten} \\
9_{ten}\overline{)108_{ten}} \\
9 \\
\hline
18 \\
18 \\
\hline
\end{array}
$$

Problem Divide 11101101 by 1101.

ANSWER

Check

$$
\begin{array}{r}
10010_{two}R11_{two} \\
1101_{two}\overline{)11101101_{two}} \\
1101 \\
\hline
1110 \\
1101 \\
\hline
11
\end{array}
\qquad
\begin{array}{r}
18_{ten}R3_{ten} \\
13_{ten}\overline{)237_{ten}} \\
13 \\
\hline
107 \\
104 \\
\hline
3
\end{array}
$$

Machine Arithmètique (1647–53, Blaise Pascal, France), the first real calculating machine. (Courtesy of International Business Machines)

Notice that this problem checks in base-ten, even down to the remainder. There is no guessing in these problems.

As you can see, base-two arithmetic is not any harder (and sometimes easier) than base-ten. The problem for humans is that it takes too many digits to express the numbers. But binary numbers and their arithmetic are very easy to build into a computer.

PROBLEM SET 9–2–2

1. What are the advantages and disadvantages of base-two arithmetic?

Add the following in base two and check in base ten.

2. 11_{two}
$+ 10_{two}$

3. 101_{two}
$+ 110_{two}$

4. 1001_{two}
$+ 1111_{two}$

5. 10111_{two}
$+ \ 1011_{two}$

6. 100111_{two}
$+ 110011_{two}$

7. 1001011_{two}
$+ 1101101_{two}$

Subtract the following in base two and check in base ten.

8. 11_{two}
$- 10_{two}$

9. 101_{two}
$- \ 11_{two}$

10. 110_{two}
$- 101_{two}$

11. $\quad 1101_{\text{two}}$
$\quad\quad - 1010_{\text{two}}$

12. $\quad 10001_{\text{two}}$
$\quad\quad - 1110_{\text{two}}$

13. $\quad 1011011_{\text{two}}$
$\quad\quad - 101101_{\text{two}}$

Multiply the following in base two and check in base ten.

14. $\quad 11_{\text{two}}$
$\quad\quad \times 10_{\text{two}}$

15. $\quad 101_{\text{two}}$
$\quad\quad \times 100_{\text{two}}$

16. $\quad 1101_{\text{two}}$
$\quad\quad \times 101_{\text{two}}$

17. $\quad 1001_{\text{two}}$
$\quad\quad \times 1001_{\text{two}}$

18. $\quad 10001_{\text{two}}$
$\quad\quad \times 1100_{\text{two}}$

19. $\quad 111001_{\text{two}}$
$\quad\quad \times 10001_{\text{two}}$

Divide the following in base two and check in base ten (including remainders).

20. $10_{\text{two}}\overline{)100_{\text{two}}}$

21. $11_{\text{two}}\overline{)1001_{\text{two}}}$

22. $101_{\text{two}}\overline{)110111_{\text{two}}}$

23. $1101_{\text{two}}\overline{)1110101_{\text{two}}}$

24. $101_{\text{two}}\overline{)101011011_{\text{two}}}$

25. $1011_{\text{two}}\overline{)10110111_{\text{two}}}$

9–3 FLOW CHARTS

Question   What are programs and flow charts?

ANSWER   A **program** is a plan for solving a problem. If the problem is hard, the program will usually be long and complex. If the problem is easy, the program will usually be short and simple.

Since computers do not think, we have to tell the computer exactly what to do. This is why we must make a careful plan, or program, of what to tell the computer to do.

It usually helps to put our program in diagram or picture form. This is called a **flow chart**. Flow charts show just what is to be done and in what order. They also show what decisions have to be made.

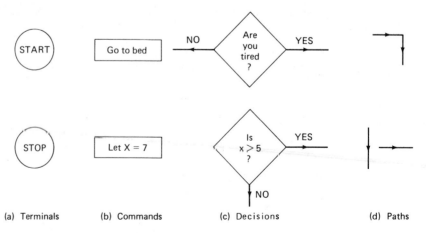

(a) Terminals   (b) Commands   (c) Decisions   (d) Paths

Examples of the symbols used in flowcharts are shown in the figure. The basic parts are as follows:

    a. **Terminals**. These tell us where a program is to stop and start. They are usually drawn with circles.

    b. **Commands**. These are the basic instructions of the program: do this, do that, and so on. These commands usually begin with a verb. They are usually drawn with rectangles.

    c. **Decisions**. Here is where our program hits a fork in the road, and we must make a choice. We are asking a question. If the answer is "no," we go one way; if the answer is "yes," we go in another direction. These are drawn with diamonds.

    d. **Paths**. These are arrows that tell us where to go and what to do next.

Problem   Draw a flow chart for playing a record.

ANSWER   We will start with this simple example (see the accompanying figure). We start with START. We then put each of the usual steps in a box: turn on stereo, take album from collection, and so on.

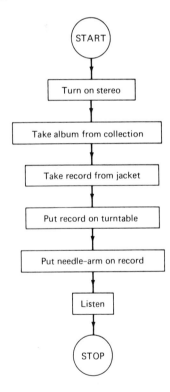

In this flow chart there are no decisions to be made. We follow the arrows from step to step. Now we will see a program with decisions to be made.

Problem   Write a flow chart for waking up in the morning.

ANSWER

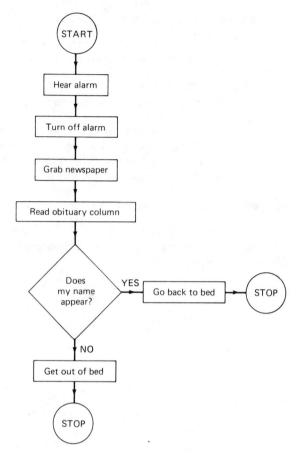

We again begin with START. Then we have four commands in a row, that we do in that order. Finally, we get to the question, "Does my name appear?" This will be answered yes or no. Now there are *two* arrows from the question diamond. The "yes" branch looks to the right and the box "Go back to bed." The "no" branch leads down to "Get out of bed." Then both paths come to STOP.

When making a flow chart we must be careful to make sure that all the paths out of all the decision diamonds lead somewhere. The next figure shows two flow charts. On the left, the person forgot the "yes" branch. On the right, this has been corrected to have both a "yes" and "no" path.

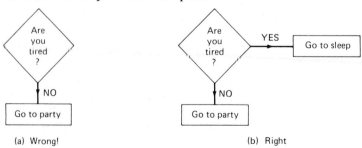

Problem Draw a flow chart for serving in tennis.

ANSWER In tennis, a player gets two serves. Usually, he or she hits the first one hard; if that one isn't good, he or she slows up on the second serve.

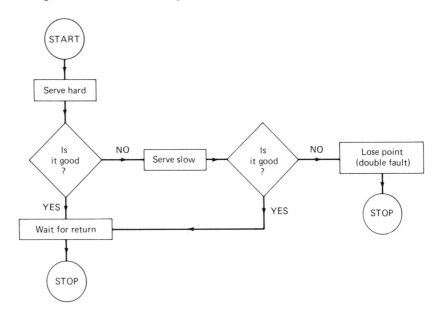

The figure shows how the tennis serve can be diagrammed in a flow chart. First, we serve hard. Now, we quickly must answer a question, "Is it good (in bounds)?" If the answer is "yes," we follow the path down, which tells us to wait for the return shot.

If the answer is "no," this means our serve was bad (a fault). So we follow the path to the right, which tells us to serve slow. Again, we ask ourselves, "Is it good?" If the answer is "yes," we go to the instruction to wait for return. If the answer is "no," this means we have double faulted. So we lose the point.

Notice that it is possible for two paths to lead into the same instruction. In the flow chart there are two separate paths into the box, "Wait for return." This is because this instruction comes after both the first serve and the second serve.

Problem Draw a flow chart for taking a shower.

ANSWER The flow chart shows that we have two simple commands at the beginning: turn on the water, and feel water. Now we get the usual "fiddling" most people do with the temperature.

If the water is too hot, we turn up the cold and feel the water again. If the water is too cold, we turn up the hot and feel the water again. This part of the flow chart is called a **loop**, since we can keep going around and around, feeling the water and readjusting it, until the water is just right.

Then we can go on and wash our body and hair if necessary. Finally, we turn off the water.

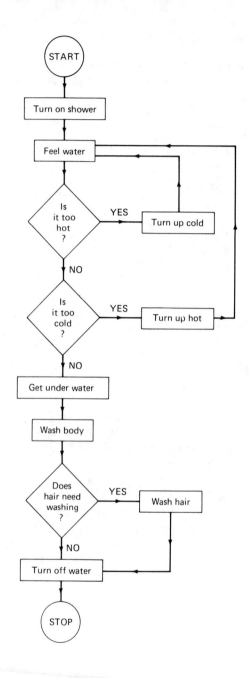

PROBLEM
SET 9-3-1

1. Define or discuss:
   (a) Program.   (b) Flow chart.   (c) Loop.   (d) Terminals.
   (e) Commands.   (f) Decisions.   (g) Paths.

Draw a flow chart for the following situations.

540   Ch. 9   /   Computers

2. Starting a car.

3. Taking a bath.

4. Being asked for a date.

5. Asking for a date.

6. Dressing for a date or important affair.

7. Catching a train.

8. Finding a library book.

9. Buying shoes.

10. Changing a flat tire.

11. Choosing food from a restaurant menu.

12. Cooking an egg.

13. Catching a football.

14. Making a phone call.

15. Looking for someone to dance with at a mixer.

16. Planning a picnic.

Flow Charts in Mathematics Problems

Question   How do we use flow charts to help us program a mathematics problem?

ANSWER   Even the simplest mathematics problem requires some plan. We can use flow charts to draw out our plan: what steps we take, what decisions we have to make.

Problem   Draw a flow chart to add 2 + 3.

ANSWER   This flow chart is as simple as we can get.

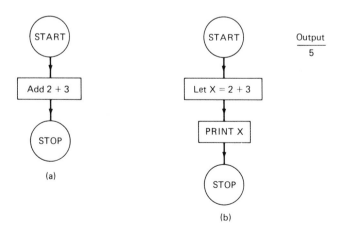

We have two possible ways to do the problem 2 + 3. Part (a) seems to show the easiest way. But this might not work with a computer. Why?

A computer is dumb. If you tell the computer to add 2 + 3, it will add 2 + 3 and get 5. However, it will not *tell* us the answer is 5 unless we ask it to. (Remember, computers take nothing for granted.)

To make sure that the computer will tell us the answer, we have to say that $X = 2 + 3$; then we say print X. This is shown in part (b). We have also shown the output that the computer would print: 5.

Problem   Write a flow chart to compute the area of a circle with a radius of 6.5.

SOLUTION: Recall from Chapter 3 that the formula for the area of a circle is $A = \pi r^2$, where $\pi = 3.14$. We will have to tell the computer what π and r are. Then we will give it the area formula. Finally, we will have the answer printed (see the accompanying flow chart).

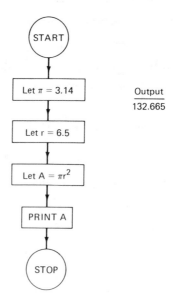

Output
132.665

Next to the flowchart we give the output (answer) that the computer would print: 132.665.

Problem   Draw a flow chart to find the mean of 10 numbers.

ANSWER   Recall from Chapter 7 that the mean is the sum of all the numbers divided by 10. To do this problem, we must first find out what the 10 numbers are. Then we add them and divide by 10. Finally, we print the mean.

In the flow chart shown, the first command, READ $A_1, A_2, \ldots, A_9, A_{10}$, is our way of getting the computer to find out what the 10 numbers are. Recall, READ is how we get numbers into the computer, and PRINT is how we get the answers out.

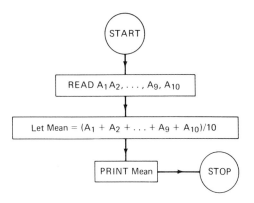

Problem   Write a flow chart to compute compound interest for different situations.

ANSWER   The formula for compound interest is given in Formula 8–1–3, $P_n = P_0(1 + r)^n$. We will need to give the computer the following data: P_0 (initial deposit), r (rate of interest), and n (years left in bank). We use a READ command to get these data in.

We then give the computer the formula to compute P_n with the data. Finally, we have the computer PRINT the answer for us. After this, we can go back and start the program over again with new data.

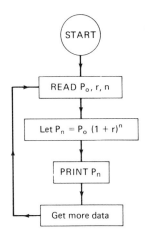

In the flow chart shown, at the bottom, we loop the program back to the READ statement to start again. Notice that there is no STOP statement. When there are no more data, the program automatically stops.

So far, none of these programs has had a decision. The next one does.

Problem   At Jennifer's Pet Palace, the hourly wage rate is $4 per hour. If employees work over 40 hours they get time and a half for the extra hours. Write a flow chart to compute their weekly wages.

ANSWER To work the problem, we need to have only one thing as input: the number of hours (H) that an employee worked that week. Also, if they work overtime, the rate is time and a half, or $6 per hour.

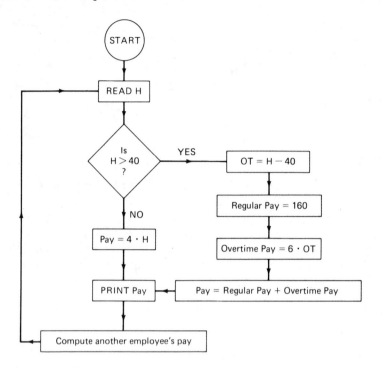

Let us look at the given flow chart. First we read in the number of hours (H). Now we have to see if the worker worked any overtime. We ask the question, "Is $H > 40$?" If the answer is "no," there was no overtime. So we take the path down. Here we simply find the pay as the rate of $4 times the hours worked H. We print this pay, and then go back for another worker.

If the answer to "Is $H > 40$?" is "yes," then there is overtime and we go out to the right. We first compute the number of overtime hours, OT, which is $H - 40$. The regular pay is the rate of $4 times 40 regular hours, or $160. The overtime pay is $6 times the overtime hours, OT. The total pay is the regular pay plus the overtime pay. We print this pay and go back for another worker.

As an example, suppose that Pat worked at Jennifer's for 45 hours last week. We will follow this through the flow chart. First, we read in that $H = 45$. Now we check to see if $H > 40$. The answer is yes, so we go to the right.

First, we calculate the overtime, OT = 45 − 40 = 5 hours. Now we compute the regular pay: $4 · 40 = $160. Next we compute the time and a half overtime: $6 · 5 = $30.

Finally, we add the regular and overtime pay to get the total pay, which is $160.00 + 30.00 = $190.00. We print this and go back for another worker.

Problem Write a flow chart to compute what an insurance company will pay on a fire damage.

ANSWER Recall from Chapter 8 that we must worry about something called *coinsurance*. If the house is insured for less than 80% of its value, the insurance company only pays a proportion of the loss. Also, the company will not pay more than the amount of the insurance.

We will need to read in three bits of information:

$$V = \text{value of house (cost to rebuild)}$$

$$I = \text{amount of insurance}$$

$$L = \text{amount of loss}$$

The flow chart is shown; it is exactly the same as the procedure given in Section 8.4, except it is in picture form. We will go through the flow chart with the following example.

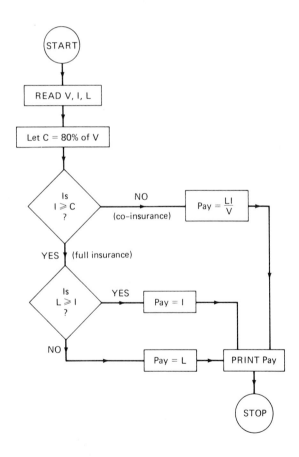

Problem Paul and Terri own a house that would cost $31,000 to be rebuilt, but they only have $22,000 of insurance on it. They have a $4000 fire. How much will the insurance company pay?

ANSWER Referring to the given flow chart, we first read in the data: V (value) = 31,000, I (insurance) = 22,000, L (loss) = 4000.

Next we compute C = 80% of V, which is $(0.80)(31,000) = \$24,800$. We now ask, "Is I (22,000) $\geqslant C(24,800)$?" The answer is "no," so we go out to the right. (This means that they are under coinsurance.)

At the right is Formula 8–4–1 for the insurance payment. We get

$$\text{Pay} = \frac{LI}{V} = \frac{(4000)(22,000)}{31,000} = \$2838.71$$

The final step is to print this answer, $2838.71, and STOP.

Problem Write a flow chart for a program to write a table of the numbers from 1 to 1000 with their squares and cubes.

ANSWER Printing the numbers from 1 to 1000 along with their squares and cubes is a simple problem. Almost any 11-year-old child could do it (given enough time and paper). However, writing a step-by-step program (or flow chart) to tell a dumb computer how to do it is something else. We use a loop.

The flow chart is shown, and we will follow through this program doing exactly what the instructions say. We start with $N = 1$; then $S = N^2 = 1$, and $C = N^3 = 1$. We print those three numbers.

Now we ask if $N = 1000$. The answer is "no," so we go down and let the (new N) = the (old N) + 1. Since N was 1, this makes $N = 2$. Loop back and get $S = N^2 = 4$ and $C = N^3 = 8$. We print these three numbers.

Now we ask again if $N = 1000$. The answer is "no," so we go down and let the (new N) = the (old N) + 1. So now $N = 3$. We loop back again and get $S = N^2 = 9$ and $C = N^3 = 27$. Then we print these three numbers.

An old-time slide rule.

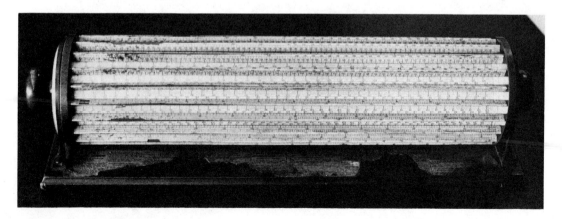

Again, we ask if $N = 1000$. The answer is still "no," so we go down. By now, you should see the pattern: we make N one bigger; find N^2 and N^3; print N, N^2, and N^3; and then check again to see if $N = 1000$.

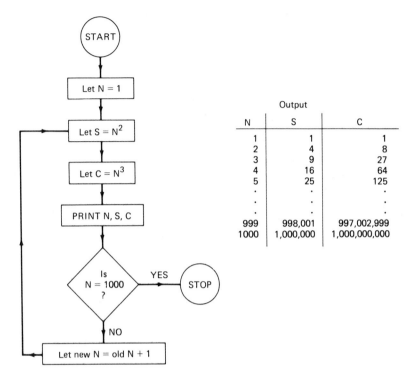

N	S	C
	Output	
1	1	1
2	4	8
3	9	27
4	16	64
5	25	125
.	.	.
.	.	.
.	.	.
999	998,001	997,002,999
1000	1,000,000	1,000,000,000

Finally, we will get to a point where $N = 999$ and then becomes 1000. Now $S = N^2 = 1,000,000$ and $C = N^3 = 1,000,000,000$. We print these three numbers. Now when we ask if $N = 1000$, the answer is "yes." So we go to the right and STOP.

PROBLEM SET 9-3-2

Draw the flow charts to work the following problems. Be sure to read in any needed data and print out all answers.

1. Find the area of several rectangles (see Formula 3-2-1).

2. Find the volume of several cylinders (see Formula 3-4-4).

3. Find the APR (true interest of a loan) (see Formula 5-2-2).

4. Find a person's tax table income (see Formula 5-4-1).

5. Compute the value of a depreciating car (see Formula 5-5-1).

6. Find the expected value of an experiment (see Formula 6-4-1). Let N = 5.

7. Find the standard deviation of a group (see Formulas 7-2-2 and 7-2-3). Let N = 5.

8. Find the cost of electricity per month (see Formula 8-3-3).

9. Find the cost of insuring a frame house in a class C community (see Tab. 8–5).

10. In a certain state, the state income tax is computed as follows:
   (a) Multiply the number of dependents by $1000.
   (b) Subtract this from your total income.
   (c) If this difference is negative, your tax is $0.
   (d) If the difference is positive, your tax is 2.5% of the difference.
   Draw a flow chart to find and print the tax for people in this state.

11. Draw a flow chart to find and print the state income tax in your state.

12. Draw a flow chart to determine which of three numbers is largest.

13. Draw a flow chart to determine a person's medical deduction on the federal income tax (see Section 5.4).

14. Draw a flow chart to determine how much the insurance company pays on an automobile accident when the car policy has $100 deductible. (Remember that the insurance won't pay more than the value of the car.)

With the following flow charts, you are to play computer. Follow the directions exactly and print out answers until you come to STOP or run out of data.

15.

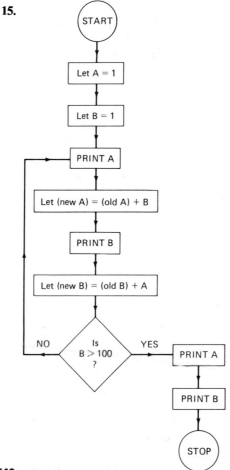

16.

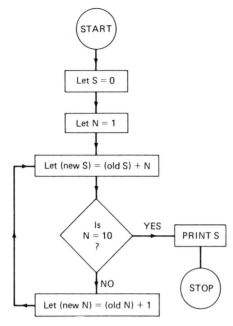

17.

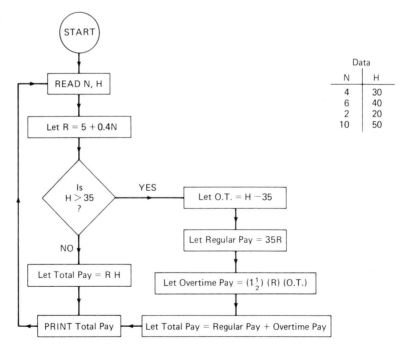

Data	
N	H
4	30
6	40
2	20
10	50

9–4 BASIC

Question   How do we actually talk to the computer?

ANSWER   We have to use a computer language. There are at least 200 different computer languages. Some are designed for scientists, and some are designed for business use. Some require a lot of instructions, and some require very few instructions.

The language we will study is probably the easiest of all to learn. It is called **BASIC** (Beginner's All-purpose Symbolic Instruction Code). BASIC is easy to learn because the instructions look like ordinary English and mathematics.

As we saw in the flow charting sections, the basic operations that a computer performs are the following:

1. Arithmetic (add, subtract, multiply, divide, and take exponents)
2. Input (Read)
3. Output (Print)
4. Decisions (Which number is bigger?)

We will see how all four of these are translated from English into BASIC.

BASIC *Arithmetic*

Question   How do we write formulas and numbers in BASIC?

ANSWER   Arithmetic in BASIC looks very similar to ordinary arithmetic. The arithmetic symbols in BASIC are

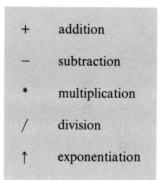

+	addition
−	subtraction
*	multiplication
/	division
↑	exponentiation

Notice that multiplication (*) and exponentiation (↑) look a bit strange, but this is to avoid confusion with other symbols.

Before every arithmetic statement we use the English word LET. The variables are either a single letter (such as X or A) or a single letter followed by a single digit (such as M5 or P2).

Table 9–1 shows some examples of translating from mathematics into BASIC. On the left is the usual mathematical statement. On the right is the same statement in BASIC. Notice that all BASIC statements are on one line across and have all capital letters. Also, notice that each BASIC statement can only have a single variable on the left of the equal sign.

Table 9–1 Examples of Mathematics-to-BASIC Translations

Mathematics statement	BASIC statement
$a = 6$	LET A = 6
$y = 10x + 1$	LET Y = 10*X + 1
$p = 6a - 5b$	LET P = 6*A − 5*B
$r = \dfrac{x + y}{7}$	LET R = (X + Y)/7
$u = t^5$	LET U = T↑5
$A = \pi r^2$	LET A = 3.14*R↑2
$P_n = P_0(1 + r)^n$	LET P = P0*(1 + R)↑N
$s = a^2 + b^2 + c^2$	LET S = A↑2 + B↑2 + C↑2
$m = \dfrac{a_1 + a_2 + a_3}{3}$	LET M = (A1 + A2 + A3)/3
$x = \dfrac{2}{5}a^2b^3$	LET X = (2/5)*A↑2*B↑3
$\text{APR} = \dfrac{24I}{B(n + 1)}$	LET A = 24*I/(B*(N + 1))

Table 9–1 shows some valid examples of BASIC statements. Table 9–2 is different. Every statement has some small mistake. Many of the mistakes would be overlooked by a human, but not by a computer.

Table 9–2 Common Errors in BASIC

BASIC statement (incorrect!)	Error (and possible correction)
A = 10	Forgot LET LET A = 10
LET X = $4\frac{1}{2}$	No mixed numbers allowed LET X = 4.5
LET A = (HI + LO)/2	Can't have two-letter variables LET A = (H + L)/2
LET C = D(X − Y)	Forgot * for multiplying LET C = D*(X − Y)
LET 2*X = 4 + Y	Can have only the variable on the left LET X = (4 + Y)/2

The student using any computer language (such as BASIC) must force himself or herself to be very precise. The computer will not overlook little mistakes as a friend or professor might.

One last special notation should be given. This is for square roots.

Mathematics	BASIC
$y = \sqrt{x}$	LET Y = X↑.5

Notice that $\sqrt{x}$ is just like $x^{0.5}$, or X↑.5 in BASIC. This is the way it is usually written in algebra.

PROBLEM SET 9–4–1 Complete the following table, converting between usual mathematics and BASIC.

	Usual mathematics statement	BASIC statement
1.	$x = 7$	
2.	$y = a + b$	
3.	$z = 6x + 7y + 1$	
4.	$a = b^5$	
5.	$s = 32t^2$	
6.	$E = mc^2$	
7.	$m = \dfrac{b_1 + b_2 + b_3 + b_4}{4}$	
8.	$y = \sqrt{a}$	
9.	$x = a^3 + b^3$	
10.	$V = 4\pi r^3/3$	
11.		LET A = 912
12.		LET X = 8*G
13.		LET Y = X↑4
14.		LET R = (A + B)/6
15.		LET K = A/4 + 3
16.		LET G = A1↑2 + A2↑2
17.		LET C = (A↑2 + B↑2)↑.5

Find the errors in the following statements (and then make a possible correction).

18. LEET X = 1

19. A = 6 + B

20. LET = 4

21. LET X = 6A + B

22. LET A = BOB + SUE

23. LET 2A = 6 − 3/X

Input and Output

Question   How do we read data into the computer?

ANSWER   Very simply. We use a READ statement. If we want the computer to read in the variables A, B, and C, we write

READ A, B, C

Below this, we put in the actual data for A, B, and C with a DATA statement. For example, consider the statements

READ A, B, C
DATA 6, 5.2, 107.91

These tell the computer to make $A = 6$, $B = 5.2$, and $C = 107.91$. We might have the statements

READ X, Y
DATA 10, 20, 400, 200, 40, 50

Here the data are read in pairs. First, the computer makes $X = 10$ and $Y = 20$, and then goes through the program. Then it starts all over again with the next two, $X = 400$ and $Y = 200$. Finally, on the last time through, the computer sets $X = 40$ and $Y = 50$. Notice that the data are taken in pairs.

Here are a few simple rules we have to follow with the READ/DATA pair:

1. After READ, we list all the variables to be read.

2. After DATA, we list all the data to be used.

3. The numbers must be separated by commas in the DATA statement.

4. Only decimals can be used (no fractions).

5. The number of data terms must be a multiple of the number of variables. (Examples: If we say READ X, Y, then after DATA there should be 2, 4, 6, 8, 10, and so on, numbers. If we say READ A, B, C, then after DATA there should be 3, 6, 9, 12, and so on, numbers).

Question How do we print our answers?

ANSWER Very simple, also. We use a PRINT statement. If we want the computer to print the value of R, we write

$$\text{PRINT R}$$

If we want to print the values of A, B, and C, we write

$$\text{PRINT A, B, C}$$

Question How do we tell the computer that a program is done?

ANSWER We put the statement END at the end. With these five statements (LET, READ, DATA, PRINT, and END) we are ready to write some simple programs.

Problem Write a program to find the area of the given rectangle.

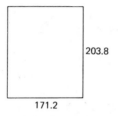

171.2

ANSWER We recall that the formula for the area of a rectangle is $A = l \cdot w$. We give the flow chart and the BASIC program for this. The instructions in a flow chart box match up to the BASIC statements in the program. The READ and DATA are two BASIC statements for the "Read l and w" box.

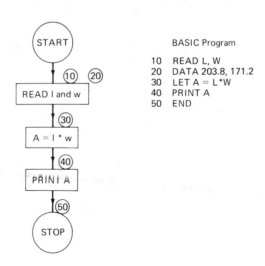

BASIC Program

```
10   READ L, W
20   DATA 203.8, 171.2
30   LET A = L*W
40   PRINT A
50   END
```

Notice that each BASIC statement has a number: 10, 20, 30, and so on. Every statement has to have a number. The numbers do not have to be 10, 20, 30, and so on. Any number up to five digits will do.

Question How do we write a program to find the areas of several rectangles, as shown in the figure?

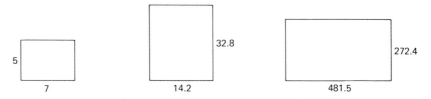

ANSWER The flow chart and program are as shown. They look much like the preceding flow chart and program. The only difference is the box, "Get another rectangle," in the flow chart. The matching BASIC statement is GO TO 10. This will automatically send the program back to statement 10 READ L, W for another rectangle.

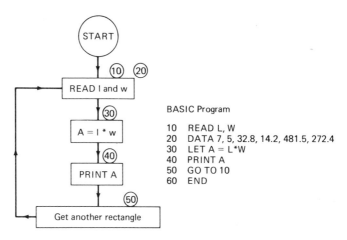

BASIC Program

```
10   READ L, W
20   DATA 7, 5, 32.8, 14.2, 481.5, 272.4
30   LET A = L*W
40   PRINT A
50   GO TO 10
60   END
```

The data will be read in two at a time until they are all used. Then the program will stop. Notice how the GO TO command creates a loop so that the program can go back and start over.

Problem Write a BASIC program to compute compound interest for different situations.

ANSWER We drew the flow chart for this previously. Now we simply translate the flow chart into BASIC along with the DATA statement and the data themselves.

The flow chart and program are shown for calculating compound interest in two situations: one where $1000 is deposited at 6% for 20 years, and the other where $3500 is deposited at 6.5% for 15 years. Notice how these data are put in the DATA statement in triples. We could easily put more situations into the data statement.

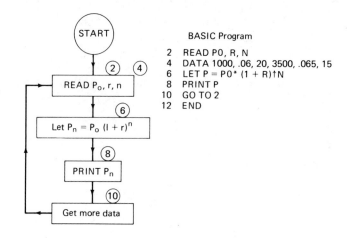

BASIC Program

```
 2   READ P0, R, N
 4   DATA 1000, .06, 20, 3500, .065, 15
 6   LET P = P0* (1 + R)↑N
 8   PRINT P
10   GO TO 2
12   END
```

Problem   Write a program to find the means and standard deviations of the following scores from two classes, each with 10 students.

$$\text{Class A:} \quad 65, 87, 93, 52, 86, 81, 77, 73, 69, 60$$
$$\text{Class B:} \quad 45, 69, 53, 82, 41, 56, 62, 93, 36, 100$$

ANSWER   We will first draw the flow chart, which uses the rules for finding mean and standard deviation from Chapter 7. Then we will give the program.

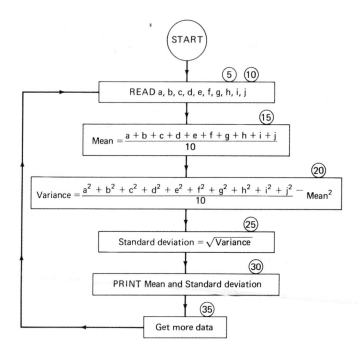

BASIC Program

```
 5  READ A, B, C, D, E, F, G, H, I, J
10  DATA 65, 87, 93, 52, 86, 81, 77, 73, 69, 60, 45, 69, 53, 82, 41, 56, 62, 93, 36, 100
15  LET M = (A + B + C + D + E + F + G + H + I + J)/10
20  LET V = (A↑2 + B↑2 + C↑2 + D↑2 + E↑2 + F↑2 + G↑2 + H↑2 + I↑2 + J↑2)/10 − M↑2
25  LET S = V↑.5
30  PRINT M, S
35  GO TO 5
40  END
```

The flow chart follows the usual procedures (in Chapter 7) exactly. After we get the data, we compute the mean; then we compute the variance, using Formula 7–2–2; finally we take the square root to get the standard deviation. We print the mean and standard deviation, and then go back for more data.

The BASIC program is given just below the flow chart. It follows the flow chart closely. There are a few changes: we have to change all the variables to single capital letters; we have to put our fractions all on one line, using parentheses if necessary. We use V↑.5 for $\sqrt{V}$. Otherwise, the program and the flow chart are the same.

Question   Can we have the computer print anything besides numbers?

ANSWER   Yes. We can have the computer print words in addition to numbers.

> To have the computer print words, we put the words to be printed inside quotation marks: "WORDS."

Problem   How would we have the computer print the answer to an area problem by saying, "the area is _____ square meters."

ANSWER   To print the word parts of the sentence, we use quotation marks. For the number part (the area itself) we just use the variable *A* (or whatever area was called). The BASIC statement would read as follows:

PRINT "THE AREA IS", A, "SQUARE METERS"

This statement will tell the computer to print the words "The area is." Then it will print the actual value of the area, which is called *A*. Finally, the computer will print the words "square meters." This is exactly what we want.

Problem   How can we print headings for tables?

ANSWER   We again use quotation marks. For each variable we want printed, we put its heading in quotation marks in a PRINT statement. For example, in the mean–

standard deviation problem, we could put headings in by inserting the following statement at the very beginning.

3 PRINT "MEAN", "STANDARD DEVIATION"

This statement will print these two headings in a line with the future answers. We put it at the very beginning before READ so that it will only be printed once, and not as part of the loop.

As a summary, let us again refer to a table of BASIC statements (Table 9–3) with little mistakes. Then we will point out the mistakes and correct them.

Table 9–3

BASIC statements (incorrect!)	Error (and possible corrections)
READ A, B, C DADA 5, 6, 7	Misspelled DATA DATA 5, 6, 7
READ 9, 10, 20	Must have data in DATA statement READ X, Y, Z DATA 9, 10, 20
READ X, Y, Z DATA 5, 10, 30, 50, 90, 120, 200	Data are *not* in multiples of 3 DATA 5, 10, 30, 50, 90, 120
READ A, B DATA 1020	Possibly forgot commas in data DATA 10, 20
PRINT YOUR TAX IS T	Forgot quotation marks PRINT "YOUR TAX IS" T

PROBLEM SET 9–4–2

Write the READ/DATA pairs of statements to read the following sets of data into the computer.

1.

x	y
16	20
20	50
24	102
28	190

2.

a	b	c
1.2	1.3	1.4
8.5	10.1	16.3
2.0	3.1	7.9

3.

l	h	w
2.5	34	4.9
106	254	199
9.9	10.3	1.5

4.

p	q	r	s
2	20	30	62
14	13	12	11

Find the errors in the following statements. (Then make a possible correction.)

5. REED A, B, C

IBM's 3250 graphics display system used here in the design of a new automobile. (Courtesy of International Business Machines)

6. READ A, B, C, D
DATA 5, 6.1, 7.9, 8.2, 5.4, 9.7

7. READ 6, 5, 2

8. READ X, Y
DATA 2, 7, 4, $6\frac{1}{2}$

9. READ XYZ
DATA 5, 4, 7

10. PRINT 5

11. PRINT THE ANSWER IS X

12. PRINT "THE COST IS C

13–21. Write a BASIC program for each of the flow charts that you drew in Problems 1–9 of Problem Set 9–3–2.

22. Draw a flow chart to compute the volume of many rectangular boxes.

23. Write a program in BASIC to find this volume in the following cases:

Length	Width	Height
16	19	25
6.5	8.2	10.3
151	209	175

A car's gas mileage is simply the number of miles driven divided by the number of gallons of gas used.

24. Draw a flow chart to compute the gas mileage of several cars.

25. Write a program in BASIC to find the gas mileage in the following cases:

Miles driven	Gas used
225	14.2
306	19.4
172	9.3
277	13.2

Becky's Balloon Factory employs six people. They each have a gross pay. From this, 2.5% state tax is deducted, 15% federal tax is deducted, and 8% retirement fund is deducted.

26. Draw a flow chart to compute and print the gross pay, state tax, federal tax, retirement contribution, and net pay (after the deductions) for each employee.

27. Write a program in BASIC for this flow chart with the following data: gross pay, $1200; $900; $650; $775; $810; $865.

The following is a formula for how many calories a person must cut from their diet everyday: $C = \dfrac{500 \cdot W}{T}$, where C = calorie cutback per day, W = weight to be lost (in pounds), and T = time to lose weight (in weeks).

28. Draw a flow chart to compute this for several cases.

29. Write a program in BASIC to do this in the following situations.

Weight to be lost (lb)	Time (weeks)
20	15
70	50
10	12
15	16
22	52

You are to play computer with the following BASIC programs. Execute each command just as the computer would. (Be sure to print answers.)

30. 5 READ A, B
10 DATA 3, 4, 5, 12
15 LET C = (A*A + B*B)↑.5
20 PRINT C
25 GO TO 5
30 END

31. 100 READ R, H
200 DATA 10, 20, 100, 50, 30, 40
300 LET V = 3.14*R*R*H
400 PRINT "THE VOLUME IS", V
500 GO TO 100
600 END

32. 10 PRINT "NUMBER", "SQUARE", "CUBE"
20 READ N
30 DATA 1, 2, 3, 4, 5, 6, 7, 8, 9, 10
40 LET S = N↑2
50 LET C = N↑3
60 PRINT N, S, C
70 GO TO 20
80 END

Decisions

Question How do we ask the computer to make decisions, such as "Is $X > 40$?"

ANSWER In flow charts, we use the decision diamond. In BASIC, we use the IF/THEN statement. Consider the question, "Is $X > 40$?" In BASIC this might be written as follows:

$$\text{IF X} > 40 \text{ THEN } 80$$

This statement says to the computer:

1. If $X > 40$ (yes), then go to statement 80.
2. If X is not > 40 (no), then just go on to the next statement.

> All IF/THEN statements have the same form:
>
> 1. If the answer is "yes," we jump to the statement numbered after the THEN.
>
> 2. If the answer is "no," we just go on to the next statement below.

We have six possible symbols in IF/THEN statements.

BASIC symbol	Meaning
=	Equal to
< >	Not equal to
>	Greater than
<	Less than
> =	Greater than or equal to
< =	Less than or equal to

Example 1 IF $A = 20$ THEN 65   means

        (a) If A is equal to 20, then jump to statement 65.

        (b) If A is not equal to 20, then just go to the next statement.

Example 2 IF H $> = 40$ THEN 5   means

        (a) If H is greater than or equal to 40, then jump to statement 5.

        (b) If H is not greater than or equal to 40, then just go to the next statement.

Problem Draw a flow chart and write the BASIC program to determine which of two different numbers is bigger.

ANSWER This is a very simple problem, but a good place to start. All it requires is one comparison. The flowchart and program are shown. In the flow chart, we read in A and B. We then ask, "Is $A > B$?" If the answer is "yes," A is bigger, we go down and print A. If the answer is "no," B is bigger, we go to the right and print B. Then we go back for more data.

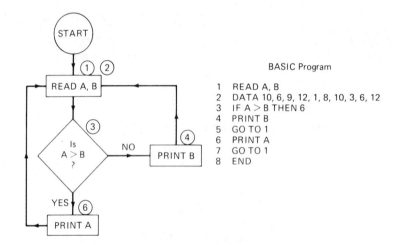

On the right is the BASIC program. First we READ A, B. Then we have some sample data in the DATA statement. Now we say, IF A $>$ B THEN 6. So, if A $>$ B, the computer will jump to statement 6 and PRINT A. If A is not $> B$, the computer will just go on to the next statement, which is PRINT B (since B is larger). Then both go back to 1, and read another pair, A and B.

Problem Write a BASIC program for the flow chart to make a table of numbers, the squares and cubes, from 1 to 1000.

ANSWER The flow chart is shown again and the BASIC program. The flow chart was already explained in Section 9.3. Let us look at the program on the right. Statements 1, 2, 3, and 4 are direct translations into BASIC. Statement 5 is the way we translate the question. We say IF N $= 1000$ THEN 8. So, if the answer is "yes," we jump to statement 8 and stop.

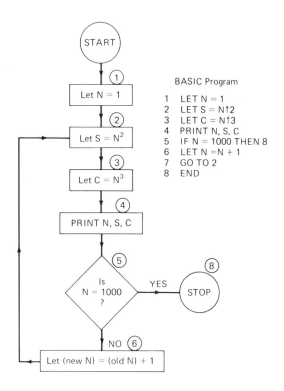

BASIC Program

```
1   LET N = 1
2   LET S = N↑2
3   LET C = N↑3
4   PRINT N, S, C
5   IF N = 1000 THEN 8
6   LET N = N + 1
7   GO TO 2
8   END
```

If the answer is "no," we go down to the next statement. In the flow chart, we say "Let (new N) = (old N) + 1." In BASIC, we write this as LET N = N + 1. This tells the computer to let N be the old N plus 1 more. Notice that we don't have to say "new" and "old." Once this is done, we loop back to statement 2 and start again.

Traffic control center in Tokyo. (Fred Ward/Black Star)

Question What does LET N = N + 1 mean?

ANSWER As a mathematical statement, $n = n + 1$ makes no sense at all. But in BASIC, it does have a meaning. The variables on the right are always old. The one on the left is always new. Sometimes, the old value is erased, and the new one is put in.

Consider the statement LET N = N + 1. Suppose that $N = 6$. When the computer sees LET N = N + 1, it will take the N on the right, which is 6, and add 1 to get 7. It will now erase the old 6, and call $N = 7$.

Consider the statement LET X = 5*X. Suppose that $X = 4$. When the computer sees LET X = 5*X, it will take the X on the right, which is 4, and multiply it by 5 to get 20. It will now erase the old 4, and call $X = 20$.

Problem Slinki's Gift Shop has the following credit policy for the balance due on a credit purchase.

    (a) If the balance is less than or equal to $33.33, the monthly finance charge is $0.50.

    (b) If the balance is over $33.33, the monthly finance charge is 1.5% of the balance.

Draw a flow chart and write a BASIC program for this problem with the following balances: $2.50, $75.00, $8.10, $3.45.

ANSWER The flow chart and program are as shown. The flow chart should seem easy. We have a decision diamond for "Is balance $(B) \leqslant 33.33$?" If the answer is "yes," we go to finance charge is $0.50. If the answer is "no," we go to finance charge is 1.5% of the balance. In either case, we print the finance charge, and go back for more customers.

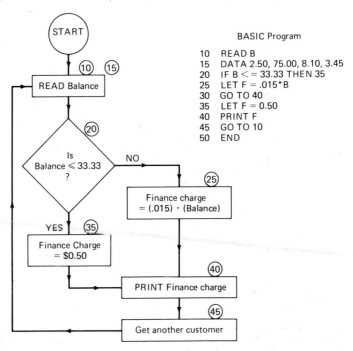

BASIC Program

```
10   READ B
15   DATA 2.50, 75.00, 8.10, 3.45
20   IF B < = 33.33 THEN 35
25   LET F = .015*B
30   GO TO 40
35   LET F = 0.50
40   PRINT F
45   GO TO 10
50   END
```

The BASIC program is given to the right of the flow chart. Let us look at it. The READ/DATA pair at the beginning should be clear. This is how we get the balance (B) into the machine.

Now we translate "Is balance ≤ 33.33?" into BASIC. This then becomes IF B < = 33.33 THEN 35. This means if the answer is "yes," we jump straight to statement 35, which is LET F = 0.50. This is just like in the flow chart. After statement 35, we say PRINT F, and GO TO 10 returns us back to READ B.

Suppose that B is not ≤ 33.33. Now the IF/THEN statement will just move us down to the next statement below, 25 LET F = .015*B. This also is just like in the flow chart. After this, we want to go to the print statement. This is why we have the GO TO 40 statement so that we can jump over statement 35 and go straight to statement 40. Otherwise, the computer would accidentally do statement 35 instead of skipping it.

Problem   Write a BASIC program for the flow chart for Jennifer's Pet Palace.

ANSWER   Most of the instructions in the flow chart will be fairly easy to translate to BASIC statement. The tricky part will be the question, "Is $H > 40$?" Here, we have to have the two separate paths (the yes path and the no path). The flow chart and program are shown. The statements in the flow chart all have their addresses on them.

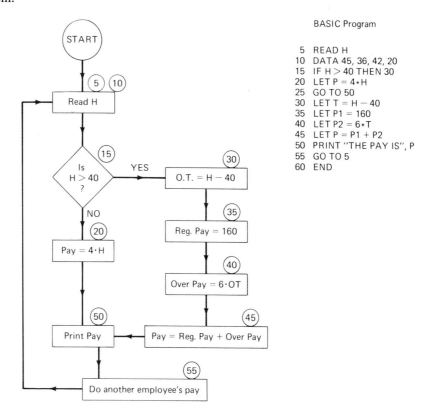

BASIC Program

```
 5  READ H
10  DATA 45, 36, 42, 20
15  IF H > 40 THEN 30
20  LET P = 4∗H
25  GO TO 50
30  LET T = H − 40
35  LET P1 = 160
40  LET P2 = 6∗T
45  LET P = P1 + P2
50  PRINT "THE PAY IS", P
55  GO TO 5
60  END
```

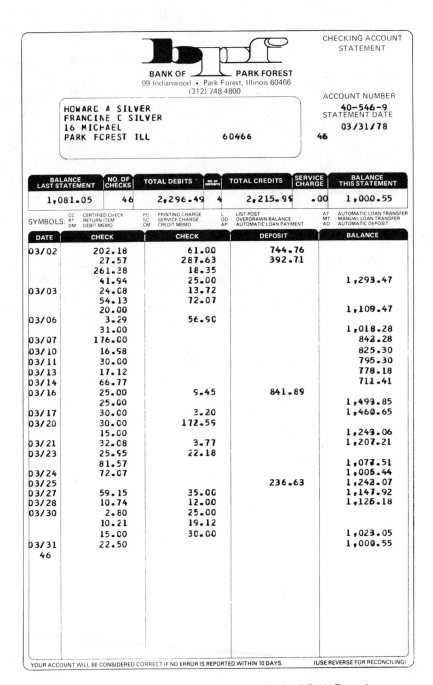

A computerized bank statement. (Courtesy of Bank of Park Forest)

Statements 5 and 10 are straight translations of the READ box of the flow chart. Statement 15 is IF H > 40 THEN 30. This is like the decision diamond of the flow chart. If the answer is "yes" (H > 40), we jump to statements 30, 35, 40, 45, and 50. These are the statements for overtime work.

If the answer is "no," we go on to the next statement (20) for the regular pay. Then we jump to statement 50 to print the pay. Finally, statement 55 says go back to the READ statement and start again.

PROBLEM SET 9-4-3 Translate the following flow-chart decisions into a BASIC IF/THEN statement.

1.

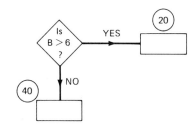

2.

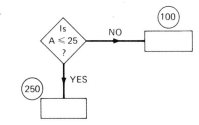

3.

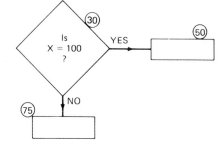

4.

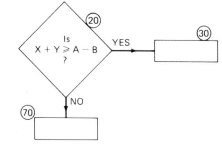

Find the errors in the following BASIC statements, and make a possible correction.

5. IS A > B THEN 50

6. IF X < = 100

7. IF X = 60 THEN

8. IF B + 6 THEN 100

9. IF M = N THAN 25

You are to play computer with the following BASIC programs. Execute each command just as the computer would. (Be sure to print answers.)

10.   20   LET R = 1
         40   PRINT R
         60   LET R = 2*R
         80   IF R < 200 THEN 40
       100   END

11.   1   LET S = 0
         2   LET N = 1
         3   LET S = S + N
         4   IF N = 10 THEN 7
         5   LET N = N + 1
         6   GO TO 3
         7   PRINT S
         8   END

12.   10   READ H, N
         20   DATA 40, 6, 20, 4, 32, 7
         30   LET R = 6 + .5*N
         40   IF H > 30 THEN 80
         50   LET P = H*R
         60   PRINT P
         70   GO TO 10
         80   LET T = H − 30
         90   LET P1 = 30*R
       100   LET P2 = 1.5*T*R
       110   LET P = P1 + P2
       120   GO TO 60
       130   END

13.   5   READ A, B, C
        10   DATA 1, 2, 5, 6, 4, 5, 2, 10, 4
        15   IF A > B THEN 35
        20   IF C > B THEN 50
        25   PRINT B, "IS BIGGEST"
        30   GO TO 5

```
35  IF C > A, THEN 50
40  PRINT A, "IS BIGGEST"
45  GO TO 5
50  PRINT C, "IS BIGGEST"
55  GO TO 5
60  END
```

14. Write a BASIC program for the flow chart drawn in Problem 10 of Problem Set 9–3–2.

BASIC (9–4)
binary system (9–2)
bit (9–2)
commands (9–3)
computer (9–1)
control (9–1)
decisions (9–3)
END (9–4)
flow chart (9–3)
GO TO (9–4)
IF/THEN (9–4)
input (9–1)

LET (9–4)
loop (9–3)
magnetic core (9–2)
memory (9–1)
output (9–1)
paths (9–3)
PRINT (9–4)
program (9–3)
READ/DATA (9–4)
retrieve (9–1)
store (9–1)
terminals (9–3)

REVIEW EXERCISES

1. Give a brief history of the computer.

2. Discuss what the computer can and cannot do.

3. Write 1011011_{two} in base ten.

4. Write 89_{ten} in base two.

Perform the indicated operations in base two and then check answers in base ten.

5. 1101101_{two}
$+ 1010110_{two}$

6. 1001001_{two}
$- 110111_{two}$

7. 10001_{two}
$\times \ 101_{two}$

8. $1101_{two}\ \overline{)1001110_{two}}$

9. Draw a flow chart for paying a bill by check.

10. Draw a flow chart for making a left turn in a car on a busy street.

11. Draw a flow chart for finding the hypotenuse of a right triangle (see Formula 3–6–1).

12. Draw a flow chart to find the cost of insuring a brick house in a class A community (see Tab. 8–5).

13. Follow the directions in the given flow chart. You are to play computer. (Use a hand calculator if necessary to execute the instructions.)

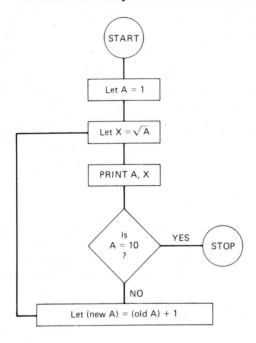

Complete the following table.

	Usual mathematical statement	BASIC statement
14.	$y = 3x + 4$	?
15.	$c = \sqrt{a^2 + b^2}$	?
16.	?	LET G = A↑3
17.	?	LET M = (A + B + C)/3

18. Write a READ/DATA pair for the following data.

x	y	z
5	7	6.2
3.8	−1	4.5
−2	0	4

Find the mistakes in the following BASIC statements. Then make a possible correction.

19. A = 10

20. LET BOB = 2

21. READ 5, 7, 9

22. READ A, B, C
DATA 6, 5, $4\frac{1}{2}$

23. IS A > 6 THEN 40

24. IF A < = 10 GO TO 25

The formula to convert °F to °C is given by $°C = \frac{5}{9}(°F - 32)$.

25. Draw a flow chart to program this for various temperatures.

26. Write a BASIC program to read in and convert the temperatures given below:

Temp. (°F)
0
10
20
30
40

In a certain department store, the employees are paid $3.10 per hour for every hour that they work up to 40 hours. If they work over 40 hours, they still only get paid for 40 hours.

27. Draw a flow chart to read in the number of hours an employee worked (*H*) and to print out his or her pay (*P*).

28. Write a BASIC program to read in the hours, shown below, and to print out the pay.

Hours
16
37
43
20
50

ANSWERS
TO
SELECTED
EXERCISES

SOME OF THE
ANSWERS MAY
VARY *SLIGHTLY*
BECAUSE OF THE
DIFFERENT WAYS
TO ROUND OFF
WHILE DOING THE
PROBLEM

Problem Set 1–1–1 (page 12)

2. $\frac{3}{4}$ **3.** $\frac{1}{3}$ **5.** $\frac{1}{6}$ **7.** $\frac{3}{2}$ **9.** $\frac{69}{10} = 6\frac{9}{10}$ **10.** $\frac{3}{4}$

11. $\frac{6}{7}$ **13.** $\frac{11}{12}$ **15.** $\frac{13}{100}$ **17.** $\frac{289}{180} = 1\frac{109}{180}$

19. $15\frac{3}{4}$ **21.** $13\frac{1}{15}$ **23.** $37\frac{7}{24}$ **25.** $\frac{2}{5}$ **27.** $\frac{1}{8}$ **29.** $\frac{1}{24}$

31. $\frac{7}{72}$ **33.** $\frac{11}{100}$ **34.** $11\frac{1}{4}$ **35.** $12\frac{1}{10}$ **37.** $4\frac{1}{3}$ **39.** $5\frac{1}{77}$

40. $18\frac{1}{5}$ **41.** $120\frac{9}{10}$ **43.** $4\frac{11}{12}$

Problem Set 1–1–2 (page 17)

2. $\frac{5}{2}$ **3.** $\frac{32}{5}$ **5.** $\frac{50}{11}$ **7.** $\frac{227}{12}$ **8.** $1\frac{4}{7}$ **9.** $5\frac{2}{3}$

11. $7\frac{1}{5}$ **13.** $13\frac{3}{13}$ **14.** $\frac{1}{6}$ **15.** $\frac{24}{35}$ **17.** $\frac{1}{4}$ **19.** $\frac{1}{100}$

20. 15   **21.** $14\frac{22}{25}$ **23.** $12\frac{1}{3}$ **25.** 112   **26.** $\frac{55}{3}$ **27.** $\frac{8}{15}$

29. 8   **31.** $\frac{7}{39}$ **32.** $2\frac{13}{40}$ **33.** $10\frac{4}{5}$ **35.** $1\frac{1}{20}$ **37.** $\frac{42}{81}$

38. $1\frac{2}{3}$ cups sugar, $3\frac{1}{8}$ cups flour, $1\frac{1}{4}$ tsp salt, and so on

39. $16\frac{11}{12}$ ft   **41.** $4\frac{1}{2}$ in.   **43.** 240   **45.** 5   **47.** $266\frac{2}{3}$ **49.** $3\frac{3}{4}$

Problem Set 1–2–1 (page 20)

1. 26.21   **2.** 83.228   **3.** 1547.117   **5.** 201.28   **7.** 233.827

9. 87.42   **10.** 6.389   **11.** 0.141   **13.** 10.26   **15.** 63.747

17. $231.18

Problem Set 1–2–2 (page 25)

1. 3847.0726   **2.** 2143.428   **3.** 74.087   **5.** 0.0000028   **7.** 812.5

9. 17.07   **11.** 0.778   **13.** 250.1   **15.** 41.62   **17.** 28.41   **19.** 17.273

21. 51.389   **23.** $437.50   **25.** 10.62   **27.** 2.385   **29.** 163.58

31. 2.26¢   **33.** 4.29

Problem Set 1–3–1 (page 31)

1. 0.500   **3.** 0.143   **5.** 0.692   **7.** 1.600   **9.** 3.143   **10.** $\frac{2}{5}$

11. $\frac{39}{100}$ **13.** $\frac{1}{4}$ **15.** $\frac{1}{5000}$ **17.** $\frac{21}{400}$ **19.** 29%   **21.** 7%

23. 99.44%   **25.** 0.05%   **27.** 102%   **28.** 0.32   **29.** 0.91   **31.** 0.0625

33. 0.041   **35.** 0.00001   **37.** 50%   **39.** 14.3%   **41.** 69.2%

43. 160%   **45.** 314.3%   **46.** $\frac{8}{25}$ **47.** $\frac{91}{100}$ **49.** $\frac{1}{16}$

51. $\frac{41}{1000}$ **53.** $\frac{1}{100000}$

Problem Set 1–4–1 (page 34)

2. $+7.20$ **3.** -8 **5.** $+3$ **7.** $+100$ **9.** -14 **11.** $+9$ **13.** -5

15. $+22.52$ **17.** $-\dfrac{1}{6}$ **19.** $-\dfrac{37}{60}$ **21.** 0.8 mi backward   **23.** Lose 7 lb

Problem Set 1–4–2 (page 36)

1. $+7$ **3.** $+7$ **5.** -1.85 **7.** $-\dfrac{1}{20}$ **9.** $-\dfrac{31}{12}$

Problem Set 1–4–3 (page 39)

1. $+32$ **3.** -156 **5.** 34.98   **7.** $-3\frac{3}{4}$ **9.** $\frac{11}{18}$ **10.** -6

11. -4 **13.** -7.69 **15.** -2.93 **17.** $4\frac{3}{8}$

Problem Set 1–5–1 (page 40)

2. 6^4 **3.** 8^{10} **5.** $(0.75)^5$ **7.** 1024   **9.** 1.134225   **11.** 0.4096

13. 64,000   **15.** 2116   **17.** 2143.5733   **19.** 607.75

Problem Set 1–5–2 (page 43)

2. 7.15×10^4 **3.** 8.7×10^7 **5.** 6.2×10^1 **7.** 1.5×10^{-4}

9. 2×10^8 **11.** 1.4×10^9 **13.** 7.25×10^{-6}

14. $6.18 \times 10^7 = 61,800,000$ **15.** $7.05 \times 10^2 = 705$

17. $8.1 \times 10^{12} = 8,100,000,000,000$ **19.** $3.34175 \times 10^{-7} = 0.000000334175$

20. 9.3×10^7 **21.** 3.3×10^6 **23.** 3×10^{10} **25.** 4.7×10^9

27. 1.06×10^{11} **29.** 7×10^{-10} **31.** 5×10^{-7}

Problem Set 1–5–3 (page 46)

Exact square roots: **2.** 3.1623   **3.** 5.3852   **5.** 43.5890   **7.** 3.5384

9. 1.4142   **11.** 6.5955   **13.** 7.7460

Problem Set 1–6–1 (page 54)

1. 5   **3.** 66.8   **5.** 1618.157   **7.** 4   **9.** 17.9   **11.** -589.15

13. 375.42   **15.** 752.61   **17.** 84   **19.** 28,087.773   **21.** 0.00002891

23. 5   **25.** 5.0699301   **27.** 2.7306273   **28.** 19.155694   **29.** $-47,253.55$

Problem Set 1–6–2 (page 59)

1. 42   **3.** 18.36   **5.** -49.833 **7.** 1.9628385   **9.** -1.1011905

11. -0.1163654 **13.** 1.3114176   **15.** 33.189055   **17.** $900.22   **19.** $2224

Problem Set 1–6–3 (page 64)

1. 25 **3.** 10,000 **5.** 1,378,584.9 **7.** 2.571841 **9.** 129.74634

11. 5998.3849 **13.** 1383.9126 **15.** 5377.2888 **17.** 7.5498344

19. 42.237424 **21.** 0.27946377 **23.** 1.3076697

Review Exercises, Chapter 1 (page 64)

1. $\frac{71}{120}$ **2.** $8\frac{7}{30}$ **3.** $\frac{11}{45}$ **4.** $7\frac{5}{16}$ **5.** $\frac{3}{28}$ **6.** 8 **7.** $\frac{21}{26}$

8. $1\frac{1}{3}$ **9.** $\frac{29}{4}$ **10.** $8\frac{6}{7}$ **11.** 256.016 **12.** 17.087 **13.** 25.515

14. 7.61 **15.** 14.176 **16.** 35.18 **17.** 0.00000146 **18.** 0.08

19. 17.381 **20.** 0.643 **21.** $\frac{9}{20}$ **22.** 31.6% **23.** 0.001 **24.** 27.3%

25. $\frac{11}{25}$ **26.** -7 **27.** -11.13 **28.** $\frac{3}{4}$ **29.** 0.8 **30.** 25.2

31. $-\frac{1}{12}$ **32.** 10 **33.** -8 **34.** 32 **35.** $(1.3)^4$ **36.** 2.45×10^{10}

37. 8.2×10^{-7} **38.** 7 **39.** 7.81025 **40.** 0.0363178 **41.** 22.720695

42. 0.10880391 **43.** 35.553388 **44.** 1.8771375 **45.** 1.9404639

46. 144.58 **47.** 32 **48.** 16.878049

Problem Set 2–1–1 (page 70)

2. $5a + 5b$ **3.** $7x + 7y$ **5.** $10p - 10q$ **7.** $6a + ra$ **9.** $xy - xz$

11. $10a + 10b + 10c$ **13.** $x^2 + xy - 12x$ **14.** $9a$ **15.** $6x$

17. $6(x - 2y)$ **19.** $t(s - 7)$ **21.** $3(2a + b - 3c)$

Problem Set 2–1–2 (page 77)

1. 49 **3.** $\frac{13}{8}$ **5.** 78 **7.** $\frac{7}{10}$ **9.** 40 **11.** 10 **13.** 1

15. 1.462 **17.** $\frac{143}{6} = 23.833$ **19.** 6 **21.** $-\frac{10}{3}$ **23.** $-\dfrac{3}{(3 + t)}$

25. 15.5 **27.** 13

Problem Set 2–1–3 (page 79)

2.

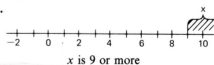

x is less than 7

3.

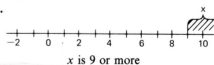

x is 9 or more

5.

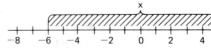

x is more than -6

7.

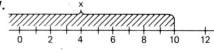

10 is more than x

9.

x is between −4 and 4
(including end points)

11.

x is between 10.1 and 10.3
(including end points)

Problem Set 2–2–1 (page 81)

1. $p - 3$ **3.** $13 \cdot p$ **5.** $(\frac{9}{10}) \cdot D$ **7.** $\frac{1}{3} \cdot A$ **9.** $0.14 \cdot D$

11. $10 + 2F$ **12.** $67 = p - 6$ **13.** $\frac{1}{5} \cdot S = 650$ **15.** $D = 0.20 \cdot 95$

17. $A = l \cdot w$ **19.** $M = 0.80 \cdot P$ **21.** $G = 2 \cdot B - 7$ **23.** $E + M < D$

Problem Set 2–2–2 (page 86)

1. $10,666.67   **3.** $23.33   **5.** 20   **7.** $2941.18   **9.** 22.64   **11.** 220

13. Nigel: 70,000; Clarence: 105,000

Problem Set 2–3–1 (page 89)

1. 0.9   **3.** 110.3   **5.** 30.6%   **7.** 0.44%   **9.** 3216.7

11. 1875   **13.** 118.4%

Problem Set 2–3–2 (page 94)

1. 3939   **3.** 20%   **5.** 28.6%   **7.** 153,846.15   **9.** 21.9%   **11.** 32,500

13. 51.9%   **15.** 57.1%   **17.** 530   **19.** 10,600   **21.** 16.5%

23. 154.5%   **25.** 29,787.23   **27.** 288

Problem Set 2–4–1 (page 97)

2. $28.13 \dfrac{\text{money saved}}{\text{week}}$ **3.** $\dfrac{5}{7} \dfrac{\text{money from Joe}}{\text{money from Ralph}}$ **5.** $0.87 \dfrac{\text{liberals}}{\text{conservative}}$

7. $144.6 \dfrac{\text{g of protein}}{\text{dollars}}$ **9.** $14.7 \dfrac{\text{mi}}{\text{gal}}$ **11.** $1.24 \dfrac{\text{games won}}{\text{games lost}}$

13. $1.61 \dfrac{\text{km}}{\text{mi}}$ **15.** $1.05 \dfrac{\$ \text{ cost in 1978}}{\$ \text{ cost in 1977}}$

Problem Set 2–4–2 (page 102)

2. 262   **3.** 80.7   **5.** 19.7   **7.** 46.55   **9.** 39.37   **11.** 565

13. 32.3   **15.** 5,100,000   **17.** 4.5 per 1000

Problem Set 2–5–1 (page 114)

	mm	cm	m	km	in.	ft	yd	mi
2.	457	45.7	0.457	×	<u>18</u>	1.5	0.5	×
3.	400	40	<u>0.4</u>	×	15.75	1.31	0.44	×
4.	×	×	606	0.606	×	1980	<u>660</u>	0.375
5.	<u>35</u>	3.5	0.035	×	1.38	0.115	×	×
7.	×	427	4.27	×	168	<u>14</u>	4.67	×
9.	1.7×10^7	1.7×10^6	17,000	<u>17</u>	×	×	18,530	10.56

	mg	g	kg	MT	oz	lb	ton
10	<u>500</u>	0.5	0.0005	×	0.0176	0.0011	×
11.	227,000	227	0.227	×	<u>8</u>	0.5	×
13.	7.5×10^7	75,000	<u>75</u>	0.075	2640	165	0.0825

	ml	L	cup	pt	qt	gal
15.	475	0.475	<u>2</u>	1	0.5	0.125
17.	×	3800	×	8000	4000	<u>1000</u>
19.	11,400	11.4	48	24	<u>12</u>	3

21. 1.78 m   **23.** 1.38 in.   **25.** 505.54 km   **27.** 65.9 kg   **29.** 0.0176 oz

31. 113.5 g   **33.** 0.01 L; 0.042 cup   **35.** 4.75 L   **37.** 2.25 gal; 8.55 L

39. 0.75 L   **41.** 0.112 MT   **43.** 400 ml

Review Exercises, Chapter 2 (page 116)

1. $5a - 15$ **2.** $6x - x^2 - xy$ **3.** $a(7 + m)$ **4.** $b(a + b - 2)$

5. 13   **6.** $\frac{31}{6}$ **7.** $\frac{24}{7}$ **8.** $\frac{51}{2}$ **9.** $-\frac{13}{2}$ **10.** $\frac{16}{9}$

11.
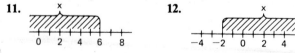

12.

13.
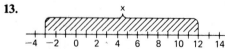

14. $(0.05) \cdot P$ **15.** $225 = \left(\frac{2}{11}\right) \cdot i$ **16.** 279   **17.** 8.53   **18.** 144.5

19. 2083.3   **20.** 31.5%   **21.** 193.60   **22.** 45.6%   **23.** 1090.91

24. 142.9 gal   **25.** 351,120   **26.** 16,363.64   **27.** 1.88 m

28. 53.6 kg   **29.** 750 g   **30.** 5.53 gal   **31.** 22.36 m/sec   **32.** 1.5 km

Problem Set 3–1–1 (page 123)

2. 23°   **3.** 66°   **5.** 114°   **7.** 122°   **9.** 14.4%   **11.** 162°   **13.** 342°

15.   **17.** 40°   **19.** 15°

Problem Set 3–1–2 (page 130)

2. 21   **3.** 12   **5.** 30.4   **7.** 56   **9.** 54.95   **11.** 3140   **13.** 84.78

15. 64.25   **17.** 53.55   **18.** 54.67 ft   **19.** $9.35 for 55 ft

21. 14   **23.** 265.6 m   **25.** 6.06   **27.** 63.7 yd

Problem Set 3–2–1 (page 138)

2. 60   **3.** 110.04 m^2 **5.** $5\frac{5}{6}$ mi^2 **7.** 704.7   **9.** $48\frac{3}{4}$ **11.** 3.42 m^2

13. 3346.455 cm^2 **15.** 271.58 km^2 **17.** 508.68 cm^2 **19.** 176.625 m^2

20. 10.71   **21.** 22.36   **23.** 13.58   **25.** 30.95 cm   **27.** 7.98 ft

29. 11.29

Problem Set 3–2–2 (page 143)

1. 16   **3.** 66.73   **5.** 82.1   **7.** 187.44

Problem Set 3–3–1 (page 150)

	in.^{2}	ft^2	yd^2	mi^2	acre	cm^2	m^2	km^2
1.	28,800	200	22.2	×	×	186,000	18.6	×
2.	6000	41.67	4.63	×	×	38,700	3.87	×
3.	×	×	2,420,000	0.78	500	×	2,020,000	2.02
5.	69.75	0.484	0.054	×	×	450	0.045	×
7.	930,000	6450	713	×	×	6,000,000	600	×

8. 21,600 ft^2 **9.** 58.5 ft^2 **11.** $8\frac{3}{4}$ in.^{2} **13.** 32 **15.** 0.16 acres

16. 2350 ft^2 **17.** 11.75 lb **19.** 261.1 yd^2 **21.** 1319.8 ft

23. 89,496.5 ft^2 **25.** $7856 **27.** 19,166 in.^{2} = 133.1 ft^2

29. $7.00 **30.** 1065 in. = 88.7 ft = 29.6 yd **31.** 31.1 yd

33. 397.29 ft^2 **35.** $551.75 **37.** 181.78 ft^2 **39.** 25 tiles

41. 400 in.^{2} **43.** 1 package (0.93 exactly) **45.** 113 in.^{2} **47.** 201 in.^{2}

Problem Set 3–4–1 (page 160)

2. 420 **3.** 232.5 **5.** 770.044 **7.** 4.284 **9.** 3617.28 cm^3

11. 261,666.7 ft^3 **13.** 49.08 **15.** 2016 **16.** 6.49 **17.** 2 cm

19. 10.62 mm **21.** 17.9 **23.** 22.5

Problem Set 3–4–2 (page 170)

1. 66,501.01 **3.** 108.47 mm^3 **5.** 37.68 **7.** 1507.2

	ft^3	in.^{3}	yd^3	cm^3	m^3
9.	35	60,480	1.296	990,000	0.99
11.	378	653,184	14	1.07 × 10^7	10.7
13.	56,500	9.76 × 10^7	2090	1.6 × 10^9	1600

15. 4.5 ft^3 **17.** $10.50 **19.** 528 ft^3 **21.** 11,488.2 cm^3

22. 379.7 ft^3 **23.** 85 bags **25.** 1071.8 ft^3 = 39.7 yd^3 **27.** 16.28 oz

29. 245,000 L **31.** 4.65 oz **33.** 4.36 × 10^5 m^3 **35.** 1.08 × 10^{12} km^3

36. 25.12 in.^{3} **37.** 50.24 in.^{3} **39.** 78.5 in.^{3} **41.** 56.52 in.^{3}

43. 169.56 in.^{3} **45.** 153.86 in.^{3}

Problem Set 3–5–1 (page 179)

2. $x = 9$; $y = 7.5$; $t = 12.5$; $z = 10$ **3.** $x = 2.73$

5. $x = 0.43$; $y = 0.71$; $z = 0.29$; $t = 0.29$; $w = 0.21$ **7.** 20.74

9. 14.3 in.

Problem Set 3–5–2 (page 183)

1.

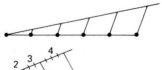

3.

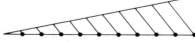

Problem Set 3–5–3 (page 186)

1. 160   **3.** 8   **5.** 367.5   **7.** $\sqrt{2} \approx 1.4$ **9.** 52 m   **11.** 69%

13. $\sqrt{1.30} \approx 1.14$ **14.** Decrease 4 times   **15.** Increase 8 times

17. Increase 4 times   **19.** Increase 32 times   **21.** Increase 2 times

Problem Set 3–6–1 (page 193)

2. 5   **3.** 13   **5.** 12.53   **7.** 1.41   **9.** 10.20   **11.** 10.95   **13.** 9.95

15. 128.06   **17.** 88.55 km/hr   **19.** 64.4 km/hr   **21.** 282.8 mi

23. 5.14   **25.** A to B directly   **26.** 43.5 mi   **27.** 33.6 mi   **29.** 29.0 mi

31. 6.7 mi   **33.** 8.49 in.   **35.** 17.68 in.   **37.** 9.9 ft

39. $h = 8.66$; area = 43.3

Review Exercises, Chapter 3 (page 197)

1. 31.1°   **2.** 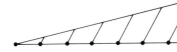   **3.** 17.22   **4.** 72   **5.** 100.48

6. 100.26   **7.** 44.28   **8.** 5   **9.** 3738

10. 6050   **11.** 452.16   **12.** 14,346.5

13. 16.26   **14.** 5   **15.** 84.17 m²

16. 24,000 cm²   **17.** 356 ft²   **18.** 39.6 yd²

19. $475.20   **20.** 2.45 m²   **21.** 622.16

22. 8478   **23.** 259.6   **24.** 3108.6   **25.** 108.8   **26.** 1436.03

27. 0.05 m³   **28.** 1762.26 ft³   **29.** 395.64 ft³   **30.** 2967.3 gal

31. $x = 19.125$; $y = 21.25$ **33.**

35. 58.99   **36.** 6.4   **37.** 17.32   **38.** 10.6 in.   **39.** 112.5 in.²

Problem Set 4–1–1 (page 208)

2. $A = 3$; $B = 6\frac{2}{3}$; $C = -2$; $D = -4\frac{3}{4}$; $E = 0$; $F = 4\frac{1}{2}$; $G = -6\frac{1}{2}$

3.

5. $A = (10, 4000)$; $B = (0, 3000)$; $C = (-10, 4500)$; $D = (-20, 3600)$
$E = (-27, 1400)$; $F = (-18, 0)$; $G = (-16, -2400)$; $H = (3, -3300)$
$I = (16, -1700)$; $J = (23, 0)$

7.

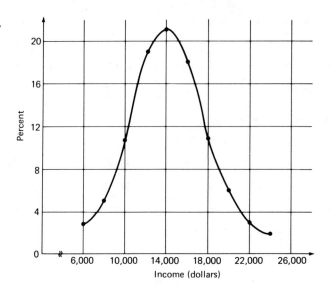

Problem Set 4–1–2 (page 211)

2. 96° to 105°

3. 0 to 200
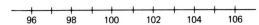

5. 0 to 20
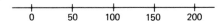

7. $8000 to $50,000

9. 0 to 10 mm
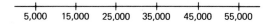

11. −10 to 100°F; −23 to 38°C

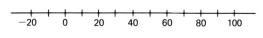

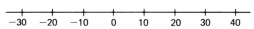

13. 0 to 15

15. 0 to 12,000

16.

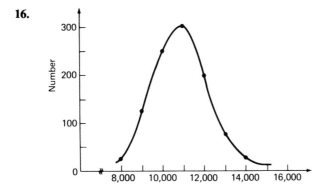

17.

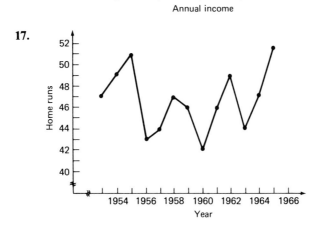

19.

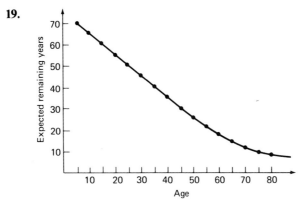

21.

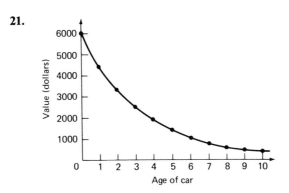

Problem Set 4–2–1 (page 216)

2.

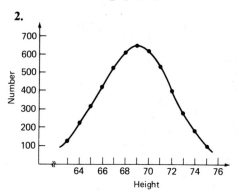

3.

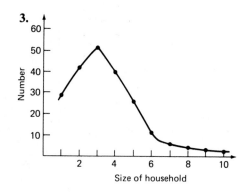

5.

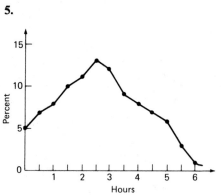

7.

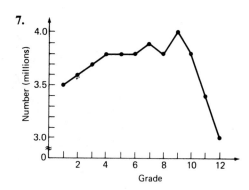

9.

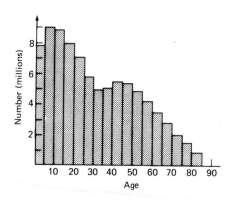

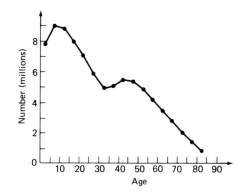

11.

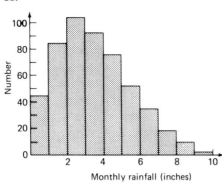

13.

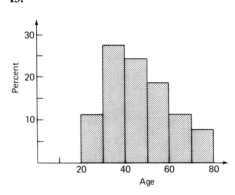

15.

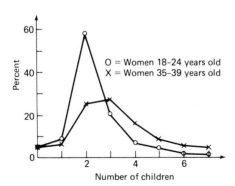

17.

Annual income	10000–12999	13000–15999	16000–18999	19000–21999	22000–24999
Number	||| 3	卌 || 7	||| 3	卌	6	|| 2

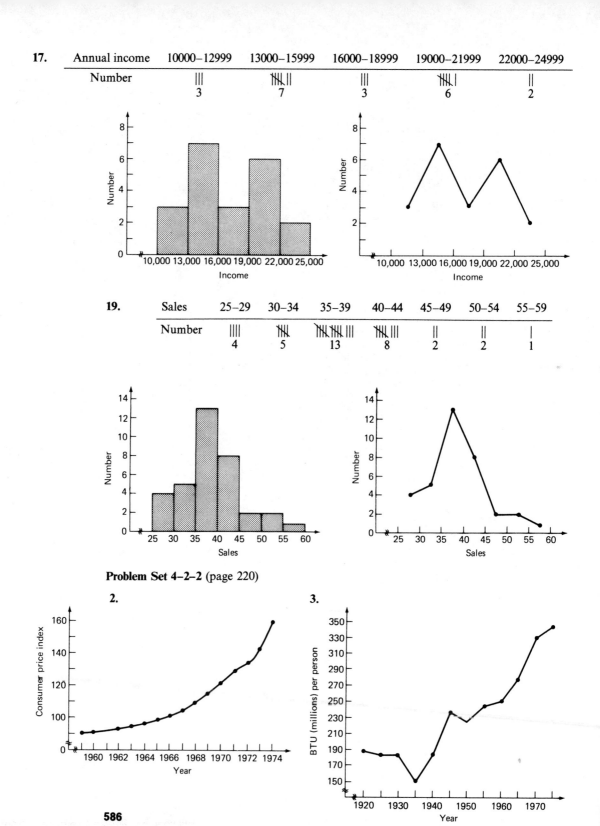

19.

Sales	25–29	30–34	35–39	40–44	45–49	50–54	55–59
Number	|||| 4	卌 5	卌 卌 ||| 13	卌 ||| 8	|| 2	|| 2		1

Problem Set 4–2–2 (page 220)

2.

3.

5.

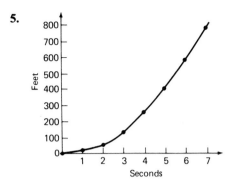

7.

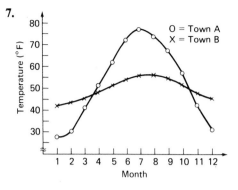

9.

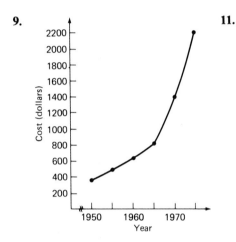

11.

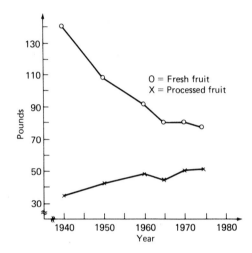

13.

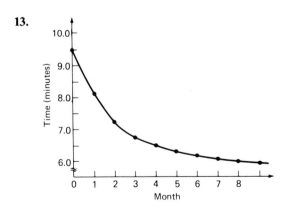

Problem Set 4–2–3 (page 225)

2.

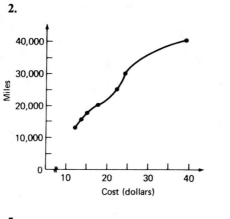

3.

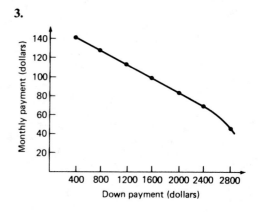

5.

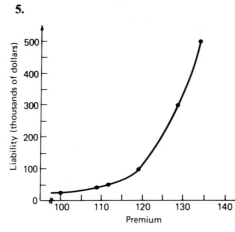

7.

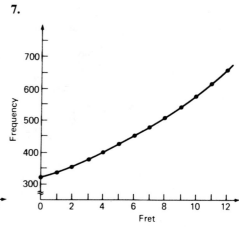

9.

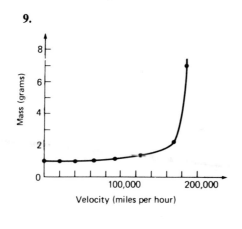

11.

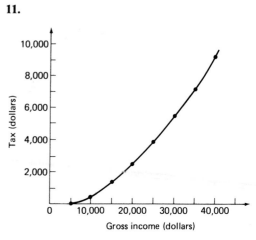

13.

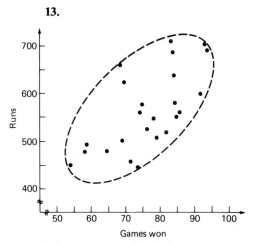

15.

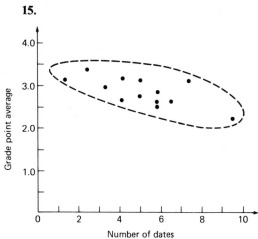

The more runs scored, the more games the teams *tended* to win.

More dates *tended* to have *slightly* lower grades.

Problem Set 4–3–1 (page 232)

2. 3   **3.** −0.2   **5.** 12   **7.** −0.6   **9.** −0.0417

10. 2.5   **11.** 4   **13.** −1.43

15. 12.4

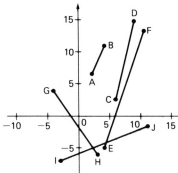

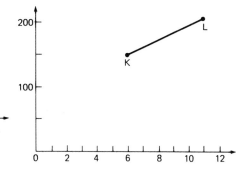

17. *m* = 4.67

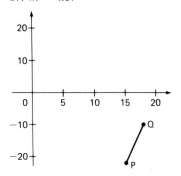

19. *m* = −250

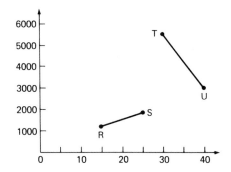

21. $m_3 = 0.055$;　　$m_4 = -0.11$;　　intersection = (17, 0.9)

23. $m_7 = 3.33$;　　$m_8 = -3.33$;　　intersection = (1960, 125)

25. $m_{11} = -5.5$;　　$m_{12} = -12.3$;　　intersection = (25, 290)

Problem Set 4–3–2 (page 240)

1. $y = 10x - 56$　　　**3.** $y = -4x + 69$　　　**5.** $y = 1.2x - 8.58$

7. $y = 2x - 3$　　　**9.** $y = -3x - 8$　　　**11.** $y = -41.7x + 2830$

13. $l_1: y = 1.7x - 12$; $l_2: y = 0.35x + 11$; intersection = (17, 17)

15. $y = 100 + x$　　　　**17.** Men: $y = 4x - 130$; women: $y = 3.5x - 110$

19. (0, \$770); (5000, \$1070); (10,000, \$1370); (20,000, \$1970)

21. $y = 0.06x + 770$

23.

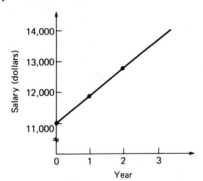

25. 11.25 yr

27. $y = -0.2x + 223$

Problem Set 4–4–1 (page 248)

2.

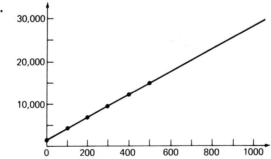

3. (50, 3250); (160, 6000); (325, 10125)　　　**5.** $m = 25$

7, 10.

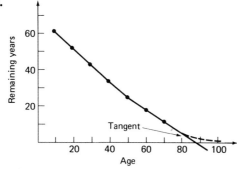

Tangent line

9. $\sqrt[3]{1400} \approx 11.2$; $\sqrt[3]{2000} \approx 12.6$ **11.** $m \approx 0.0135$

13. (15, 57.5); (23, 50.5); (37, 38); (45, 29); (65, 15)

15.

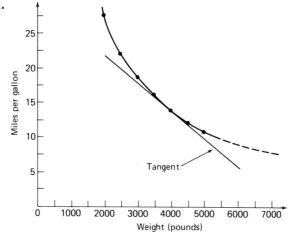

Remaining years

Tangent

Age

17. For every year older we get, we lose $\frac{2}{3}$ year of expected life (at 70 years old).

19. (2250, 25); (2800, 20); (3700, 15); (4400, 12.7)

21.

Miles per gallon

Tangent

Weight (pounds)

23. For every extra pound of car weight, the gas mileage goes *down* about 0.0035 miles per gallon.

25. (1980, 1,300,000); (1990, 2,500,000)   **27.** log scale

29. (2000, $383);
$m_{\text{tangent}} \approx \$8.88$

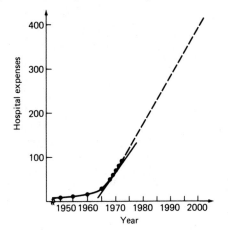

31. (2000, 0.9525);
$m_{\text{tangent}} \approx -0.00025$

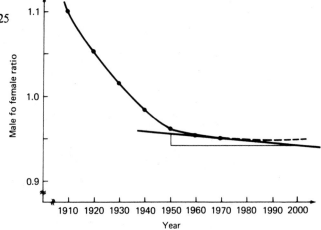

33. (2000, 3,000,000);
$m_{\text{tangent}} \approx 72,000$

2.

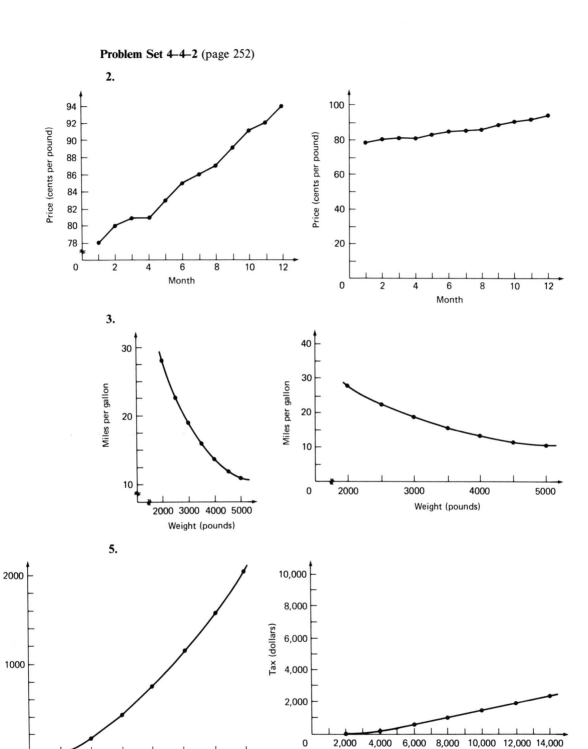

3.

5.

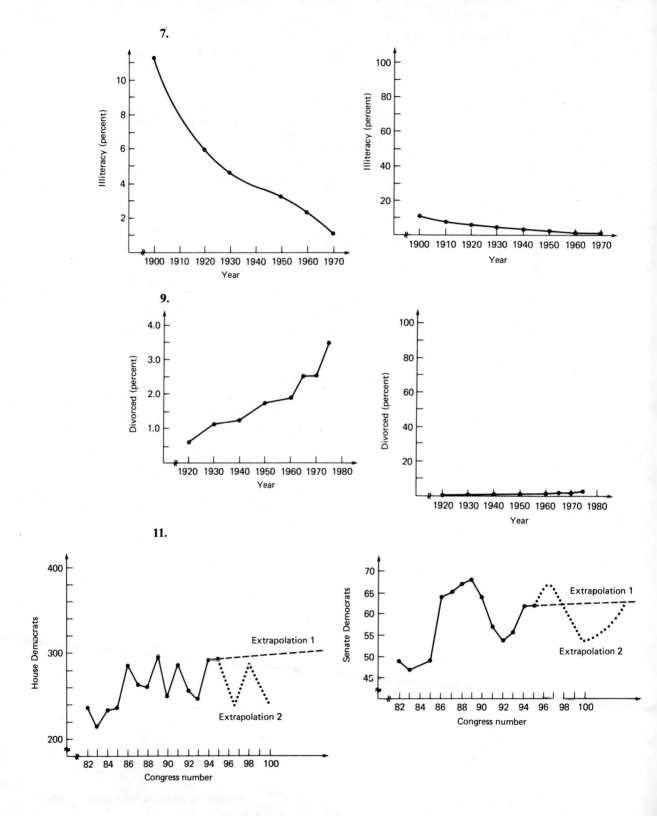

7.

9.

11.

13.

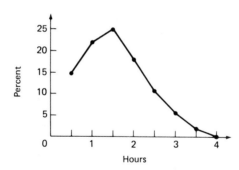

Review Exercises, Chapter 4 (page 253)

1. (4, 300) **2.** (−8, 200) **3.** (−6, −200) **4.** (10, −150)

5. 0 to 4

6. 0 to 100,000

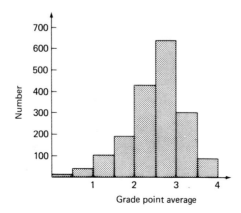

7.

8.

9.

Price	0–10	10–19	20–29	30–39	40–49	50–59	60–69	70–79	80–89	90–99
Number	卌 I	卌 III	卌 II	卌 I	II	I			I	I
	6	8	7	6	2	1	0	0	1	1

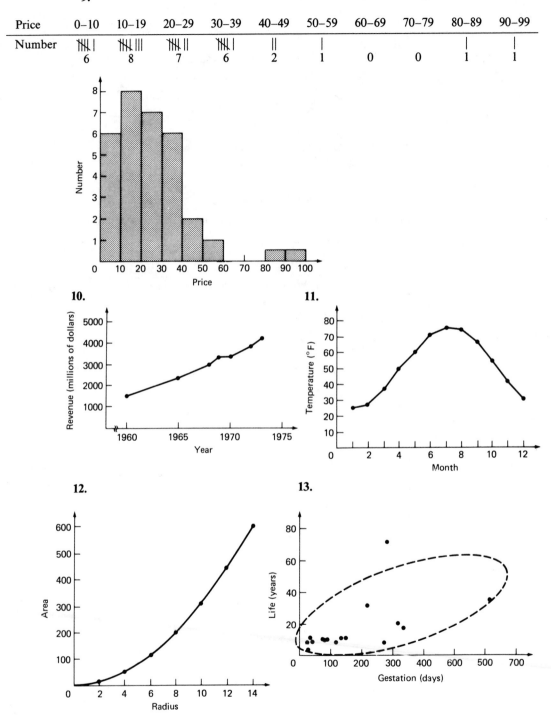

10.

11.

12.

13.

14. -0.214 **15.** 777 **16.** 0.25 **17.** -1.5 **18.** 30

19. $y = 4x + 66$ **20.** $y = 3x - 2$ **21.** $C = 0.065m + 1200$

22.

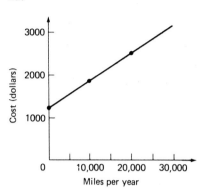

23, 25.

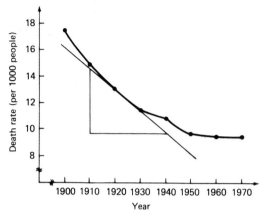

24. $(1915, 13.8)$; $(1935, 11.0)$ **26.** $m_{\text{tangent}} = -0.16$

27. In that year the death rate went down by 0.16 per 1000.

28. One graph is Problem 23. The other is as shown.

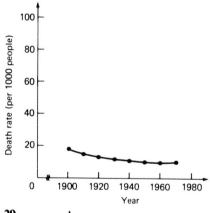

29.

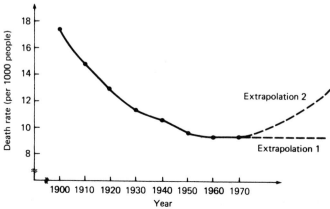

Problem Set 5–1–1 (page 260)

1. $1.45 for 28 oz (5.18¢/oz)   **3.** 3 cups for 19¢ (0.79¢/oz)

5a. $1.19 for 24 oz (4.96¢/oz)   **5b.** 88¢ for 16 oz (5.5¢/oz)

7. $4 for 30 lb (80% eaten) (16.7¢/lb)

9.

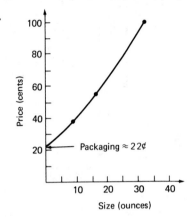

Problem Set 5–1–2 (page 263)

2. PPD = 57.4 (gP/$)   **3.** PPD = 51.4 (gP/$)   **5.** 5.6 g

Problem Set 5–1–3 (page 268)

2. 165   **3.** 144   **5.** 105   **7.** 152   **8.** 1614   **9.** 2234   **11.** 2409

13. 2700   **16.** 6 ft, $6\frac{1}{2}$ in.   **17.** 7 ft, 1 in.   **19.** 5 ft, 9 in.

Problem Set 5–1–4 (page 270)

1. 2.86   **3.** 70,000   **5.** 0.23   **7.** 10 days; 50 days; 200 days

9. Bob: 72 days; Betty: 116.7 days   **11.** 83.3 cal/day

13. 8.7 lb   **15.** 109 days

Problem Set 5–2–1 (page 277)

2. 450   **3.** 12%   **5.** $40.50   **7.** 16%   **9.** $555   **11.** 17.3%

13. $4200   **15.** $4882.50   **17.** $162.75   **19.** $800   **21.** $280

23. 22.9%   **25.** $57.60   **27.** $258   **29.** $134.44   **31.** $100

33. $728   **35.** $354   **37.** $64.75   **39.** 18.1%   **41.** 12.0%

43. 21.6%   **45.** 17.8%

Problem Set 5–3–1 (page 284)

2. 344.18   **3.** 112.20   **5.** 32,061.07   **7.** 820.25   **9.** 4.17%

11. 69.9%   **13.** 39,805.83   **15.** 3280.61   **17.** 29.3%   **19.** Decrease of 8.5%

21. 11,236.34   **23.** 300   **25.** 2096.60   **27.** 4.8%

Problem Set 5–4–1 (page 294)

2. 1616   **3.** 1635   **5.** 3983   **7.** 533   **9.** 3164   **11.** 4213

13. 877   **15.** 1864

17.

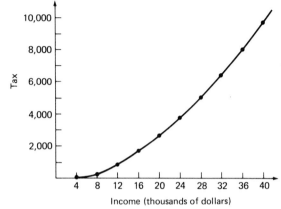

Income (thousands of dollars)

	Married joint	Married sep.	Both single, H.H.
18.	4256	4358	3278
19.	4256	4395	3299
21.	4256	4546	3403
23.	14,321		

24. 1395   **25.** 9.7%   **27.** 20   **29.** 2523   **31.** 2550   **33.** 27%

Problem Set 5–4–2 (page 299)

	2–10.		11–19.		
Problems	I.D.	DEDUCT	Tax Table Income	TAX	BALANCE OR REFUND?
2, 11.	5640	2440	21,060	2799	701 Refund
3, 12.	1000	0	14,500	2141	241 Due
5, 14.	755	0	19,609	2624	224 Due
7, 16.	3040	840	9,660	916	284 Refund
9, 18.	3990	790	14,710	1325	75 Refund

Problem Set 5–4–3 (page 302)

2. 17%; 83%   **3.** 28%; 72%   **5.** 42%; 58%   **7.** 25%; 75%   **9.** 39%; 61%

11. 375   **13.** 340   **15.** They pay $551; U.S. pays $399

17. 1455.97; 44.03 lost   **19.** 56   **21.** 675; 506.25; 11.25%

Problem Set 5–5–1 (page 308)

2. $C = 1500 + 0.06m$ **5.** Higher repair bills and tire costs

3, 7.

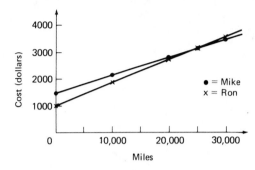

9. At 25,000 miles per year, it costs the same to run both cars. Less than 25,000, the used car is cheaper; more than 25,000, the new car is cheaper.

11. 1250; 950; 400; 100   **12.** 3360; 1140   **13.** 2520; 840

15. 1440; 480   **17.** 780; 300   **19.** 480; 120   **21.** 5000

23. 2560   **25.** $V\left(\dfrac{100 - D}{100}\right)$; $V\left(\dfrac{100 - D}{100}\right)^2$; $V\left(\dfrac{100 - D}{100}\right)^n$

Problem Set 5–5–2 (page 313)

2. 2.7¢/mi   **3.** 5¢/mi   **5.** 2.3¢/mi   **7.** 15.3 mi/gal

9. 31.1; 2.25   **11.** 18.7; 3.75   **13.** 11.9; 5.9   **14.** 60   **15.** 600

17. 504   **19.** 180   **21.** 0.42¢/mi (18 qt used in 3000 mi)

22. 0.625   **23.** 0.533   **25.** 0.475   **27.** 62.50   **29.** 60%

31. For 5000 miles per year:

Yr.	Int.	Ins.	Tax	Deprec.	Tire	Gas	Oil	Repair	Total
1	210	300	50	1000	0	195	15	30	1800
2	210	300	50	760	0	195	15	60	1590
3	210	300	50	560	0*	195	15	90	1420
4	0	300	50	400	0*	195	15	120	1080
5	0	300	50	320	34	195	15	150	1064
6	0	300	50	240	34	195	15	180	1014
7	0	300	50	200	34	195	15	210	1004
8	0	300	50	120	34	195	15	240	954
9	0	300	50	80	34	195	15	270	944
10	0	300	50	80	34	195	15	300	974

*Since tires don't wear out for 4 years.

Problem Set 5–5–3 (page 317)

1. 4220; 4642 **3.** 4955; 5451 **5.** 5465; 6012 **7.** 6846.13

9. 171.70 **11.** 3634.68 **13.** 975.32 **15.** No. It is only $25 over cost.

17. 1280 **19.** 3913

Review Exercises, Chapter 5 (page 318)

1(a). 79¢ for 16 oz (4.94¢/oz) **(b).** $2.19 for 23 oz (9.5¢/oz)

2.

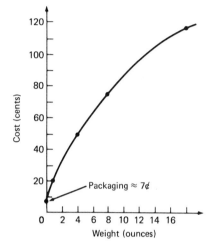

3. 49.3 gP/$ **4.** 159 lb **5.** 1901 **6.** 62,400 cal/yr; 17.8 lb/yr

7. 500 **8.** 9% **9.** 4000 **10.** 5040 **11.** 1040 **12.** 16.9%

13. 13,714.29 **14.** 13,312.50 **15.** 12,475.71 **16.** Down 0.2%

17. 2380 **18.** 3049 **19.** 716 Refund **20.** 2645 **21.** 6148.40

22. 2948.40 **23.** 23,971.60 **24.** 3577 **25.** 723 **26.** 25%

27. 75% **28.** 150 **29.** 37.50 **30.** $C = 0.055m + 1100$

31. (5000, 1375); (10,000, 1650); (25,000, 2475) **32.** 5625; 1875

33. 4200; 1425 **34.** 3150; 1050 **35.** 12.2 mi/gal **36.** 5.97¢/mi

37. 336 **38.** 0.73¢/mi **39.** 23.83 **40.** 3644.20 **41.** 4008.62

Problem Set 6–1–1 (page 327)

3. Maybe ├──────↑──────┤ **5.** Unlikely ├──↑──────────┤
  0   1   0   1

12. {1, 2, 3, 4, 5, 6} **13.** {2, 3, 4, 5, 6, 7, 8, 9, 10, 11, 12}

15. {get across safely, get hit by car, fall in manhole, and so on.}

17. {love it, like it, ho-hum, dislike, hate it}

19. {honest fast service, honest slow service, dishonest slow service, dishonest fast service}

21. {J♣, J♠, Q♣, Q♠, K♣, K♠}

23. { } impossible   **25.** {all clubs, spades, and diamonds}

27. {red cards}   **29.** {all cards}   **31.** { } impossible

33. {a, b, c, d}   **35.** {a, b, f}

Problem Set 6–1–2 (page 332)

2. No; does not add up to 1.   **3.** No; does not up to 1.   **5.** Yes

7. No; has a negative probability   **9.** Yes

10. {Democrat wins, Republican wins, third party wins}

11. {married, single, divorced, widowed}

13. {0, 1, 2, 3, etc., children}   **15.** {1, 2, 3, 4, etc.}   **17.** 0.73

19. 83.7%   **21.** 0.18   **22.** 77/78   **23.** 0.997   **25.** 99.99%

27. 0   **29.** 1/79   **31.** 999,999/1,000,000

Problem Set 6–2–1 (page 336)

2. $\frac{3}{8}$; $\frac{5}{8}$ **3.** $\frac{23}{129}$; $\frac{106}{129}$ **5.** 0.982; 0.018   **7.** 1; 0

9. 29.7%; 70.3%   **11.** $\frac{1}{3}$ **13.** $\frac{1}{52}$ **15.** $\frac{5}{26}$ **17.** $\frac{29}{30}$

19. $\frac{11}{15}$ **21.** 8.3%   **23.** $\frac{8}{47}$

Problem Set 6–2–2 (page 339)

2. 80,808   **3.** 36   **5.** 863,040   **7.** 8064   **9.** 840   **11.** 6720

Problem Set 6–2–3 (page 341)

2. 77.2%   **3.** 0.331   **5.** 56.0%   **7.** 0.00022   **8.** Age 40: 0.9241359; age 60: 0.7698698; age 80: 0.2626372   **9.** 72

11.

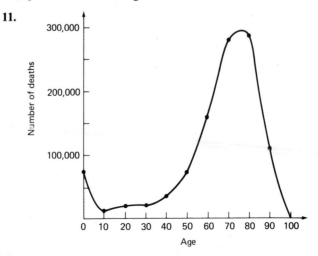

13. There are much fewer 97-year-olds than 25-year-olds to use as a base.

15. 3×10^{-9}

17.

Number of heads	Tally	Number	Probability
0	|||	3	0.06
1	||||| ||||| ||	12	0.24
2	||||| ||||| ||||| ||||	19	0.38
3	||||| ||||| |||	13	0.26
4	|||	3	0.06

19.

Event	Tally	Number	Probability
At least one six	||||| ||||| ||||| |||||		21	0.42
No sixes	||||| ||||| ||||| ||||| ||||| ||||	29	0.58
		50	

Problem Set 6–2–4 (page 348)

2. $\frac{1}{5}$ **3.** $\frac{3}{5}$ **5.** $\frac{1}{2}$ **7.** $\frac{1}{21}$ **9.** $\frac{100}{101}$

10. 3 to 7 for; 7 to 3 against   **11.** 3 to 2 for; 2 to 3 against

13. 11 to 9 for; 9 to 11 against   **15.** 1 to 49 for; 49 to 1 against

17. Bet 3 on A = bet 7 on *not A*   **19.** Bet 2 on C = bet 1 on *not C*

21. Bet 1 on E = bet 4 on *not E*   **23.** Bet 1 on G = bet 39 on *not G*

Problem Set 6–3–1 (page 351)

2. Independent   **3.** Neither   **5.** Mutually exclusive

7. Independent   **9.** Neither   **11.** Independent

Problem Set 6–3–2 (page 358)

2. $\frac{1}{2}$ **3.** $\frac{1}{4}$ **5.** $\frac{1}{16}$ **7.** $\left(\frac{1}{2}\right)^{10}$ **9.** $\left(\frac{1}{2}\right)^{100}$ **11.** 72%   **13.** 15%

15. 65%   **17.** 27.36%   **19.** 0.06%   **21.** 0.672   **23.** $\left(\frac{1}{6}\right)^3 = 0.0046$

25. 0.421   **27.** 0.32   **29.** 0.36   **31.** $\frac{1}{5}$ **33.** 19 to 1

Problem Set 6–3–3 (page 367)

2. 0.25   **3.** 0.235   **5.** 0.0005   **7.** 0.0027   **9.** 0.078   **11.** 0.143

13. 0.3   **15.** 0.65

17. M = married; L = goes to law school

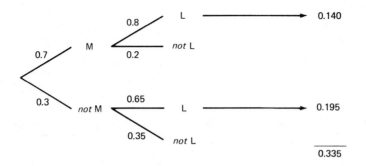

18. 0.335

19. *P*(winning) = 0.495

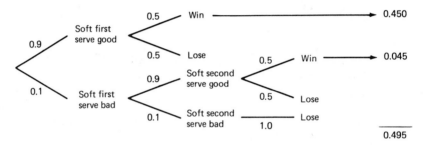

Problem Set 6–4–1 (page 374)

2. 119   **3.** 300   **5.** −$0.17;   unfair   **7.** $9.47;   same as on red

9. 75¢ (unfair)   **11.** $120   **13.** 100   **15.** Scientific   **17.** 0.4

19. 0.0025   **21.** 0.0975   **23.** 30

Review Exercises, Chapter 6 (page 376)

1. Very, very unlikely   **2.** Likely (if group is popular)

3. {war, lasting peace, uneasy truce, and so on}

4. {accepted, rejected, rainchecked, and so on}

5. {getting nothing, getting a fraction of $20, getting all $20, getting more than $20}

6. {hearts, diamonds, spades}

7. {J♣, Q♣, K♣}   **8.** {all clubs, all face cards}

9. No: does not add up to 1   **10.** Yes: all between 0 and 1, add to 1

11. 0.57   **12.** 13/15   **13.** 0.038   **14.** 10/3,000,000   **15.** 725,760

16. 0.55   **17.** 7 to 3   **18.** 6/11   **19.** Bet 15 on *A* = bet 85 on *not A*

20. Independent   **21.** Not independent   **22.** $(\frac{1}{2})^4$ **23.** 0.0144

24. 0.1344   **25.** 0.8656   **26.** 3/20   **27.** 2/19   **28.** 6/380   **29.** 0.7

30.

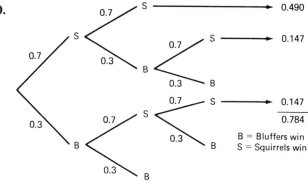

31. 0.784   **32.** 230   **33.** 67¢   **34.** 25

SOME OF THE ANSWERS MAY VARY *SLIGHTLY* BECAUSE OF THE DIFFERENT WAYS TO ROUND OFF WHILE DOING THE PROBLEM

Problem Set 7–1–1 (page 386)

2. 46.46   **3.** 471.8   **5.** 17.57   **7.** 656.8   **9.** 60.75   **11.** 50.67

13. 25.67   **15.** 64   **17.** 59.5   **19.** 72   **21.** 54.47   **23.** 35.19

25. 43.92   **27.** 1.79   **29.** 25,250

Problem Set 7–1–2 (page 391)

2. 45   **3.** 317   **5.** 18.0   **7.** 643   **9.** 61.5   **11.** 52.5   **13.** 27

15. 64   **17.** 60.5   **19.** 70.5   **21.** 54.5   **23.** 32   **25.** 43

27. 2   **29.** 23,950

Problem Set 7–2–1 (page 402)

Range is first; σ is second.   **2.** 66; 17.70   **3.** 1150; 413.29

5. 21.4; 4.92   **7.** 447; 125.76   **9.** 36; 13.21   **11.** 51; 18.10

13. 73; 26.29   **15.** 26; 7.45   **17.** 42; 15.23   **19.** 43; 15.09

21. 26; 5.75   **23.** 81; 17.68   **25.** 59; 14.81   **27.** 4; 1.07

29. 41,000; 11,095.70

Problem Set 7–3–1 (page 404)

2. 656   **3.** 917   **5.** 273,200   **7.** 399,000   **9.** 500   **11.** 764

13. 2   **15.** 8500 to 12,500   **17.** 6500 to 14,500   **19.** 4500 to 16,500

Problem Set 7–3–2 (page 412)

2. $+1.4$ **3.** -1.8 **5.** -0.77 **7.** -0.5 **9.** 365   **11.** 97.1%

13. 24.2%   **15.** 86.4%   **17.** $+1.3$ **19.** -2.3

	z	$F(z)$ (%)	% below x	% above x
21.	$+0.5$	69.1	69.1	30.9
23.	-0.7	24.2	24.2	75.8
25.	$+0.3$	61.8	61.8	38.2
27.	$+0.9$	81.6	81.6	18.4
29.	$+1.2$	88.5	88.5	11.5

	z_A	z_B	$F(z_A)$ (%)	$F(z_B)$ (%)	% between A and B
31.	-0.1	$+0.7$	46.0	75.8	29.8
33.	-0.7	$+1.0$	24.2	84.1	59.9
35.	-0.5	$+1.5$	30.9	93.3	62.4
37.	-0.6	$+1.7$	27.4	95.5	68.1
39.	21.2%				

Problem Set 7–4–1 (page 418)

2. 68.4%   **3.** 55.3%   **5.** 50.5%   **7.** 245   **9.** 102

11. 9.8% (yes); 90.2% (no)

Problem Set 7–4–2 (page 426)

	E	N	p	CI
2.	63	95	0.663	0.097
3.	247	610	0.405	0.040
5.	75	918	0.082	0.018
7.	88.2	420	0.21	0.040
9.	217	247	0.88	0.041

11. 816 **13.** 494 **15.** 0.058 **17.** 0.115 **19.** 3.5%

21. $p = 12.9\%$; CI = 7.3% **23.** $p = 88\%$; CI = 5.3%

25. Bean: 55%; Box: 45% **27.** No

Problem Set 7–5–1 (page 432)

2. That is just an average; a new lawyer will probably earn less

3. Phony accuracy **5.** Faster than what, popcorn kernels?

7. Twice as much as what, peanut butter?

9. Perhaps natural breads build bodies in 20 or 50 ways; perhaps Ding Dong Bread hurts bodies in 14 ways

Review Exercises, Chapter 7 (page 433)

1. 44 **2.** 166.71 **3.** 15,576.47 **4.** 41 **5.** 159 **6.** 15,100

7. 68 **8.** 220 **9.** 17,900 **10.** 19.19 **11.** 56.73 **12.** 4690.86

13. 2254 **14.** 3152 **15.** 3292 **16.** -0.44 **17.** -2.75

18. 130 **19.** 91.9% **20.** 38.2% **21.** $+0.4$ **22.** -1.1

23. -0.3; $+0.7$; 61.8%; 37.6% **24.** $+1$; $+1.5$; 15.9%; 9.2%

	E	N	p	CI
25.	201	253	0.79	0.051
26.	143	622	0.23	0.034
27.	45	368	0.122	0.034

28. 278 **29.** 8.2% **30.** 41%; 6%

31. Missing base; sales may have doubled: gone from 1 to 2

Problem Set 8–1–1 (page 448)

2. $30 **3.** $78 **5.** 7.1% **7.** 4.8% **9.** 1909.09 **11.** 1.092

13. 2.382 **15.** 8 yr **17.** 7^+ yr **19.** $7\frac{1}{4}\%$ **20.** 5465.30

21. 638 **23.** 11,422.05 **25.** 12 **27.** 7% **29.** $4231.10

31. 795.90 **33.** $777.60 **35.** $900,000 **37.** $24,732,000

39. 7^+ yr **41.** $4\frac{1}{2}\%$

43.

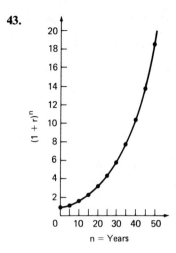

Problem Set 8–1–2 (page 458)

	DT	TT	QT
2.	72	115.2	144
3.	36	57.6	72
5.	18	28.8	36
7.	13.1	20.9	26.2
9.	11.1	17.7	22.2
11.	9.6	15.4	19.2
13.	8	12.8	16
15.	6	9.6	12
17.	3.6	5.8	7.2

18. 13.1; 26.2   **19.** 4 yr   **21.** 16 yr   **23.** $790   **25.** 364

27. 18%   **29.** 6%   **31.** 22.5   **33.** 45   **35.** 25.7   **37.** 42.4

39. 120   **41.** 32.7   **43.** 180   **45.** 18   **47.** 65.5   **49.** 90

51. 20   **53.** 30   **55.** 80   **57.** 65.5   **59.** 22.5   **61.** 80   **63.** 120

Problem Set 8–2–1 (page 468)

2. 12.578   **3.** 23.276   **5.** 18.287   **7.** 28   **9.** $7\frac{3}{4}\%$ **10.** 16,870

11. 37,693.50   **13.** 543.69   **15.** 25 yr   **17.** $6\frac{3}{4}\%$ **19.** $206,798

21. $6973.60   **23.** $72,555   **25.** $3863.49

27. $4413.72/yr; $367.81/mo; $84.88/wk; $12.09/day

29. 7^- yr   **31.** 16 yr

Problem Set 8–3–1 (page 474)

2. $10,000 or more   **3.** $12,500 or more   **5.** $44,000 or less

7. $28,000 or less   **9.** $52,000 or less   **11.** $30,500 or more

	Price	% Down	Down payment	Mortgage
12.	33,000	15	4,950	28,050
13.	43,000	20	8,600	34,400
15.	22,500	25	5,625	16,875
17.	52,000	25	13,000	39,000
19.	43,333.33	15	6,500	36,833.33
21.	27,000	10	2,700	24,300

23. $10,080   **25.** $36,000   **27.** Yes   **29.** $6800   **31.** $11,800

33. $6620   **35.** $16,550

Problem Set 8–3–2 (page 482)

1. $185.28   **3.** $257.60   **5.** $130.00   **7.** $35,599   **9.** 9%

10. $5800   **11.** $23,200   **13.** $4200   **15.** $304.29   **16.** $27,838.43

17. $34,798.04   **19.** $266.65   **21.** $207.40   **23.** $47,000

25. $367.16   **27.** 14,100

Problem Set 8–3–3 (page 487)

2. 2.39%   **3.** $880   **5.** 43,571.43   **7.** $55,925.93   **9.** 22%   **11.** 31.5%

Problem Set 8–3–4 (page 496)

	M	Y	I	E
2.	164.60	1975.20	1750	225.20
3.	288.00	3456	2880	576
5.	190.07	2280.79	2147	133.79
7.	163.69	1964.24	1766.75	197.49

	P_0	r (%)	P_1	P_5	P_{20}
8.	25,000	5	26,250	31,900	66,325
9.	32,000	$4\frac{1}{2}$	33,440	39,872	77,184
11.	20,701	$5\frac{1}{2}$	21,840	27,057	60,406
13.	25,000	$4\frac{1}{2}$	26,125	31,150	60,000
15.	39,009	5	40,960	49,776	103,491

16. $800   **17.** $633.60   **19.** $1257.20   **21.** $1890.40

23. (a) Yes; (b) 7000; (c) 28,000; (d) 235.20; (e) 2520;
(f) 302.40; (g) 36,575; 43,610; 54,355; (h) 945;
(i) 866.25; (j) 39,685.10

25. (a) Yes; (b) 4400; (c) 39,600; (d) 333.04; (e) 3762;
(f) 234.48; (g) 46,200; 56,144; 71,676; (h) 1364;
(i) 1435.28; (j) 51,091.04

27. (a) Yes; (b) 12,250; (c) 36,750; (d) 302.45; (e) 3399.38;
(f) 230.02; (g) 51,205; 61,054; 76,097; (h) 1274;
(i) 1495.48; (j) 55,559.14

Problem Set 8–4–1 (page 502)

2. 1905.12   **3.** 330.24   **5.** 417.09   **7.** 1071.36   **8.** 1100

9. 0   **11.** 1400   **13.** 3400

Problem Set 8–4–2 (page 507)

2. 414.50; 145.08   **3.** 281.90; 98.67   **5.** 421.00; 147.35

7. 437.56; 153.15   **9.** 557; 194.95

11.

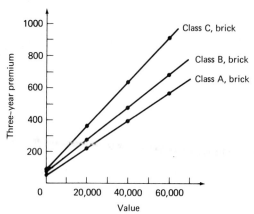

	80% of value	Insurance payment
13.	24,000	12,000
15.	20,400	4313.73
17.	28,800	24,000
19.	13,200	14,000
21.	39,200	44,000

23. $4285.71

Problem Set 8–4–3 (page 511)

2. 160.25 **3.** 303.05 **5.** 148.50 **7.** 337.81 **9.** 30,303

11. 63,091 **13.** 15,060 **15.** 28,400 **17.** 40,850 **19.** 24,000

21. 35 **23.** 212.40 **25.** 47,319 **27.** 40,789 **29.** 47,297 **31.** 40,770

33.   **35.**

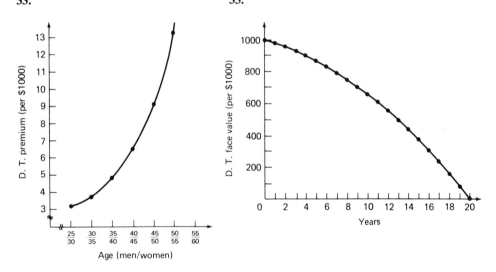

Problem Set 8–4–4 (page 516)

2. 682.80 **3.** 341.40 **5.** 296.55 **7.** 15,191 **9.** 30,349

11. 4620 **13.** 19,750 **15.** 4368 **17.** 20,160 **19.** 50

21. 155,000 **23.** 35 **24.** 593.10 **25.** 308.41 **27.** 5370

29. 576.00 **31.** 152.64 **33.** 8700 **35.** 60,606 **36.** 17,704

37. 97.55 **39.** 1437.45 **41.** 853.50

43.

45. 222.00   **47.** 51,720; 3271.17; 2640   **49.** 23,460; 14,293.38; 12,280

Review Exercises, Chapter 8 (page 518)

1. 35.75   **2.** 5.7%   **3.** 950   **4.** 3492   **5.** 2583.98   **6.** 14.5 yr

7. $7\frac{1}{4}\%$ **8.** 14.4 yr   **9.** 9 yr   **10.** 7.2   **11.** 14.4%   **12.** 6%

13. 255.15   **14.** 103,650   **15.** 2439.32   **16.** 11^- yr   **17.** $6\frac{1}{4}\%$

18. 3162.11   **19.** $46,000 or less   **20.** 9250; 27,750

21. 23.3%; 32,200   **22.** 35,000; 28,000   **23.** 233.74   **24.** 34,783

25. 192.36   **26.** 9800   **27.** 39,200   **28.** 335.94   **29.** 2.15%

30. 12%   **31.** 18.58   **32.** 8400   **33.** 33,600   **34.** 276.53

35. 2940   **36.** 378.36   **37.** 52,332; 65,226   **38.** 921.60   **39.** 47,622

40. 585   **41.** 1071.36   **42.** 750   **43.** 400   **44.** 480.40

45. 376.00   **46.** 32,000; 27,500   **47.** 25,600; 3281.25

48. 21,600; 24,000   **49.** 294.75   **50.** 173.25   **51.** 39,002

52. 494.25   **53.** 18,850   **54.** 13,620   **55.** 49,930   **56.** 7290

57. 11,925

Problem Set 9–2–1 (page 531)

2. 5_{ten} **3.** 12_{ten} **5.** 8_{ten} **7.** 127_{ten} **9.** 427_{ten}

11. 11001_{two} **13.** 1100000_{two} **15.** 110110_{two} **17.** 100000_{two}

19. 1001111_{two}

Problem Set 9–2–2 (page 535)

2. 101_{two} $(2 + 3 = 5)$ **3.** 1011_{two} $(5 + 6 = 11)$ **5.** 100010_{two} $(23 + 11 = 34)$

7. 10111000_{two} $(75 + 109 = 184)$ **8.** 1_{two} $(3 - 2 = 1)$ **9.** 10_{two} $(5 - 3 = 2)$

11. 11_{two} $(13 - 10 = 3)$ **13.** 101110_{two} $(91 - 45 = 46)$ **14.** 110_{two} $(3 \times 2 = 6)$

15. 10100_{two} $(5 \times 4 = 20)$ **17.** 1010001_{two} $(9 \times 9 = 81)$

19. 111001001_{two} $(57 \times 17 = 969)$ **20.** 10_{two} $(4 \div 2 = 2)$

21. 11_{two} $(9 \div 3 = 3)$ **23.** 1001_{two} $(117 \div 13 = 9)$

25. 10000_{two} $R111_{two}$ $(183 \div 11 = 16$ R7$)$

Problem Set 9–3–1 (page 540)

2. Starting a car

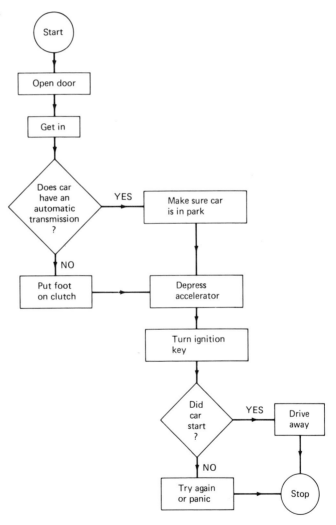

3. Taking a bath

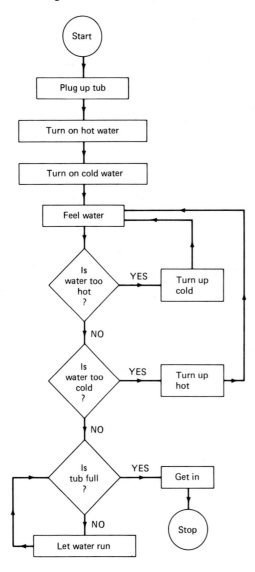

5. Asking for a date

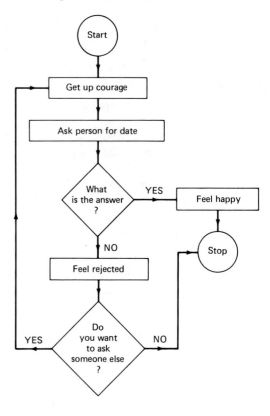

7. Catching a train

9. Buying shoes

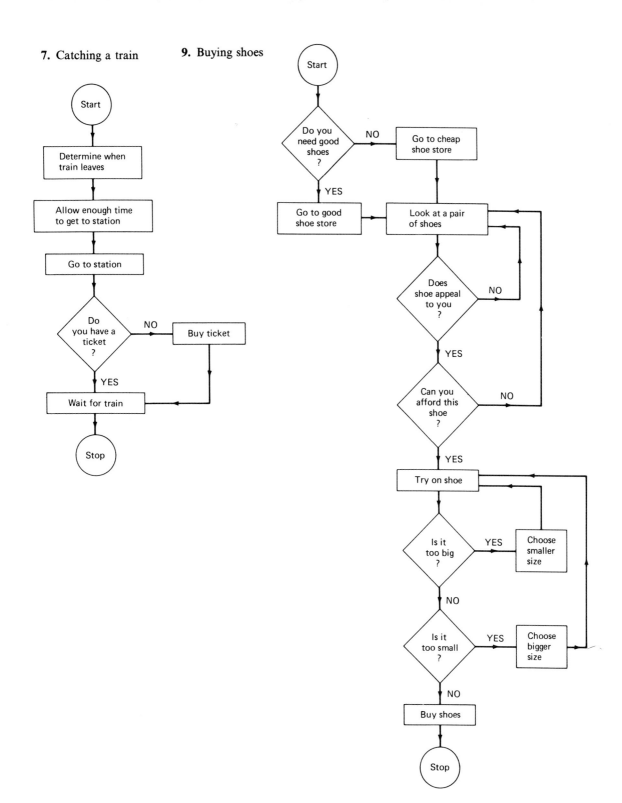

11. Choosing food from a restaurant menu

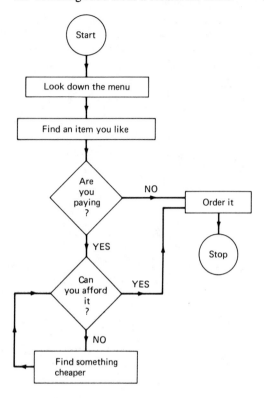

13. Catching a football

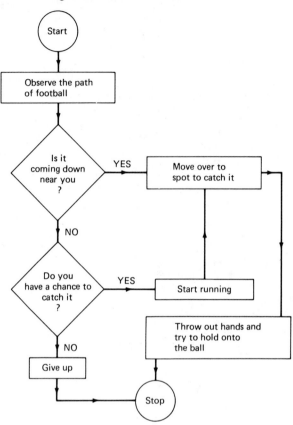

15. Looking for someone to dance with at a mixer

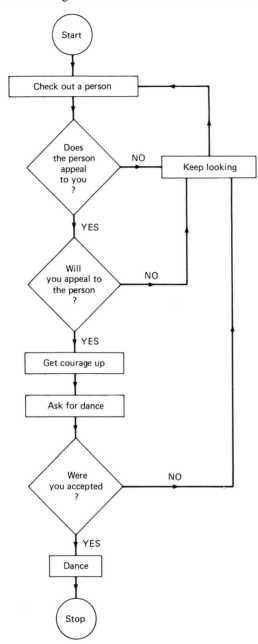

Problem Set 9–3–2 (page 547)

1.

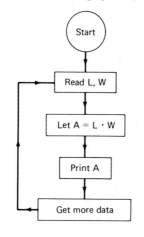

3.

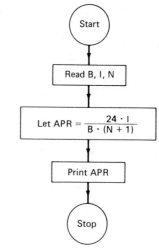

5.

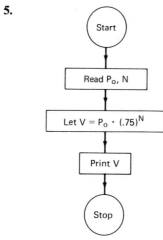

7.

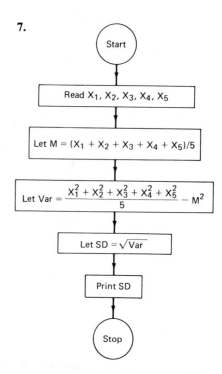

9. **13.**

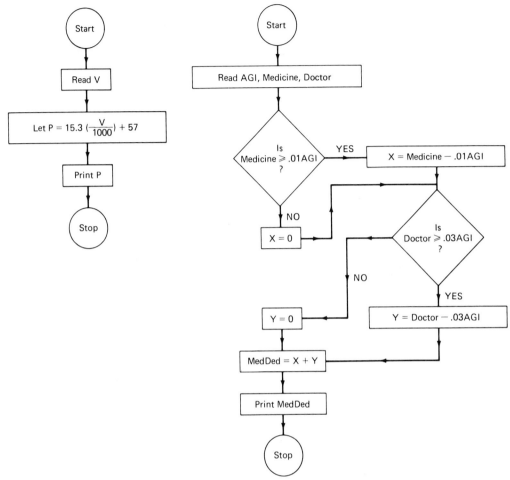

15. 1, 1, 2, 3, 5, 8, 13, 21, 34, 55, 89, 144

17. 198; 314.50; 116; 517.50

Problem Set 9–4–1 (page 552)

1. LET X = 7   **3.** LET Z = 6*X + 7*Y + 1   **5.** LET S = 32*T↑2

7. LET M = (B1 + B2 + B3 + B4)/4   **9.** LET X = A↑3 + B↑3

11. $a = 912$ **13.** $y = x^4$ **15.** $k = \dfrac{a}{4} + 3$ **17.** $c = \sqrt{a^2 + b^2}$

19. LET A = 6 + B (forgot LET)   **21.** LET X = 6*A + B (forgot *)

23. LET A = 3 − 3/(2*X) (only A can be on the left)

Problem Set 9–4–2 (page 558)

1. READ X, Y
DATA 16, 20, 20, 50, 24, 102, 28, 190

3. READ L, H, W
DATA 2.5, 3.4, 4.9, 106, 254, 199, 9.9, 10.3, 1.5

5. READ A, B, C (misspelled READ)

7. READ A, B, C (need variables in READ)
DATA 6, 5, 2

9. READ X, Y, Z (forgot commas)
DATA 5, 4, 7

11. PRINT "THE ANSWER IS" X (need " marks)

13. 10 READ L, W
20 DATA
30 LET A = L*W
40 PRINT A
50 GO TO 10
60 END

15. 10 READ B, I, N
20 DATA
30 LET A = 24*I/(B*(N + 1))
40 PRINT A
50 END

17. 10 READ P, N
20 DATA
30 LET V = P*0.75↑N
40 PRINT V
50 END

19. 10 READ X1, X2, X3, X4, X5
20 DATA
30 LET M = (X1 + X2 + X3 + X4 + X5)/5
40 LET V = (X1↑2 + X2↑2 + X3↑2 + X4↑2 + X5↑2)/5 − M↑2
50 LET S = V↑.5
60 PRINT S
70 END

21. 10 READ V
20 DATA
30 LET P = 15.3*V/1000 + 57
40 PRINT P
50 END

23. 10 READ L, W, H
20 DATA 16, 19, 25, 6.5, 8.2, 10.3, 151, 209, 175
30 LET V = L*W*H
40 PRINT V
50 GO TO 10
60 END

25. 10 READ M, G
20 DATA 225, 14.2, 306, 19.4, 172, 9.3, 277, 13.2
30 LET M1 = M/G
40 PRINT M1
50 GO TO 10
60 END

27. 10 READ G
20 DATA 1200, 900, 650, 775, 810, 865
30 LET S = .025*G
40 LET F = .15*G
50 LET R = .08*G
60 LET N = G − S − F − R
70 PRINT G, S, F, R
80 GO TO 10
90 END

29. 10 READ W, T
20 DATA 20, 15, 70, 50, 10, 12, 15, 16, 22, 52
30 LET C = 500*W/T
40 PRINT C
50 GO TO 10
60 END

31. 6280, 1570000, 113040

Problem Set 9–4–3 (page 567)

1.   IF B > 6 THEN 20   **3.**   IF X = 100 THEN 50
40 # # # # # #   75 # # # # # # #

5. IF A > B THEN 50 (IF, not IS)

7. IF X = 60 THEN 100 (forgot statement number)

9. IF M = N THEN 25 (misspelled THEN)

11. 55   **13.** 5 IS BIGGEST; 6 IS BIGGEST; 10 IS BIGGEST

Review Exercises, Chapter 9 (page 569)

3. 91   **4.** 1011001_{two} **5.** 11000011_{two} ($109 + 86 = 195$)

6. 10010 ($73 − 55 = 18$)   **7.** 1010101_{two} ($17 \times 5 = 85$)   **8.** 110_{two} ($78 \div 13 = 6$)

9.

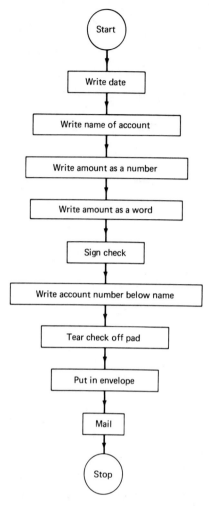

10.

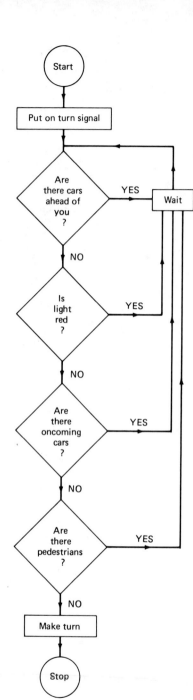

11.

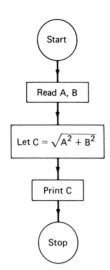

12.

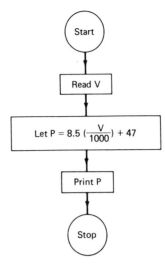

13. 1, 1.0000
2, 1.4142
3, 1.7321
4, 2.0000
5, 2.2361
6, 2.4495
7, 2.6458
8, 2.8284
9, 3.0000
10, 3.1623

14. LET Y = 3*X + 4

15. LET C = (A↑2 + B↑2)↑.5

16. $g = a^3$

17. $m = \dfrac{a + b + c}{3}$

18. READ X, Y, Z
DATA 5, 7, 6.2, 3.8, -1, 4.5, -2, 0, 4

19. LET A = 10 (forgot LET)

20. LET B = 2 (illegal variable)

21. READ X, Y, Z (forgot DATA statement)
DATA 5, 7, 9

22. READ A, B, C (no fractions allowed)
DATA 6, 5, 4.5

23. IF A > 6 THEN 40 (IF, not IS)

24. IF A < = 10 THEN 25 (THEN, not GO TO)

25.

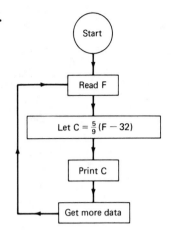

26. 10 READ F
20 DATA 0, 10, 20, 30, 40
30 LET C = (5/9)*(F − 32)
40 PRINT C
50 GO TO 10
60 END

27.

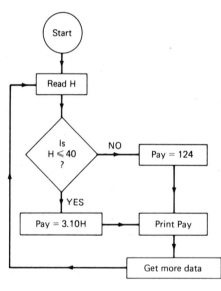

28. 10 READ H
20 DATA 16, 37, 43, 20, 50
30 IF H < = 40 THEN 60
40 LET P = 124
50 GO TO 70
60 LET P = 3.10*H
70 PRINT P
80 GO TO 10
90 END

INDEX